# Mathematics
## Using the Basic Skills

**Len Farrow** MA MSc

Tresham College, Kettering

**Susan Llewellyn** MA

**Stanley Thornes (Publishers) Ltd**

First published in 1987 by:
Stanley Thornes (Publishers) Ltd
Ellenborough House,
Wellington Street,
CHELTENHAM GL50 1YD

Reprinted 1988
Reprinted 1992
Reprinted 1995

British Library Cataloguing in Publication Data

Farrow, L.
Mathematics: using the basic skills.
1. Mathematics—1961-
I. Title II. Llewellyn S.
510 QA39.2

ISBN 0-85950-677-0

Typeset by Tech-Set, Gateshead, Tyne & Wear.
Printed and bound in Great Britain at The Bath Press, Avon.

**Mathematics**

Using the Basic Skills

*Also available from Stanley Thornes and Hulton:*

S. Llewellyn and A. Greer — MATHEMATICS THE BASIC SKILLS THIRD EDITION

Ewart Smith — EASIER EXAMPLES IN ARITHMETIC

# Contents



## Part A

# Part B

# Introduction

## A Message to the Teacher

Over the last few years there has been much discussion about making the teaching of mathematics relevant to people's lives and work. This book attempts to do this.

In part A you will find a series of practical applications of basic mathematics. Each chapter contains an introduction to the topic, a series of practical questions about the topic, a student project and some suggestions as to how other subjects can be related to the topic. Several of the projects require the student to carry out his or her own investigations, either directly, or by using information supplied by you.

Part B contains exercises in the basic mathematical skills needed for the chapters of part A. These are intended to be used for student practice as and when necessary. Some students may not need to use them at all. Each of these chapters contains an end test which could equally well be used as an initial diagnostic test.

We have tried to include all the topics required for the major examinations in arithmetic or numeracy and in the foundation level of GCSE, plus a few extras. As the book has been designed around projects emphasis has been placed on the parts of the subject which are useful at work and in everyday life.

## A Message to the Student

However simple our jobs or lives may be, most of us use numbers sometimes. This book has been written to show you how mathematics can be brought into many different everyday situations at home and at work. Some of the projects are included to show you how numbers can be used to solve a problem. Other chapters show how numbers can help you to make decisions about life – decisions which many of you will have to make at some time in the future.

Part B is included to give you practice on topics which you find difficult. If you don't understand the basic ideas then you will find much of the other work difficult, so use this part to improve your basic skills.

# Acknowledgements

The authors and publishers would like to thank the Post Office and the Department of National Savings for permission to reproduce the leaflets and stationery items in chapter 3, and acknowledge the leaflets *Postal Rates* and *Wrap Up Well* and the double lined lettering style as Post Office copyrights, and the leaflets *Savings Certificates* and *Premium Savings* withdrawal forms as Department of National Savings copyright.

We also express our thanks to the following organisations for information and permission to reproduce material: John Bartholomew and Son Ltd p. 13, British Airways p. 84, British Gas plc pp. 96–9, East Midlands Electricity Board pp. 97–100, Her Majesty's Stationery Office, Crown Copyright Reserved pp. 13 and 14, National Express Ltd p. 70, Townsend Thoresen p. 82, United Counties Omnibus Co Ltd p. 70.

Thanks also to Marion Farrow who typed the manuscript.

# PART A

Chapter 1

# Cooking for a Crowd

Most recipe books assume that you are cooking for a group of four normal adults, but you sometimes need to cook for one or two, and sometimes for a crowd. To do this you need to be able to adjust recipes to suit your needs.

In this chapter you will carry out calculations which will be useful when you cook for a large number of people. This may be something you only do occasionally or you may be working in the catering trade and have to do it regularly.

## 1. Measuring Quantities

### 1.1 Units Used in Cooking

In cookery there are two standard types of measurement needed – weight and volume. Volume is usually used for liquids. The units used to measure these are given in the following table.

| | *Imperial* | *Metric* |
|---|---|---|
| Weight | ounce (oz) | gram (g) |
| | pound (lb) | kilogram (kg) |
| Volume | pint (pt) | millilitre (ml) |
| | fluid ounce (fl oz) | litre (l) |

The relationships between these units are as follows.

| | |
|---|---|
| Imperial | 1 lb = 16 oz |
| | 1 pt = 20 fl oz |
| Metric | 1 kg = 1000 g |
| | 1 l = 1000 ml |

Some cookery books give quantities in both metric and imperial units whereas others use only metric or only imperial. It is useful to be able to carry out approximate conversions from one to the other.

Some appproximate conversions are given below.

| | |
|---|---|
| 1 oz is about 28 g<br>1 lb is about 450 g | 1 kg is about $2\frac{1}{4}$ lb |
| 1 pt is about 600 ml<br>1 fl oz is about 30 ml | 1 l is about $1\frac{3}{4}$ pt |

**1.** Convert the following into ounces.
a) $\frac{1}{2}$ lb b) $\frac{1}{4}$ lb c) $1\frac{1}{2}$ lb d) 1 lb 6 oz

**2.** Write the following as fractions of a pint.
a) 10 fl oz b) 5 fl oz c) 15 fl oz
d) 30 fl oz

**3.** Convert the following into grams.
a) 4 oz b) 6 oz c) $\frac{1}{2}$ lb d) $1\frac{1}{2}$ lb
e) $\frac{3}{4}$ lb

**4.** Convert the following into ounces, giving answers to the nearest ounce.
a) 200 g b) 500 g c) 1 kg d) 1.5 kg
e) 100 g

**5.** Convert the following into millilitres.
a) 2 pt b) $\frac{1}{2}$ pt c) $\frac{1}{4}$ pt d) 4 fl oz
e) 7 fl oz

**6.** Convert the following into fluid ounces, giving answers to the nearest fluid ounce.
a) 1 l b) 500 ml c) 100 ml d) 150 ml

**7.** How many whole cakes can be made from a 1 kg bag of flour if each cake requires
a) 200 g b) 300 g c) 4 oz d) 7 oz
e) 12 oz?

## 1.2 Measuring Temperature

In cookery it is important to be able to set the oven to the correct temperature. Gas cookers are usually regulated using *gas marks*, and electric cookers are marked in *degrees Fahrenheit* (°F) or *degrees Celsius* (or centigrade) (°C). The following table shows the most commonly used oven temperatures taken from a cookery book.

| *Gas mark* | °*C* | °*F* |
|---|---|---|
| 3 | 170 | 325 |
| 4 | 180 | 350 |
| 5 | 190 | 375 |
| 6 | 200 | 400 |
| 7 | 220 | 425 |

The conversions from °F to °C are only approximate and it is worth getting to know the method of conversion in case you don't have a table handy.

To convert from °F to °C the method is to

(a) take 32 away from the °F temperature

(b) multiply answer from a) by 5

(c) divide answer from b) by 9.

} or the other way around if easier

Using algebraic notation this would be

$$C = \frac{5(F - 32)}{9}$$

This formula is also useful when listening to the weather forecast. They usually give temperatures in °C and to convert them you need to do the reverse of the above, i.e. × 9, ÷5, + 32.

**1.** Take the temperatures in °F given in the table above and convert each one into °C to the nearest degree. How do the results compare with the °C values given in the table?

**2.** Convert the following °F temperatures into °C.
a) 250 ° b) 300 ° c) 475 ° d) 80 °
e) 60 °

**3.** Convert the following °C temperatures into °F.
a) 100 ° b) 20 ° c) 26 ° d) 0 °
e) 10 °

## 1.3 Accuracy of Measurements

If you carry out a calculation using an electronic calculator the answer given may have 8 or 9 digits. This is often too accurate to be of any use. You therefore need to approximate the answers and one way of doing this is to give an answer 'to the nearest . . .', for example

2685 = 2700 to the nearest 100

19.2 = 19 to the nearest whole number

In making measurements in cookery you are unlikely to be able to weigh anything more accurately than to the nearest $\frac{1}{2}$oz or 10 g. In fact most recipes give quantities correct to the nearest 25 g. In volume measurement accuracies of $\frac{1}{2}$fl oz or 20 ml can be obtained.

NOTE: When approximating answers, if a value is exactly half-way between two possible answers always put it up rather than down.

$5\frac{1}{2}$oz = 6 oz to the nearest oz

25 g = 30 g to the nearest 10 g

---

**1.** Rewrite the following correct to the nearest oz.
a) $3\frac{1}{2}$oz b) $4\frac{1}{4}$oz c) $6\frac{3}{4}$oz

**2.** Rewrite the following correct to the nearest 10 g.
a) 12 g b) 55 g c) 68 g d) 12.5 g

**3.** Rewrite the following correct to the nearest 25 g.
a) 140 g b) 150 g c) 220 g d) 380 g

**4.** Rewrite the following correct to the nearest 20 ml.
a) 150 ml b) 205 ml c) 560 ml
d) 355 ml

---

Notice that with imperial quantities, fractions are sometimes used for parts of units, but with metric quantities decimals are used.

## 1.4 Reading Scales

Before starting to cook you need to measure out the quantities needed. This usually involves reading scales.

The diagram below shows part of the scale on a set of kitchen scales.

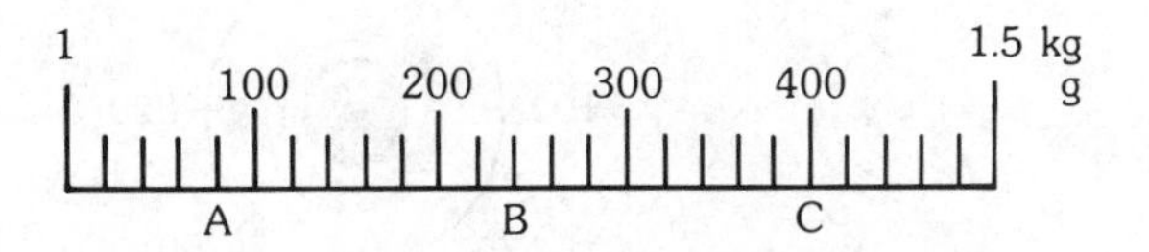


The measurements are in g and kg. As there are 5 divisions to each 100 g each small division must represent 20 g.

A therefore gives 1 kg and 80 g, i.e. 1.08 kg.

B gives 1 kg and 240 g, i.e. 1.24 kg.

C gives 1 kg and 400 g, i.e. 1.4 kg.

---

**1.** The following scale is marked in g and kg.

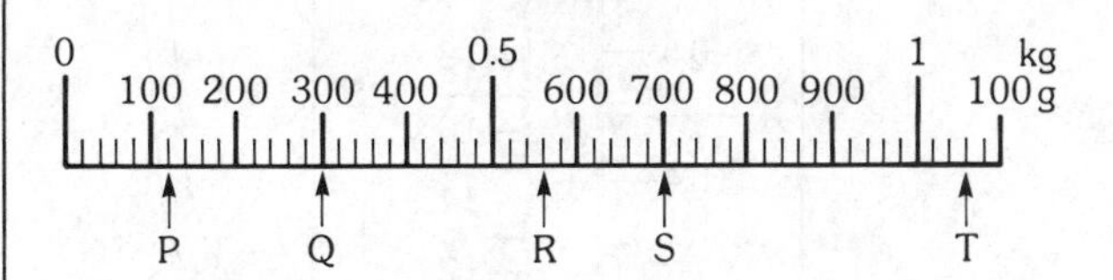


State the readings given by P, Q, R, S and T to the nearest 20 g.

**2.** The following scale is marked in oz and lb.

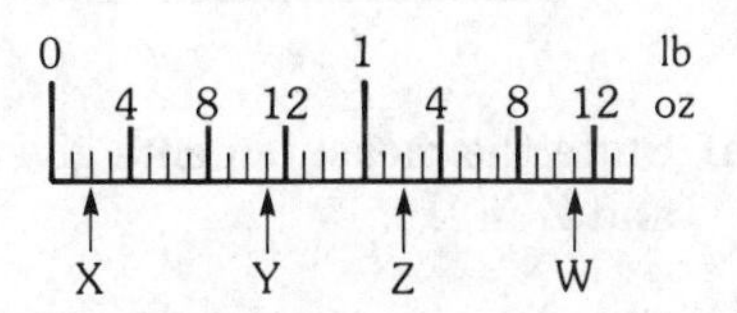


State the readings given by X, Y, Z and W to the nearest oz.

**3.** The diagram shows the scale on an electric cooker in °C.

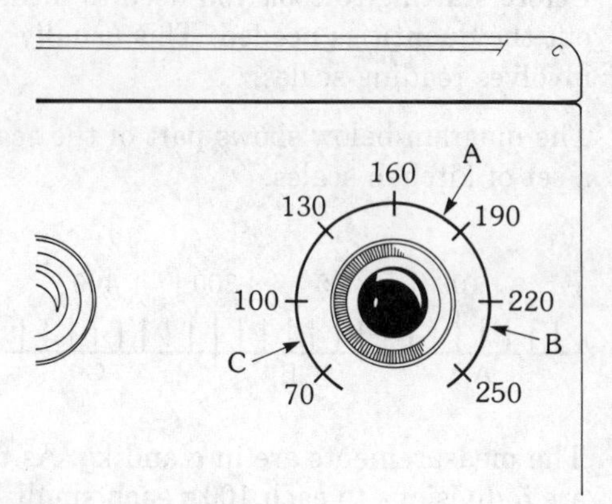


State the temperatures given by A, B and C.

**4.** The scale on a measuring jug is shown below.

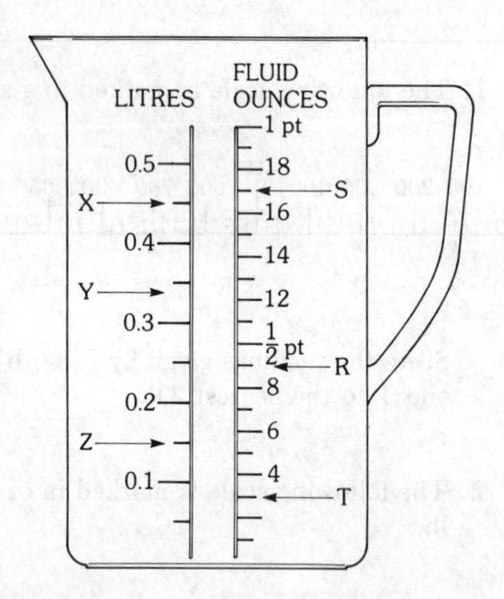


a) State the measurements given by X, Y and Z in l.

b) State the measurements given by R, S, and T in fl oz.

# 2. Increasing the Quantities

## 2.1 Proportion

Consider the simple problem, 'If 6 eggs cost 54 p, how much does 1 dozen cost?'

Unless there is a special offer on eggs by the dozen you can assume that 1 dozen (12) eggs will cost twice as much as 6 eggs, i.e. £1.08. But what about the cost of 9 eggs?

The most general way of approaching this problem is what is known as the *unitary method*.

6 eggs cost 54 p

so 1 egg costs $54\text{ p} \div 6 = 9\text{ p}$

and 9 eggs cost $9\text{ p} \times 9 = 81\text{ p}$

In practice this can be shortened by writing

$$\text{cost of 9 eggs} = \frac{54}{6} \times 9 \quad \text{or} \quad \frac{54}{6} \times \frac{9}{1}$$

and then cancelling down.

This method is particularly useful when the *unit cost* (cost of 1 egg) is not a whole number, for example if 6 eggs cost 58 p, how much do 9 cost?

$$\begin{aligned} \text{cost of 9 eggs} &= \frac{58}{6} \times \frac{9}{1} \\ &= \frac{58}{2} \times \frac{3}{1} \quad \text{(cancelling by 3)} \\ &= 29 \times 3 \\ &= 87\text{ p} \end{aligned}$$

Sometimes the final answer will not be a whole number and you will have to correct it to the nearest 1 p or £1 or other whole unit.

**1.** If 6 eggs cost 50 p find the cost of the following numbers of eggs giving answers correct to the nearest 1 p.
a) 2 doz b) 9 c) 4 d) 15 e) 8

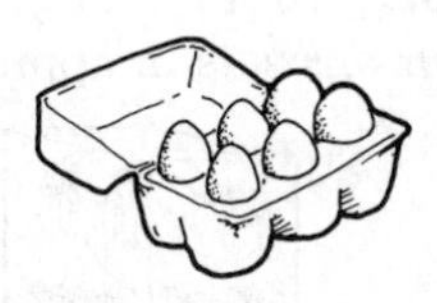

**2.** If 500 g of margarine costs 54 p find the cost to the nearest 1 p of the following amounts of margarine.
a) 100 g b) 250 g c) 150 g d) 600 g

If the cancelling down and multiplication of fractions is causing you problems then you will need to turn to part B, chapter 3 for further practice.

**3.** If 1 kg of flour costs 85 p find the cost to the nearest 1 p of
a) 200 g b) 600 g c) 150 g d) 220 g.

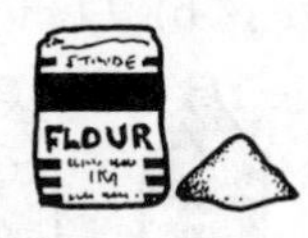


**4.** Minced beef is priced at £1.10 per lb. Find the cost of
a) 5 lb b) $\frac{1}{2}$ lb c) $1\frac{1}{2}$ lb d) $\frac{3}{4}$ lb.

**5.** A pint of milk costs 23 p. Find the cost of
a) 4 pt b) $\frac{3}{4}$ pt c) 15 fl oz d) 8 fl oz.

## 2.2 Scaling up Recipes

The basic ideas of proportion must be understood if you are going to take a recipe for 4 people and extend it to produce a meal for 6 or 9. Adapting a recipe for 4 people to feed 8 is easy – you just double everything, but how do you feed 7?

A recipe for spaghetti sauce for 4 people includes 1 lb of minced beef. For 8 people 2 lb would be needed. For 7 people you need

$$1 \times \frac{7}{4} = 1\frac{3}{4}\text{ lb}$$

**1.** To feed 4 people you need 10 oz of spaghetti. How much spaghetti would be needed for
a) 6 people b) 10 people c) 7 people
d) 18 people?

**2.** A recipe for 'chilli con carne' for 4 people includes $1\frac{1}{2}$ lb beef. How much beef to the nearest $\frac{1}{4}$ lb would be needed for
a) 8 people b) 2 people c) 10 people
d) 18 people?

**3.** To cook goulash for 4 people you need $\frac{1}{2}$ lb onions. What weight of onions would be needed for
a) 6 people b) 18 people c) 7 people
d) 12 people?

**4.** A recipe for a chocolate sponge includes
$3\frac{1}{2}$ oz flour
$\frac{1}{2}$ oz cocoa
2 eggs
4 oz sugar
4 oz margarine
To make a gâteau you need 3 layers instead of 2 and therefore $1\frac{1}{2}$ times the amounts. How much of each ingredient would be needed?

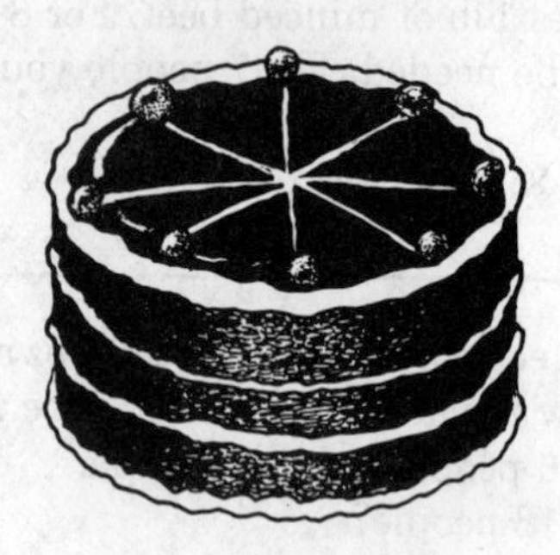

## 2.3 Getting the Mixture Right

Sometimes when you are catering for any number of people you will have to mix ingredients in a particular *ratio*. This ensures that the mixture will be the same no matter how much or how little you are making, for example a manufacturer recommends that orange squash should be made by mixing 1 part squash to 5 parts water. This could also be written as a ratio 1:5. This means that each litre of mixture should be made up of 6 parts, 1 of which should be squash and the other 5 water. Therefore the amount of squash needed for 1 l of mixture is

$$\frac{1}{6} \times 1\,\text{l} = 170\,\text{ml}$$

**1.** Concentrated apple juice should be mixed with water in the ratio 1:7. How much concentrated juice should be used to make the following amounts of mixture?
a) 20 fl oz  b) 1 l  c) $\frac{1}{2}$ l  d) 200 ml
Give your answers in appropriate units.

**2.** Crumble mixture contains flour and fat in the ratio 2:1. Ignoring the weight of other ingredients, how much fat would be needed for the following mixture weights?
a) 4 oz  b) 200 g  c) $\frac{1}{2}$ lb  d) 1 kg

**3.** A recipe for cider punch suggests mixing 2 parts cider with 3 parts lemonade and then adding small quantities of other ingredients. How much punch can be made using
a) 2 pt cider  b) 1 l cider?

**4.** For a three-fruit marmalade oranges, grapefruits and lemons are to be mixed in the ratio 3:2:1 by weight. What weights of grapefruits and lemons would be needed for
a) 6 lb oranges  b) 4 kg oranges?

# Feeding a Conference

Sanjay and Debbie belong to a local Oxfam group and they have been given the job of providing and cooking the food for a one-day conference which 30 people are expected to attend.

They have to provide

coffee mid-morning

savoury mince, potatoes, carrots and green beans for lunch, followed by rice pudding and stewed rhubarb for dessert, and coffee

tea and cake mid-afternoon

A cookery book gives the following lists of ingredients for 4 people.

**Savoury mince** 1 lb minced beef
$\frac{1}{2}$ lb onions
7 oz tin tomatoes
cornflour
salt, pepper
paprika powder

**Rice pudding** 2 oz round grain rice
1 oz sugar
1 pint milk

**Stewed rhubarb** 1 lb rhubarb
4 oz sugar

**1.** Assuming that you already have salt, pepper, paprika and cornflour, work out the quantities of other ingredients needed for the savoury mince.

**2.** Remembering that onions are usually bought to the nearest lb, tomatoes come in 7 oz and 15 oz tins, and so on, make up a shopping list for the savoury mince.

**3.** The cookery book also gives the following guidelines:

1 lb potatoes – 3 servings
1 lb carrots – 4 servings
1 lb green beans – 4 servings

Work out the quantities of these vegetables needed to the nearest lb.

**4.** Work out the quantities needed for the rice and rhubarb and add appropriate items to your shopping list from question **2.** Remember that sugar comes in kg and milk by the pt, and so on.

**5.** An 8 oz slab of fruit cake will give 8 slices. How many slabs would be needed to give 2 slices per person?

**6.** A pint of milk will supply 20 cups of coffee or tea, 6 tea bags will make a pot of tea for 15 and a 100 g jar of coffee provides 50 cups. Add milk, coffee, tea and sugar to the shopping list.

**7.** Using the quantities worked out in questions **2, 3, 4, 5** and **6** above make out and price a shopping list in local shops.

**8.** By adding 10% on to the total food cost (to cover the cost of gas, etc.) and then dividing by 30, work out the amount that each person should be charged.

**9.** If the hire of a room costs £22 and the speaker's expenses are £9 work out the total cost that each person should pay so that the conference does not run at a loss. (Round your answer up a bit to give Oxfam some profit.)

# Teacher's Notes

**1.** This project would link in well with any sort of catering or domestic science course and provide a relevant numeracy input. The work could also be included in the preparation stage for any self-catering residential course.

**2.** Links with other subject areas include

*science* nutritional content of foods, what happens in cooking, materials used for cookware

*career development* careers in catering and the food manufacturing trade

*personal development* working as a group, planning meals

*information technology* food labelling methods, stock control in a supermarket

*communications* availability of different foods, survey of quantities in which foods are sold.

**3.** Students on any course involving services to people such as catering, hotel management or caring services will find this a useful project. An interesting extension for students on caring courses would be to look at the problems involved in cooking for one or two people only. Students on courses with a retail or food production element can investigate the production and marketing of basic foodstuffs.

**4.** A visit to a food production or processing factory would fit in well with this project, as would a visit to a large supermarket or a large catering establishment.

Chapter 2

# Maps and Map Reading

If you are going somewhere which is unfamiliar to you the chances are that you will use a map to find your way. In a town you may need a plan of the town; when walking in the country you will probably use an Ordnance Survey map. Motorists often use books of maps which cover the whole country in sections. In this chapter you will carry out calculations which will help you to use a map more effectively.

## 1. Distances and Maps

### 1.1 Scale Factors

A map is a scale diagram of the real world from which you can find real distances. Somewhere on a map you will find the scale of the map, either in written form or in the form of a diagram, for example

1:50 000 or 4 miles to the inch

or

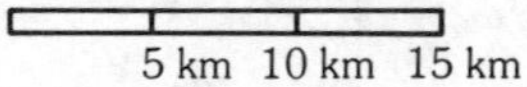


To find distances in the real world you must measure distances on the map and then convert them into real distances.

On a map where the scale is 5 miles to the inch, a distance of 2 in must represent 10 miles while a distance of 10 in will represent 50 miles.

Work through the following questions.

**1.** A map has a scale of 4 miles to the inch. Find the real distances represented by map distances of
a) $\frac{1}{2}$ in  b) 4 in  c) 10 in  d) $2\frac{1}{2}$ in.

**2.** A map has the following diagram on it to show its scale.

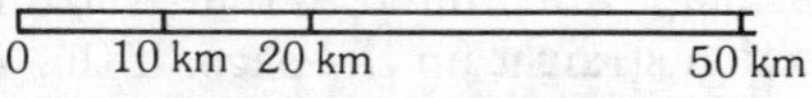


Using this diagram work out
a) how many cm on the map represents 10 km on the ground
b) the number of km represented by map distances of
i) 2 cm  ii) 5 cm  iii) 8 cm  iv) 20 cm.

**3.** A map has a scale of 1:50 000. This means that 1 cm on the map represents 50 000 cm on the ground. 50 000 cm is equivalent to $\frac{1}{2}$ km.

What real distances would be given by map distances of
a) 4 cm b) 12 cm c) 10 cm
d) 25 cm?

## 1.2 Measuring Distances

When you are travelling, the route is not usually straight, so you are faced with the problem of measuring distances along all sorts of curves on the map. To do this you can either use a special map distance measurer or just put a piece of cotton along the route, straighten it out and measure its length.

Now look at the map opposite and look for the scale. Ask the person next to you to pick two places on the map so that you can work out the real distance between them by road. Then swap over so that your partner can find a distance. Carry on like this to get some practice.

# 2. Directions and Maps

## 2.1 The Compass and Angles

Most maps are printed so that North is the direction straight up the page. If this is not the case the map should have a diagram on it showing the direction of North. The other directions are then as shown here.

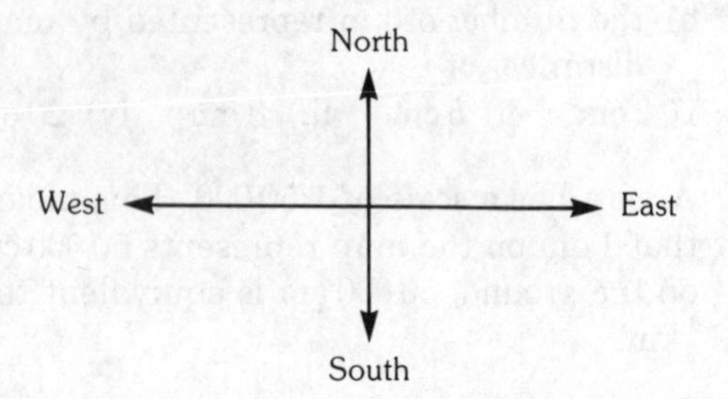


The direction of one place from another place is often described as *due* North, South, East, or West. If the direction is not one of these, a *bearing* is used to describe it. Before you can talk about bearings you need to be able to measure angles using a protractor.

Measure the following angles to the nearest degree.

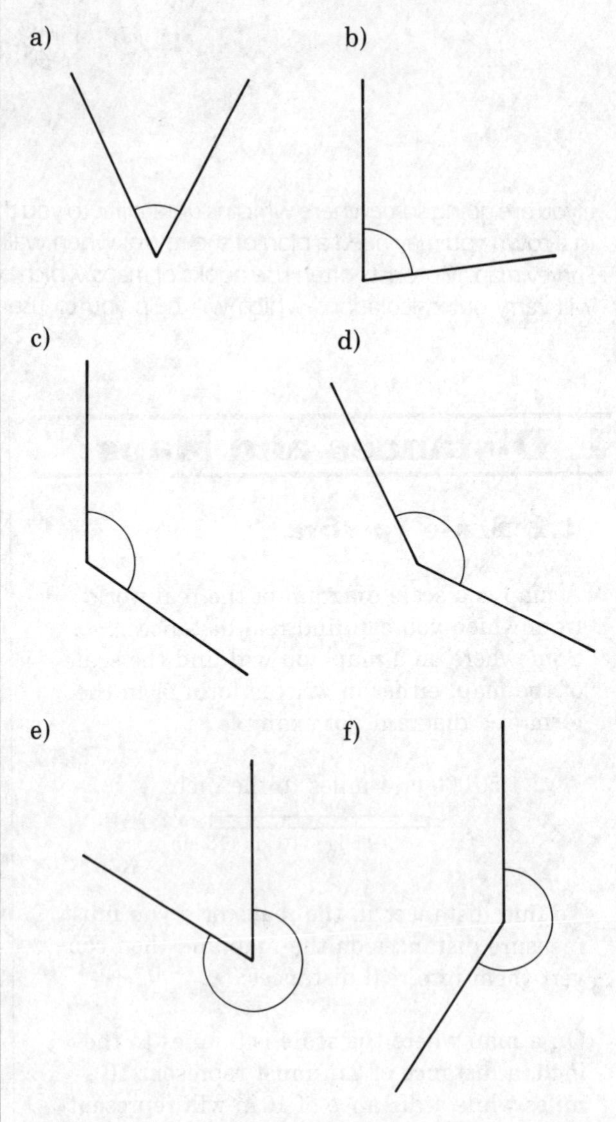


Remember that one complete turn gives 360°.

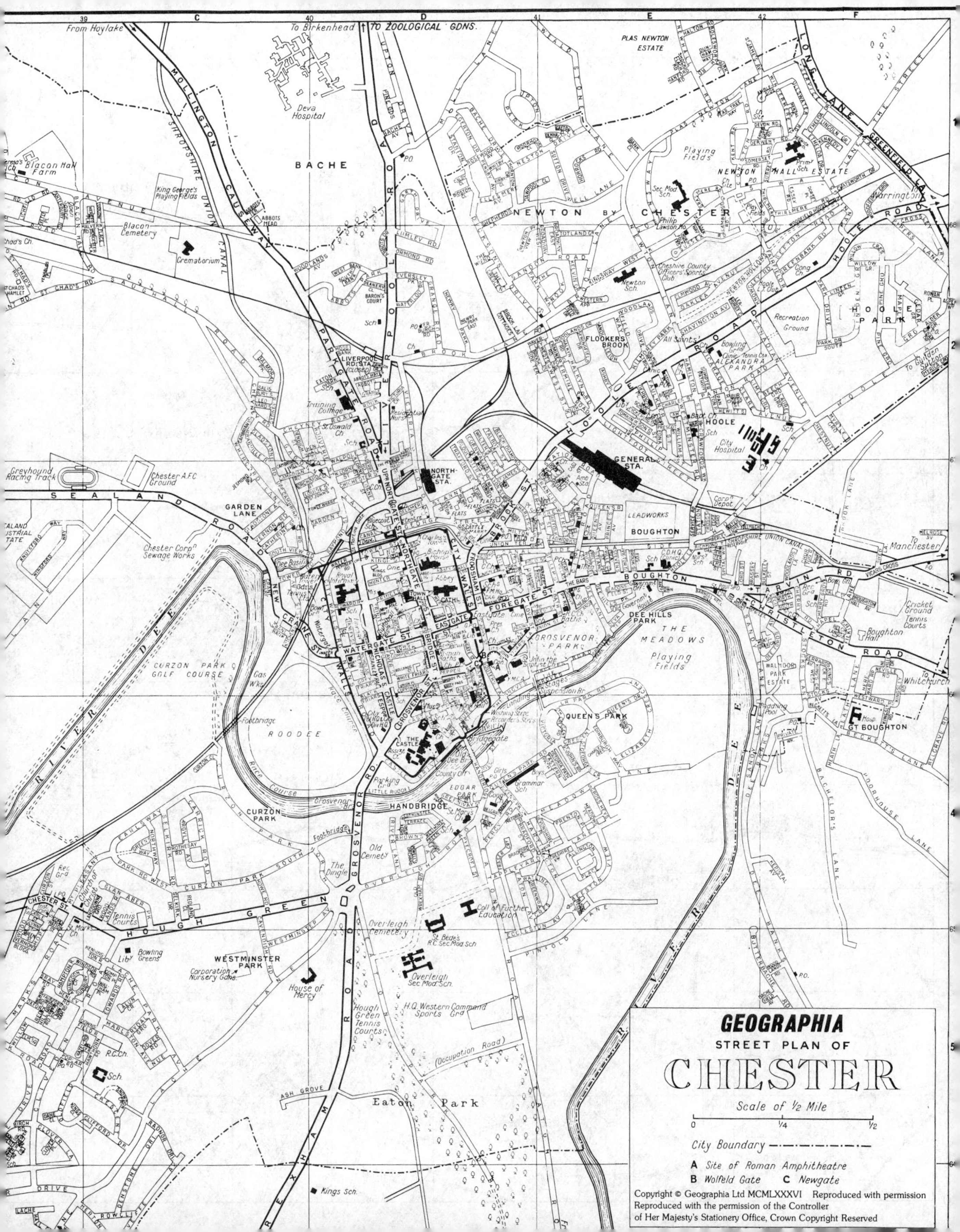
GEOGRAPHIA
STREET PLAN OF
CHESTER
Scale of 1/2 Mile
0
1/4
1/2
City Boundary
A Site of Roman Amphitheatre
B Wolfeld Gate
C Newgate
Copyright © Geographia Ltd MCMLXXXVI Reproduced with permission
Reproduced with the permission of the Controller
of Her Majesty's Stationery Office, Crown Copyright Reserved
From Hoylake
To Birkenhead
TO ZOOLOGICAL GDNS.
PLAS NEWTON ESTATE
Deva Hospital
BACHE
Blacon Hall Farm
King George's Playing Fields
Blacon Cemetery
Crematorium
MOLLINGTON
SHROPSHIRE UNION CANAL
CAUSEWAY
NEWTON BY CHESTER
NEWTON HALL ESTATE
Playing Fields
LONG LANE
GREENFIELD LA
HOOLE ROAD
HOOLE PARK
Recreation Ground
FLOOKERS BROOK
ALEXANDRA PARK
HOOLE
City Hospital
GENERAL STA.
NORTH-GATE STA.
LIVERPOOL RD. STA. (CLOSED)
Training College
Greyhound Racing Track
Chester A.F.C. Ground
SEALAND ROAD
GARDEN LANE
Chester Corpn Sewage Works
LEADWORKS
BOUGHTON
SHROPSHIRE UNION CANAL
To Manchester
TARVIN RD
CHRISTLETON ROAD
Cricket Ground
Boughton Hall
To Whitchurch
GT BOUGHTON
WALMOOR PARK ESTATE
DEE HILLS PARK
THE MEADOWS
Playing Fields
GROSVENOR PARK
QUEEN'S PARK
Suspension Br.
FOREGATE ST
EASTGATE
WATERGATE ST
NORTHGATE ST
BRIDGE ST
GROSVENOR ST
CITY WALLS
Royal Infirmary
CATHL.
THE CASTLE
RIVER DEE
CURZON PARK GOLF COURSE
ROODEE
Race Course
Grosvenor Br.
HANDBRIDGE
EDGAR PARK
CURZON PARK
Old Cemetery
The Dingle
HOUGH GREEN
CHESTER ST.
Overleigh Cemetery
Coll of Further Education
St. Bede's R.C. Sec. Mod. Sch
WESTMINSTER PARK
Corporation Nursery Gdns.
House of Mercy
Overleigh Sec. Mod. Sch.
H.Q. Western Command Sports Grd
Hough Green Tennis Courts
(Occupation Road)
Eaton Park
WREXHAM ROAD
Kings Sch.
ASH GROVE

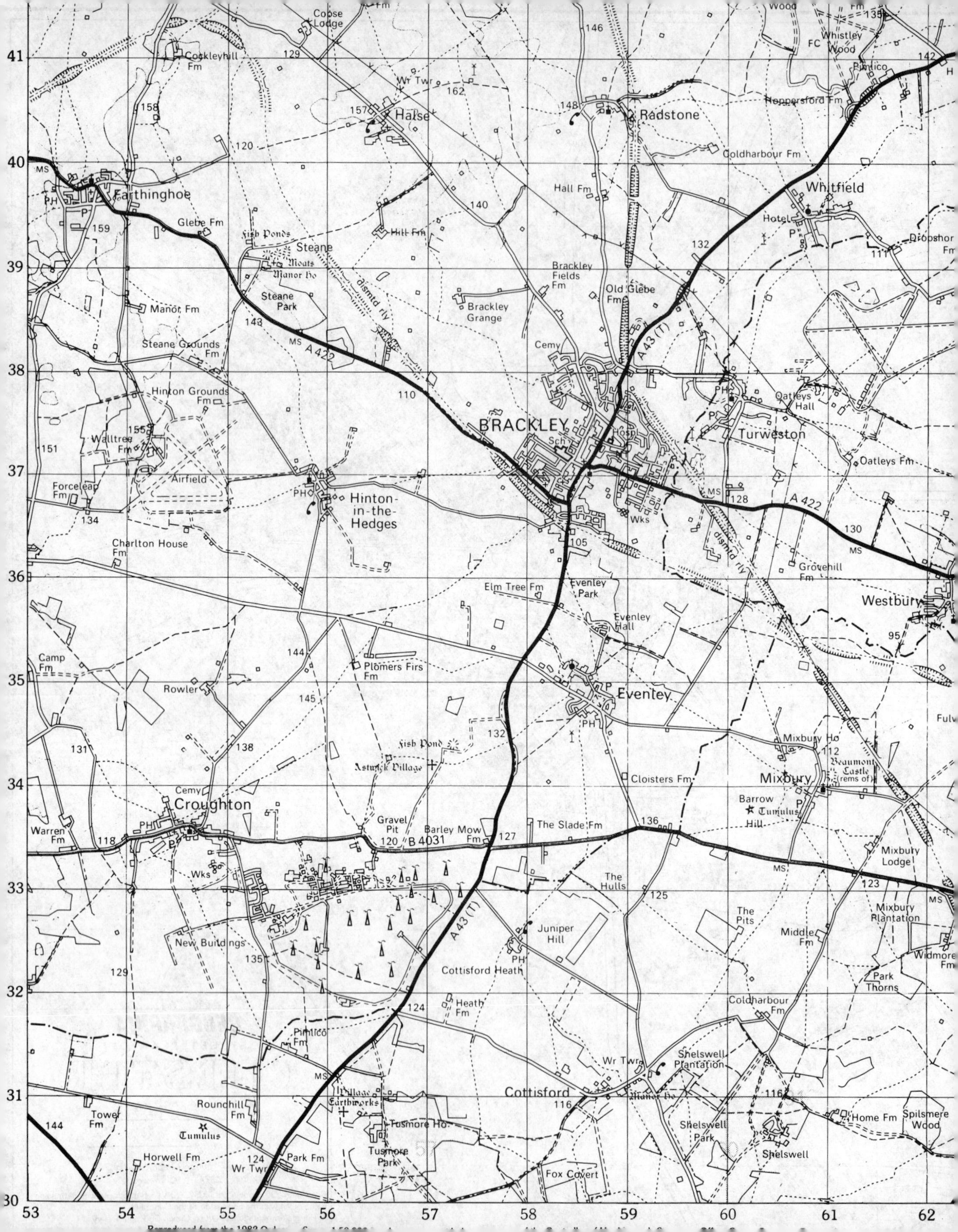

Coose Lodge
Cockleyhill Fm
129
Wr Twr
162
157
Halse
148
Radstone
146
Whistley Wood
FC
Pimlico
142
Heppersford Fm
Coldharbour Fm
158
120
MS
Farthinghoe
PH
P
159
Glebe Fm
Fish Ponds
Hill Fm
140
Hall Fm
Whitfield
Hotel
Steane
Moats
Manor Ho
Steane Park
Manor Fm
Brackley Fields Fm
Brackley Grange
Old Glebe Fm
132
111
Dropshor Fm
143
A 422
Steane Grounds Fm
dismtd rly
Cemy
A 43 (T)
Hinton Grounds Fm
110
BRACKLEY
Sch
Hosp
Oatleys Hall
Turweston
Walltree Fm
155
151
Airfield
Forceleap Fm
134
Hinton-in-the-Hedges
Oatleys Fm
128
130
Wks
105
Charlton House Fm
Grovehill Fm
Westbury
Elm Tree Fm
Evenley Park
Evenley Hall
95
Camp Fm
144
Plomers Firs Fm
Rowler
145
Evenley
132
Fish Pond
Astwick Village
131
138
Mixbury Ho
112
Beaumont Castle (rems of)
Cloisters Fm
Mixbury
Barrow
Tumulus
Hill
Croughton
Warren Fm
118
Gravel Pit
120
B 4031
Barley Mow Fm
127
The Slade Fm
136
Mixbury Lodge
The Hulls
125
123
New Buildings
135
Juniper Hill
The Pits
Middle Fm
Mixbury Plantation
Widmore Fm
129
Cottisford Heath
Park Thorns
124
Heath Fm
Coldharbour Fm
Pimlico Fm
Wr Twr
Shelswell Plantation
Village Earthworks
Cottisford
116
Manor Ho
Home Fm
Spilsmere Wood
Tower Fm
144
Rounchill Fm
Tumulus
Tusmore Ho
Shelswell Park
Shelswell
Horwell Fm
124
Wr Twr
Park Fm
Tusmore Park
Fox Covert
41
40
39
38
37
36
35
34
33
32
31
30
53
54
55
56
57
58
59
60
61
62

## 2.2 Bearings

Bearings are angles which are always measured from North in a clockwise direction. They are given as a 3-figure number even if they are less than 100°, for instance

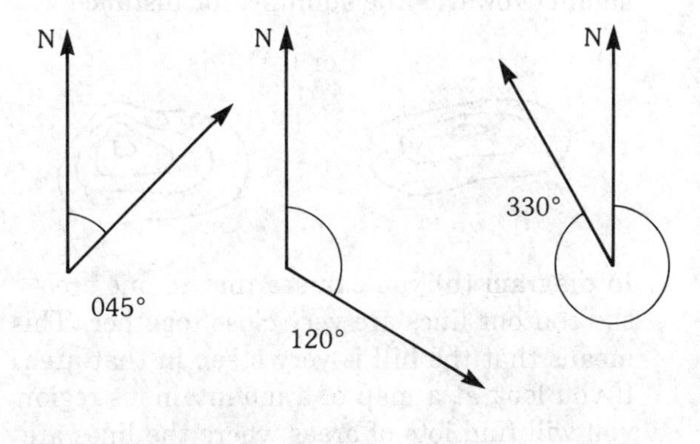


1. Draw diagrams like those above to show bearings of
   a) 075° b) 090° c) 160° d) 180°
   e) 230° f) 295°
2. Measure the bearing of A from B in each of the following cases.

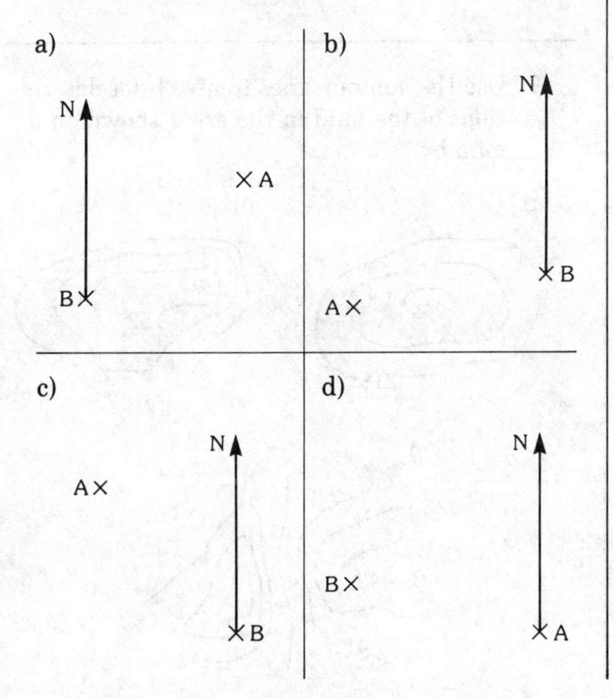


Now look at the map opposite and find the bearing of one place from another on the map. Get your partner to check your work.

Note that bearings do not depend on the scale of the map or on the distances between the places.

# 3. Finding Places on the Map

Most maps have some sort of grid system which enables you to find a place given in an index or to describe the position of a place to someone else. This is usually done through systems of numbers (Ordnance Survey) or mixtures of letters and numbers (town plans).

## 3.1 Grids and Map References

On Ordnance Survey maps such as the one shown opposite a place can be described by giving a 6-figure map reference. The first 2 figures in a map reference give the number of the line along the left-hand edge of the square containing the place in question. The 4th and 5th figures give the number of the line along the bottom edge of the square. The 3rd and 6th figures tell you how far into the square you must go in order to find the place. These are used by imagining that the sides of each square are divided into 10 smaller divisions. So 588347 means 8 imaginary divisions into the square whose left-hand edge is the line 58, and 7 imaginary divisions into the square whose lower edge is the line 34. On the map shown this would be near the road junction in Evenley.

1. Using the OS map opposite describe the places whose map references are
   a) 576335 b) 598368 c) 564307
   d) 558368.

**2.** Give the map references of the following places marked on the OS map on page 14.
a) Warren Farm near Croughton
b) Glebe Farm near Farthinghoe
c) Radstone Church
d) Brackley Grange

## 3.2 Town Plans

Streets are often identified on town plans by describing the square they are in using a combination of either two letters, or a letter and a number as on the map on page 13 (e.g. Northgate Station is in square D3). Books of maps such as motorists' maps or *A to Z*s also give a page number.

**1.** The map index for the map on page 13 includes the following information.

| | |
|---|---|
| Kingsway West | E2 |
| City Road | E3 |
| Watergate St. | D3 |
| Elizabeth Cres. | E4 |

Find these places on the map.

**2.** Find the references for the squares containing the following places.
a) Newton School
b) Cathedral
c) The Castle
d) Alexandra Park, Hoole

# 4. Ups and Downs

On a town map it is impossible to find out whether one place is higher or lower than another. On Ordnance Survey maps, however, you will find a series of lines called *contour lines* which help you to imagine the shape of the ground.

## 4.1 Contour Lines

Contour lines are curves which join places which are the same height above sea level. A hill will therefore appear as a series of closed curves inside each other getting smaller and smaller towards the summit, for instance

(a) or (b)

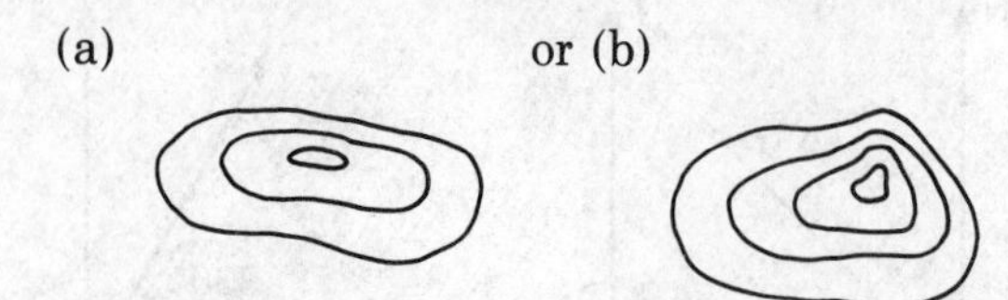

In diagram (b) you can see that in one area the contour lines are very close together. This means that the hill is very steep in that area. If you look at a map of a mountainous region you will find lots of areas where the lines are close together.

On the *Landranger* series of Ordnance Survey maps you will find that going from one contour line to the next higher one gives a height increase of 50 ft (15.24 m). Some of the contour lines have heights in m marked on them.

**1.** Use the contour lines to give brief descriptions of the land in the areas shown on a map by

a) b)

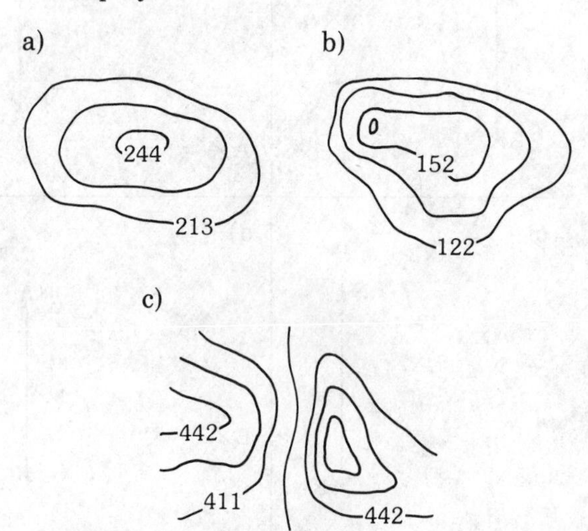

**2.** Draw a set of contour lines to show a long hill with 2 high spots, both about 280 m high.

## 4.2 Gradients

The gradient, or steepness, of a hill can be found by first measuring the distance between consecutive contour lines on the map, and then translating this into real distances. The distance must be measured at right angles to the contour lines. This can be used to find the gradient as in the following example.

On a 1:50 000 map a road crosses at right angles two contour lines which are 2 mm apart. From the scale, 2 mm represents $2 \times 50\,000 = 100$ m. The height difference between the contour lines is 15.24 m (this is standard for OS maps), i.e. the road goes up by 15.24 m for every 100 m along.

The road therefore goes up 1 m for every $100 \div 15.24$ m along, i.e. 1 m in about 7 m (to the nearest m).

The gradient of the road is therefore 1 in 7.

**1.** Find the gradients of roads where the distances between consecutive contour lines on an OS map are
a) 5 mm b) 2.5 mm c) 8 mm.
(Assume roads cross contours at right angles.)

**2.** Find the gradients of the road sections in the diagrams below assuming that the height difference between contour lines is 15.24 m. Measure the distance along the road, not just the distance between the contour lines.

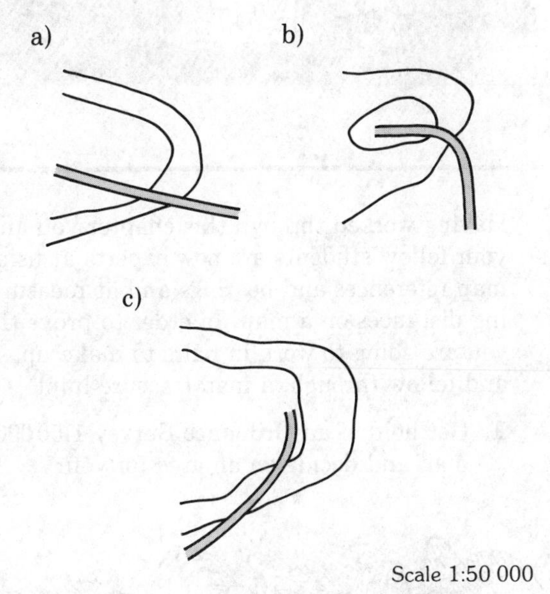


Scale 1:50 000

**3.** In some countries, and increasingly in the UK, gradients are given as percentages. The percentage is found by making the gradients found above into a fraction, and then a percentage, for instance

$$1 \text{ in } 7 \text{ gives } \frac{1}{7} \times 100 \simeq 15\%$$

Write the gradients of the roads in questions **1** and **2** above in percentage form.

Does a high percentage mean a steep road or a gentle incline?

# Planning a Treasure Hunt

Having worked through this chapter you and your fellow students are now experts at using map references and bearings and at measuring distances on a map. In order to prove this you are going to work in pairs to make up, and follow through, a map treasure hunt.

**1.** Get hold of an Ordnance Survey 1:50 000 map and decide on an area for your treasure hunt – about a quarter of the map will do.

**2.** This treasure hunt will be on roads so pick a starting point and decide on which road you are going to set out.

**3.** It's now up to you to devise a treasure hunt using map references, distances and bearings, for example 'start at 335557 and travel 2 miles starting on a bearing of 020°, change to a bearing of 280°', and so on. Decide on a finishing place and reach it using about 20 steps.

There are various ways of setting tasks for the hunt. You could expect people just to get to the correct finishing place, or you could expect them to pick up clues on the way – use your imagination.

**4.** It's now time to test out your route. Swap routes with another couple and work through their route.

There are lots of variations which you can add to this work, for instance run it as a sponsored event, go out and try the route, get people to describe the route from the map.

# Teacher's Notes

**1.** Many people accept the need for mathematics in subjects such as engineering but it is useful to be able to show students how it applies to other areas such as the humanities. This project gives students a chance to remind themselves of things they may have learnt in youth organisations and would also fit in well with a residential course.

**2.** Links with other subject areas include
*science* environmental work on the effect of pollution on an area
*personal development* planning routes, leadership qualities
*information technology* using graphics to draw simple maps
*communication* describing an area, giving directions
*social sciences* the environmental effect of urban sprawl, provision of services to an area.

**3.** This is not the sort of project which can be seen as leading directly into any particular type of employment. It does however fit in with the need to use leisure time wisely and to make the most of the environment.

**4.** Local environmental organisations such as Friends of the Earth and Conservation Corps will be happy to send speakers along to talk about the environment and a knowledge of maps and how to use them will fit in well with this.

**5.** The final assignment in this chapter has been deliberately left fairly open-ended to encourage students to use their initiative. They will of course need equipment such as maps and compasses in order to carry this out.

Chapter 3

# The Post Office

The Post Office is responsible for collecting and delivering mail. However local Post Offices also provide a variety of other services. Leaflets are to be found giving details and advice in all main Post Offices.

You can pick up holiday money at the counter, pay gas and electricity bills or find out how to claim supplementary benefit. It is well worth while finding out exactly which services the Post Offices do provide.

## 1. Postal Services

### 1.1 Expected Times of Arrival

Over 40 million letters a day are sent by post. The Post Office aims to deliver 90% of first class letters on the next day and 96% of second class letters by the third working day after collection. Post is not delivered on a Sunday. There are no collections on a Sunday.

There is a Special Delivery service for mail which has to arrive by the next day. For the price ask at the Post Office counter.

---

By what day would you expect the post to have arrived for

**1.** a second class letter posted on a Monday

**2.** a first class letter posted on a Tuesday

**3.** a second class letter posted on a Friday

**4.** a second class letter posted on a Wednesday

**5.** a first class letter posted on a Saturday

**6.** a second class letter posted on a Thursday

**7.** a first class letter posted on a Sunday

**8.** a Special Delivery letter posted on a Thursday?

## 1.2 Cost of Stamps

First class stamps cost 18 p and second class stamps cost 13 p in 1986.

How much would you pay for

**1.** stamps for five first class and two second class letters

**2.** stamps for ten second class letters

**3.** a booklet containing ten first class stamps

**4.** stamps for a dozen second class postcards

**5.** a sheet of 36 first class stamps

**6.** stamps for two dozen first class and half a dozen second class letters?

## 1.3 Weight and Size

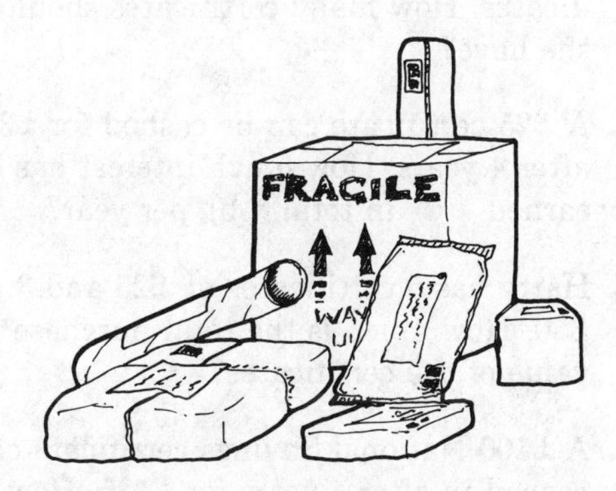


Letters, cards and packets may be sent by *letter post* but there is a limit of weight of 750 g for second class letters.

Parcels may not be more than 610 mm (2 feet) in length, 460 mm (18 inches) in width and 460 mm (18 inches) in depth. Packages in the form of a roll may not be more than 1.04 m (3 ft 5 in) for the length plus twice the diameter combined or 900 mm (2 ft 11 in) for the greatest measurement across. Larger parcels must go by *parcel post*.

Say which of the following are outside the limits for weight or size for the letter post rate, and why.

**1.** a packet measuring 2 feet by 2 feet by 1 foot

**2.** a first class package weighing 800 g

**3.** a roll that is 3 feet long and 6 inches in diameter

**4.** a parcel measuring 500 mm by 400 mm by 200 mm

**5.** a second class packet weighing 650 g measuring 620 mm by 260 mm by 130 mm

**6.** a roll sent first class weighing 900 g and measuring 700 mm long and 150 mm in diameter

## 1.4 Cost of Inland Letters

The cost of postage varies according to the weight and whether the letter or parcel is to be sent first or second class. The 1986 prices are given below.

*Inland letters and cards*

| | 60 g | 100 g | 150 g | 200 g |
|---|---|---|---|---|
| First class | 18 p | 26 p | 32 p | 40 p |
| Second class | 13 p | 20 p | 24 p | 30 p |

In 1986 it would cost 26 p to send a first class inland letter weighing 80 g.

Use the chart on page 21 to give the cost of

**1.** first class at 120 g

**2.** second class at 70 g

**3.** second class at 160 g

**4.** first class at 95 g

**5.** first class at 160 g

**6.** first class at 185 g

**7.** second class at 110 g

**8.** first class at 100 g

**9.** first class at 40 g

**10.** second class at 185 g.

### 1.5 Overseas Mail

The charge for overseas letters is different from the charge for inland letters. There are different rates for three zones and a separate rate for Europe.

Obtain from the Post Office the leaflet giving the postal rates for overseas mail. Use this leaflet to find the cost of sending

**1.** a letter under 20 g to Europe

**2.** a newspaper to the Bahamas

**3.** a postcard to Djibouti

**4.** a letter weighing 120 g to France

**5.** printed papers to the Sudan

**6.** a letter weighing 20 g to Zone C

**7.** a newspaper to Europe weighing between 60 g and 100 g

**8.** a letter under 10 g to India

**9.** a Christmas card containing less than five words of greeting (sent as printed papers) to Canada

**10.** an aerogramme to China.

Have you used air mail or surface rates?

# 2. Savings

## 2.1 National Savings Certificates

One way of saving is to buy National Savings certificates at the Post Office (see opposite).

Certificates can be bought in units of £25. If you keep the certificates for four years they *mature* and you can draw them out for more money than you started with. This extra money is called *interest* and the value of the certificates the *repayment value*. If you leave them with the Post Office for longer they will go on earning interest. The interest is tax free.

There are different 'issues' (nineteenth, twenty-first etc.) which pay different rates. For more details about these, ask at the Post Office counter. Cashing the certificates in early means you don't get such a good return for your money, though you still get some interest. Sometimes it pays to exchange matured certificates for the latest issue.

Some certificates have their best repayment value after five years rather than four.

**1.** Charlotte has £325 worth of £25 certificates. How many certificates should she have?

**2.** A £25 certificate can be cashed for £31.60 after 4 years. How much interest has been earned a) in total b) per year?

**3.** Harry has 5 certificates at £25 and 3 at £50. How much is the total purchase value of the certificates?

**4.** A £100 National Savings certificate can be cashed in after 4 years for £125. How many certificates should Keith buy to be sure of being able to draw out over £900 after 4 years?

Issue 31ST ISSUE **SPECIMEN** Purchase Date 15AUG85

**SAVINGS CERTIFICATE**

*This National Savings Certificate is issued by the Treasury under the National Loans Act 1968 and subject to the Savings Certificates Regulations 1972, as amended, and where applicable, to the Prospectus current at the date of purchase.*

SECRETARY TO THE TREASURY

A1234567

Purchase Value £50.00 Holder's Number 0320 008 12

Number of Units 2 Certificate Number BNB BDBQ 0071

Registered in the name(s) of EDNA MATHEWS

**5.** The fourteenth issue promises to pay £67 on maturity for a £50 certificate. If Elaine has three such certificates how much can she expect to draw out on maturity?

**6.** Hari has three £100 and five £25 certificates. After 4 years a £100 certificate can be cashed for £124.60 and a £25 certificate for £31.15.
Calculate a) the total repayment value after 4 years b) the total interest earned after 4 years.

**7.** A £50 certificate matures after 4 years at £67.00. What would be the repayment value of five £25 certificates after 4 years? (You get the same interest for two £25 certificates as for one £50 certificate.)

**8.** Find out the repayment value of a £25 certificate after 5 years for the latest issue.

## 2.2 National Savings Bank

The disadvantage with National Savings certificates is that you have to leave the money in the Post Office for over a year before any interest is paid. If you want to keep your money in the Post Office and be able to draw it out at any time, but still earn a small amount of interest, then you could open a National Savings Bank ordinary account.

You can open the account with £1 or more at any of over 20 000 savings bank Post Offices. You will be given a bank book and each time you *make a deposit* (put money in) or *make a withdrawal* (take money out) you fill in a form and an entry will be put into your bank book.

Withdrawal forms and deposit forms are available at all Post Offices.

National Savings Bank
NS

ACCOUNT NUMBER 852 601 352
MR. J. SMITH
SIGNATURE John Smith

Ordinary Account
MMM

*Bank book*

| Date of Deposit or Warrant etc. | Amount of Deposit (£s in words) or Method of Withdrawal | £ Th | £ H | £ T | £ U | | Initials |
|---|---|---|---|---|---|---|---|
| July 5th | Demand | = | | 3 | 0 | = | S.L. |
| | BALANCE | = | | 4 | 2 | 42 | |
| Sept 15th | FIFTY ONE POUNDS 76p | = | | 5 | 1 | 76 | R.C. |
| | BALANCE | = | | 9 | 4 | 18 | |
| Oct. 2 | Demand | = | | 4 | 0 | = | H.W. |
| | BALANCE | = | | 5 | 4 | 18 | |

*Page of a bank book*

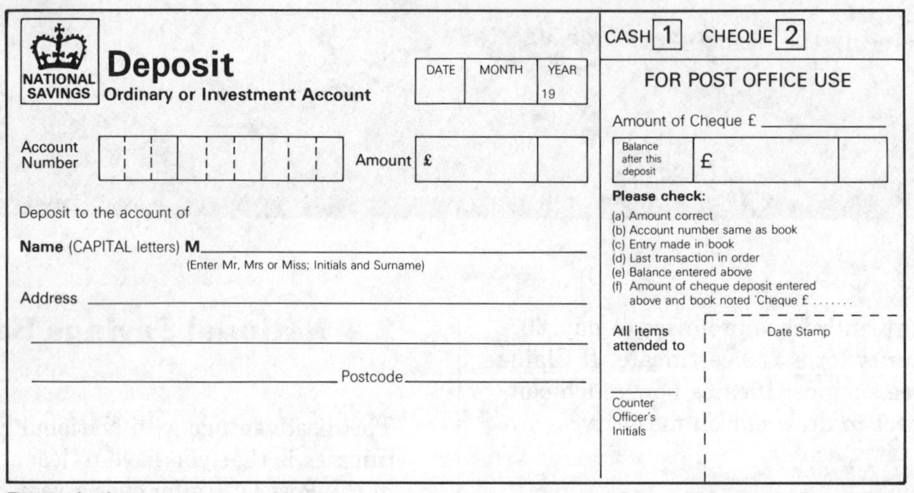

NATIONAL SAVINGS
**Deposit**
**Ordinary or Investment Account**

DATE | MONTH | YEAR 19

Account Number

Amount £

Deposit to the account of

**Name** (CAPITAL letters) **M**
(Enter Mr, Mrs or Miss; Initials and Surname)

Address

Postcode

CASH 1 CHEQUE 2

FOR POST OFFICE USE

Amount of Cheque £

Balance after this deposit £

**Please check:**
(a) Amount correct
(b) Account number same as book
(c) Entry made in book
(d) Last transaction in order
(e) Balance entered above
(f) Amount of cheque deposit entered above and book noted 'Cheque £ .......

All items attended to

Date Stamp

Counter Officer's Initials

*Deposit form*

NATIONAL SAVINGS
Ordinary Account
**Withdrawal on Demand**

DATE | MONTH | YEAR 19

Account Number

Amount £

Amount (in words)

Name (CAPITAL letters) M
(Enter Mr, Mrs or Miss)

Address

Postcode

PLEASE SIGN IN THE PRESENCE OF THE PAYING OFFICER, WHO MAY ASK FOR EVIDENCE OF IDENTITY.
I acknowledge receipt of the above amount

Signature ..........................................

FOR POST OFFICE USE

**3**

Balance after this Withdrawal £

1. Please make sure you see the signature written
2. Please check:
   (a) Signature and account number same as book
   (b) Book security featues in order
   (c) Last deposit transaction in order
   (d) Entry made in book
   (e) Balance entered above
3. Please retain the book if required to

All items attended to

Date Stamp

Paying Officer's Initials

*Withdrawal form*

The diagram opposite shows a page of a bank book.

1. On 2nd October was money put in or taken out of the account?
2. What word is used to show that money has been taken out of the account?
3. How much money was in the account immediately after the deposit of £51.76 on 15th September?
4. How much money must there have been in the account before the withdrawal of £30 on 5th July?
5. What do you think the word 'balance' means?
6. If £32.95 is deposited in the account on 5th October what will the new balance be?
7. Look at an actual withdrawal form and deposit form in your local Post Office and make two sketches showing how you would fill in these forms to show a) a deposit of £3 b) a withdrawal of £1.

## 2.3 Interest

Every so often your bank book will be sent to headquarters and interest will be added to your account. The rate in 1986 was 6% per year if more than £500 was kept in the account all year, and 3% per year otherwise. (See part B, chapter 4 for practice calculations on simple and compound interest).

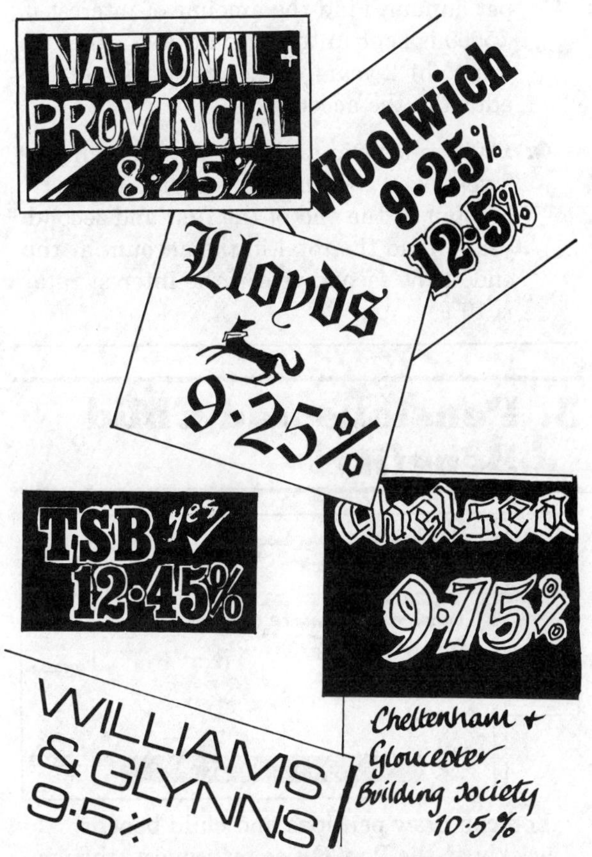


Investment accounts offer a higher rate of interest but you have to give one month's notice in writing in order to take out your money. Some accounts are *index-linked* which means that the rate of interest is related to the cost of goods in the shops. If prices rise then so will your interest. Tax has to be paid on the interest, though there is a tax-free allowance (£70 in 1986) for ordinary accounts.

**1.** If £50 is left in a savings account for 1 year at an interest rate of 3% per annum how much interest would be earned?

**2.** Give the total in the account if £1000 is kept for 1 year at 6% interest per year and the interest is added to the account at the end of the year.

**3.** An investment account pays 8% interest per annum. Find the amount of interest if £5000 is kept in the account for a) 1 year b) 2 years, if the interest is not added to the account.

**4.** £3000 is kept in an investment account for 2 years and the interest is added to the account at the end of the first and second years. Find the total in the account at the end of the second year if the interest rate is 10%.

# 3. Pensions and Child Benefit

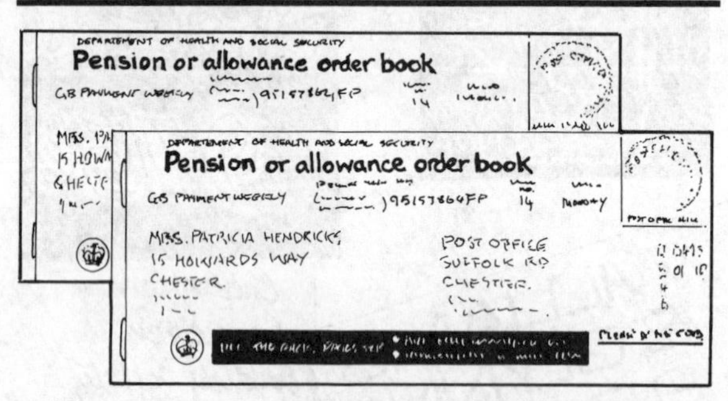


You can draw pensions and child benefit weekly at the Post Office on request, though these can also be paid monthly (every 4 weeks) into a National Savings account or bank account.

Assume child benefit is £7.10 per child per week. (The table in question **6** may be helpful for some questions.)

**1.** Anne has three children. She draws her benefit once a fortnight. How much does she draw at a time?

**2.** Anne's friend, Diane, has two children and she draws her allowance every month. What does she get per month?

**3.** Give the amount of child benefit payable for 5 weeks to a family with 4 children.

**4.** How much child benefit is paid per child per year?

**5.** Find the child benefit payable for 8 weeks to Sucharita who has three children.

**6.** What amounts of money should be in the spaces labelled a), b) and c) in the table below?

*Child benefit – £7.10 per child per week*

| | 1 child | 2 children | 3 children | 4 children |
|---|---|---|---|---|
| 1 week | £7.10 | £14.20 | £21.30 | £28.40 |
| 2 weeks | £14.20 | £28.40 | £42.60 | £56.80 |
| 3 weeks | £21.30 | £42.60 | £63.90 | £85.20 |
| 4 weeks | £28.40 | £56.80 | a)...... | £113.60 |
| 5 weeks | b)...... | £71.00 | £106.50 | £142.00 |
| 6 weeks | £42.60 | £85.20 | c)...... | £170.40 |

**7.** Anne's mother draws a weekly pension of £37.59. How much is this a) per month b) per year?

**8.** Last month Anne's mother's pension rose by 52 p per week. How much is her new pension a) per month b) per year? What is her total increase per year?

# Premium Bonds versus Savings Accounts

Dave and Liz have been saving in accounts at the Post Office. They also spent money on Premium Bonds, which instead of earning interest give them a chance of winning a cash prize. After the Premium Bonds have been held for 3 months they go into a weekly draw. The prize money is tax free.

**1.** Dave holds £300 in Premium Bonds and has £510 in an Ordinary Savings Account which pays 6% interest per year. He wins £50 on the Premium Bonds. How much money does he have in total at the end of the year?

**2.** Liz wonders if she should invest £200 in Premium Bonds or put it into a savings account paying 8% interest tax free. Would she be better off buying Bonds if she was lucky enough to win £25 during the year?

**3.** Dave's brother holds £250 in Premium Bonds but makes no win. He could have invested it at 6.75% interest. How much money has he lost in a year by buying Premium Bonds instead?

**4.** The following year Dave's brother wins £75. Does this mean that he gained or lost over the two years, assuming that he would have reinvested his interest at the end of the first year?

**5.** Dave has £3000 to invest. He considers two schemes.

a) saving in an ordinary account which pays 6% interest per year, where the first £70 of interest is tax free, but 30% tax has to be paid on the rest of the interest

b) saving with an investment account which pays 8% interest but on which tax at 30% has to be paid on the whole amount of the interest.

Compare the amount of money that he will have at the end of one year for each scheme and say which is most advantageous.

**6.** Obtain leaflets for as many different savings schemes as possible and say which you would choose and why.

# Teacher's Notes

**1.** This topic can be made very broad as the Post Office is such a valuable source of information in our community. The Post Office will be pleased to provide information packs on request. For details of educational material published, write to 'The Schools' Officer, RMM 4.2, 33 Grosvenor Place, London SW1X 7PX'.

**2.** Other subjects may be brought in as follows.

*humanities* the history of the development of the postal service, mail coaches, early stamps, first telegraph and airmail services, the Penny Post

*social studies* cost of overseas mail according to zones, difficulties of extending the postal service to less developed areas, different types of transport necessary over different terrain

*information technology* automatic sorting machines, the segregator (used for sorting letters from parcels), the Automatic Letter Facer (ALF)

*community studies* Family Income Supplement leaflets, savings, bills that can be paid at the Post Office counter, travel concessions for the elderly (details are usually available at the Post Office), pensions, child benefit

*office practice* mail offices, franking machines, record books, parcels, packing

*communications* letter writing, newspapers and printed papers etc.

**3.** The *Post Office Guide*, on sale at all main Post Offices, gives full details of all the services available.

Chapter 4

# Furnishing a Flat

If you move into a new house, or just a bedsitter you may well need to furnish and decorate it. This will almost certainly mean that you will have to measure rooms, work out areas and decide where you can fit things. This chapter and the next one will lead you through the calculations needed to do this.

How best to organise a room is a problem which you may also meet at work. An office where several people are working must be set out so that they can move about and get to equipment easily. It is far better to plan this before you start moving filing cabinets around.

## 1. Measurement of Length

### 1.1 Metric Units

The standard metric units of length are

kilometre (km) – used for distances between places

metre (m) – used for measuring large things such as a room

centimetre (cm) – used for measuring medium sized things such as a piece of paper

millimetre (mm) – used for measuring small things such as the thickness of a piece of wood.

The relationships between these units are as follows.

1 km = 1000 m

1 m = 100 cm }

1 cm = 10 mm } 1 m = 1000 mm

For this project you will only be interested in m and cm. Get a metre rule and look at these lengths to get used to them.

Carry out the following calculations.

**1.** Change these lengths from m to cm.
a) 1.6 m b) 22.5 m c) 0.8 m
d) 2.07 m

**2.** Change these lengths from cm to m.
a) 210 cm b) 85 cm c) 3500 cm
d) 6820 cm

**3.** Add together the following giving your answer in suitable units.
a) 87 cm and 113 cm b) 2.07 m and 1.65 m
c) 120 cm and 3.6 m d) 82 cm and 0.5 m

**4.** Subtract the following giving your answer in suitable units.
a) 27 cm from 162 cm
b) 3.18 m from 7.05 m
c) 120 cm from 2.9 m
d) 97 cm from 1.6 m

## 1.2 British (Imperial) Units

The standard units of length are

mile – used for distances between places

yard (yd) – used for shorter distances such as the length of a field

foot (ft) – used for short distances or large objects such as a room

inch (in) – used for medium and small objects such as a book or the thickness of a piece of wood (fractions also used).

The relationships between these units are shown below.

1 mile = 1760 yd
1 yd = 3 ft
1 ft = 12 in
} 1 yd = 36 in

Here you will be using mostly feet and inches, although yards may occasionally be useful.

**1.** Change the following measurements into inches.
a) 4 ft b) 2 ft 8 in c) 5 ft 6 in
d) $2\frac{1}{2}$ ft e) 2 yd f) $4\frac{1}{2}$ yd

**2.** Write the following in feet and inches, for example as 2 ft 4 in.
a) 16 in b) 44 in c) 30 in d) 60 in

**3.** Add together the following giving your answers in suitable units.
a) 8 in and 13 in
b) 3 ft and 2 ft 8 in
c) 2 ft 6 in and 4 ft 6 in
d) 2 ft 8 in and 3 ft 8 in

**4.** Subtract the following giving your answers in suitable units.
a) 6 in from 1 ft 10 in
b) 8 in from 2 ft 4 in
c) 3 ft 5 in from 8 ft 9 in
d) 2 ft 9 in from 6 ft 2 in

## 1.3 Changing Units

As well as being able to work in British and in metric units you need to be able to convert from one set to the other.

The basic conversion factor you need to remember is

1 in $\simeq$ 2.54 cm

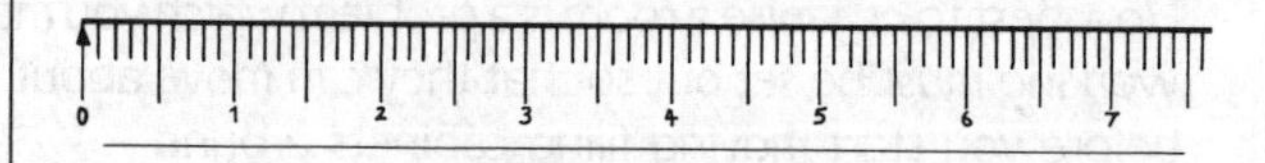


In practice when working with large objects you can use

1 in $\simeq$ 2.5 cm
then 1 ft $\simeq$ 30 cm
and 1 yd $\simeq$ 0.9 m

giving 1 m $\simeq$ 40 in

Using the above approximate conversion factors

**1.** convert the following British measurements into i) cm ii) m, in both cases to the nearest cm.
a) 10 ft b) $12\frac{1}{2}$ ft c) 6 ft 6 in
d) 5 ft 4 in e) 9 ft 10 in f) 8 ft 3 in

**2.** convert the following metric measurements into i) in ii) ft and in, in both cases to the nearest in.
a) 2 m b) 3.5 m c) 80 cm d) 4.2 m
e) 1.65 m f) 2 m 42 cm

## 1.4 Making the Measurements

Measuring length is probably one of the most important mathematical skills and is needed in various walks of life. It is also important to be able to estimate length reasonably accurately. To carry out this next exercise you will need a ruler, a metre rule and a tape measure.

**1.** Discuss with the person next to you, without using any measuring device, the lengths represented by
a) 1 cm b) 1 in c) 1 ft d) 1 m.
Check these estimates against a ruler.

**2.** Estimate the lengths of the following in suitable metric *and* imperial units.
a) your pen
b) your desk (length, width and height)
c) this book
d) your paper
e) your waist
f) your height
g) the length and breadth of the room

**3.** Measure the items in question 2 and compare the answers with your estimates. Do you find it easier to estimate in metric or imperial units?

**4.** Draw a line on a piece of paper and ask 10 people to estimate its length in cm. Find the average of the 10 estimates. Is it near to the correct measurement? Now try the same exercise with pictures like these.

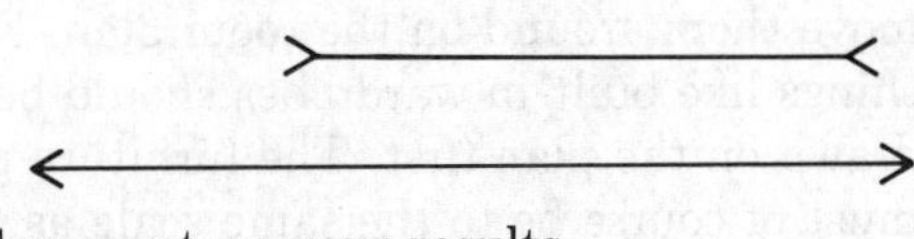

Comment on your results.

# 2. Scale Diagrams and their Use

## 2.1 The Room

In chapter 2 you had to deal with the idea of the scale or scale factor of a map. In this chapter you must choose a suitable scale and draw a scale diagram of a room. First look at an example.

A room is 4 m (400 cm) wide and 3 m (300 cm) long with a window and a door. Drawing to a scale of 1 to 100 would give a scale drawing 4 cm by 3 cm which would be rather small to work with. A scale of 1 to 10 would give a diagram too big to fit on a sheet of A4 paper. Scales of 1 to 20, giving a diagram 20 cm by 15 cm, or 1 to 50, giving 8 cm by 6 cm, would be more sensible. Your choice would depend on how much detail you wanted to put on the diagram. For the diagram below, a 1 to 80 scale has been used.

Suppose that the door is in a shorter wall, is 80 cm wide and that its left hand edge is 100 cm from the corner. The window is 200 cm wide, is on a longer wall and its left hand edge is 80 cm from the corner. For a 1 to 80 diagram you must divide each of these measurements by 80. The scale diagram might then look like this.

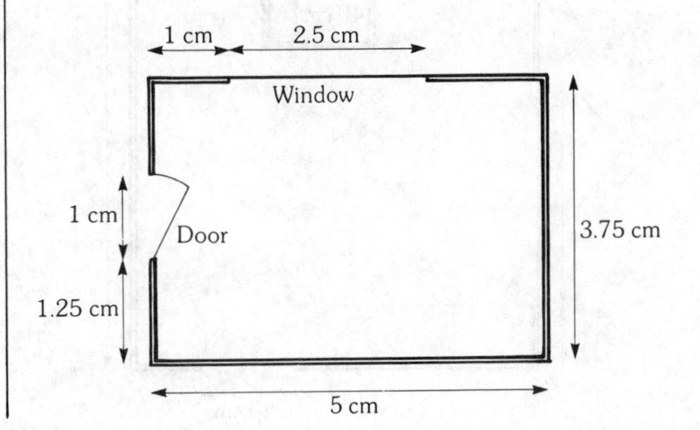

Notice how you show windows and doors (including the direction in which the door opens) on a plan.

---

Draw the following scale diagrams remembering to work out your measurements *before* you start drawing.

**1.** the room in the example above to a scale of 1 to 20. Keep this drawing for future use.

**2.** the rooms shown below to a scale of 1 to 50.

a)

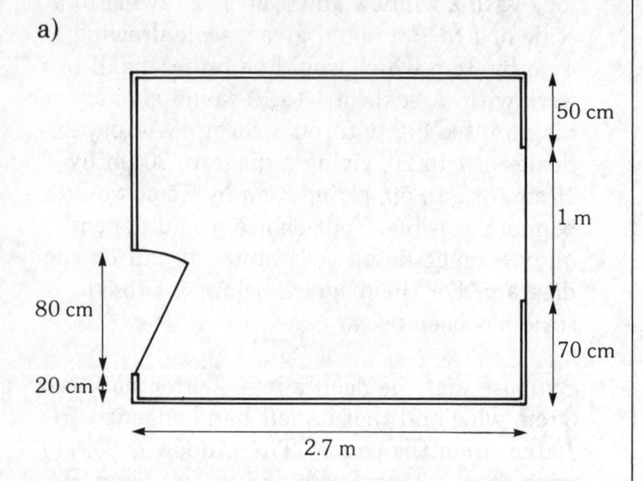


b)

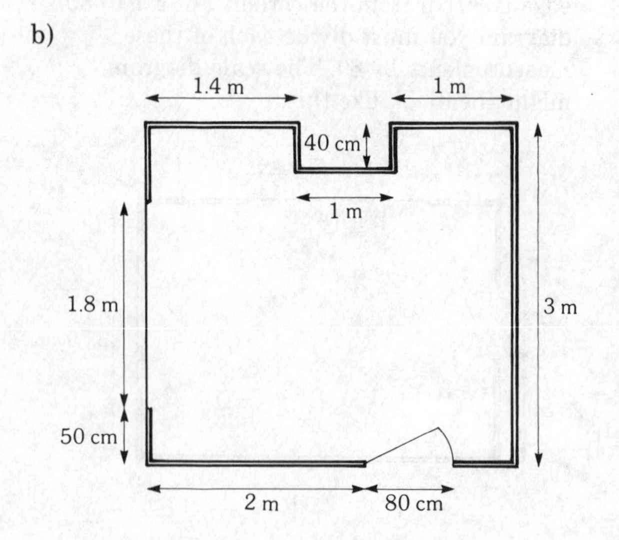


c)

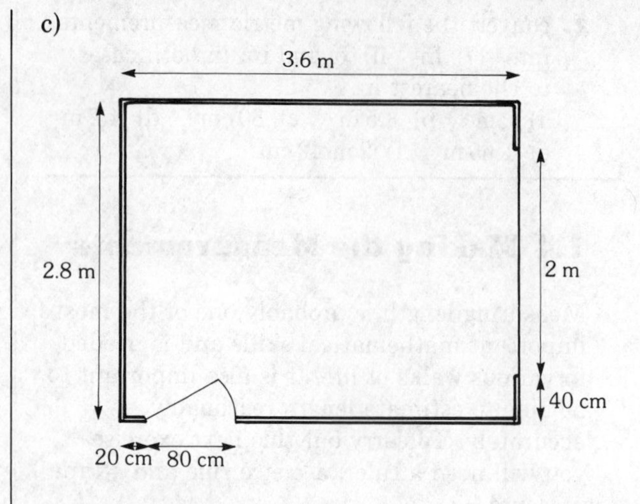


d)

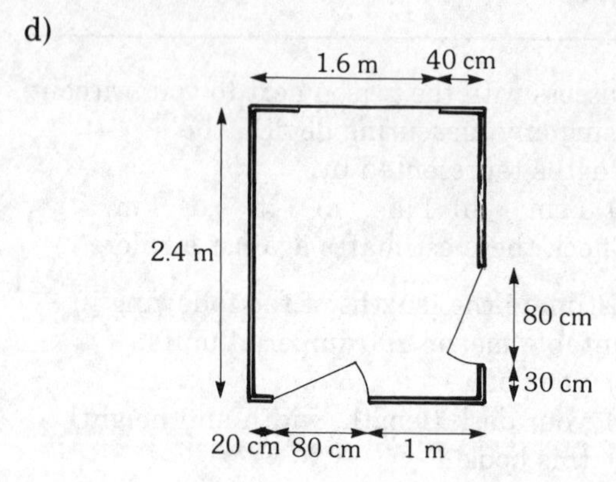


**3.** The room in which you are working to a scale of 1 to 50. First draw a rough plan and mark the actual measurements on it.

**4.** Your bedroom at home to a scale of 1 to 20.

---

## 2.2 Fitting in the Furniture

If you are trying to furnish a flat it may be useful to look first at various possible arrangements on a scale diagram. To do this you need to make cardboard plans of the pieces of furniture available so that you can move them around on the room plan. Fixed things like built-in wardrobes should be drawn on the plan first. The furniture plans must of course be to the same scale as the room plan.

Now assume that the room in question 1 opposite is to be a bedroom. Work through the following questions.

**1.** Draw on cardboard and cut out plans of the following pieces of furniture to a scale of 1 to 20.
   a) two beds each 1 m by 2 m
   b) a wardrobe 90 cm by 50 cm
   c) two chests of drawers each 40 cm by 30 cm
   d) a tall cupboard 80 cm by 50 cm

**2.** Arrange the furniture on your scale diagram in what you think is the best way. Remember to leave room for doors and drawers to open.

**3.** The wall under the window has a radiator up against it. Check that this does not cause problems with your arrangement.

**4.** Draw a sketch of the room showing where you would put the furniture. Compare this with other people's work.

**5.** Try to arrange your own bedroom furniture using a similar method.

The room you have been working with could equally well have been a small office. Copy your 1 to 20 room plan onto another piece of paper and work through the following.

**6.** The room is to be used as an office for 3 people. They will each need a desk and they must also fit in 2 filing cabinets, a table for the telephone and other oddments. The sizes are as follows.
   desk 1.2 m by 0.8 m
   filing cabinet 80 cm by 60 cm
   table 1 m by 80 cm
   
   Draw and cut out plans of each of these items to a scale of 1 to 20.

**7.** Remembering that filing cabinets need to open, telephones need wires and people need chairs and space, arrange these items on a room plan.

**8.** The three people in the room are to be Dave, Lloyd and Adrian. Because of his job Dave tends to use the telephone most. Adrian is the newcomer to the firm who usually gets the odd jobs like taking messages to other rooms. Draw the furniture positions on your plan and put in names to show where each person should sit.

**9.** Compare your solution with the person next to you.

**10.** Find a small office somewhere and look for problems in the room layout.

# Harish's Flat

Harish Mehta has moved away from home because he has got himself a job in Birmingham. He has managed to find a small flat to rent. It contains a small amount of furniture.

Harish's landlord has kindly provided him with a plan of the flat with measurements correct to the nearest 10 cm. The plan is shown below, not to scale.

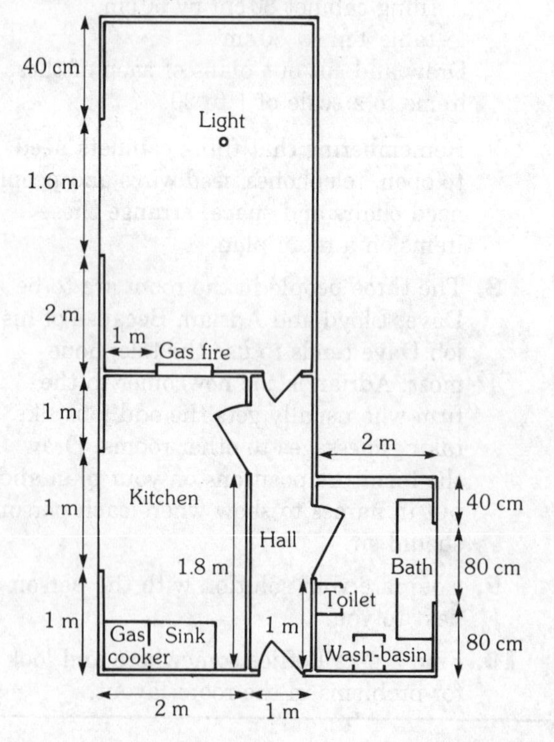


**1.** Using the measurements given draw an accurate plan of the flat to a scale of 1 to 20. You may assume that all doors are 80 cm wide.

**2.** The gas fire, cooker, sink, toilet, wash-basin and bath are fixed. Put these on your plan using measurements taken from examples in your home.

**3.** In the kitchen there is a tall moveable cupboard 80 cm by 40 cm and Harish also want to fit in a small fridge, a table 1 m by 80 cm and two small chairs. By making and rearranging cardboard plans of these mark their positions on your room plan.

**4.** The flat also contains the following.
a low cupboard 150 cm by 70 cm
two armchairs each 80 cm square
a bed 2 m by 1 m
a wardrobe 80 cm by 50 cm
a low coffee table 1 m by 60 cm
Make plans of these and think about how you could arrange them on the room plan.

**5.** Harish is fond of books and he has been given a tall bookcase 90 cm by 30 cm. Try to fit this in with the other furniture.

**6.** However much space you have you always need more! Is there anywhere where Harish could usefully fit a set of shelves on the wall?

**7.** Is there anything that you would want to fit in or change if this was your flat? If so, decide where to make the changes you want.

**8.** Finish off your plan by drawing in the positions of the various pieces of furniture. Is there any room for Harish himself?

# Teacher's Notes

**1.** This project is basically an exercise on units and scale drawing. Possible extensions could include the costing of the furniture for a flat, designing working surfaces in a kitchen and building a three-dimensional model of the flat and furniture.

**2.** Other subject areas could be brought in as follows.

*practical skills* making a model flat, designing suitable furniture and making it; cooking for a single person, looking after oneself

*information technology* computer aided design, graphics, modern office equipment

*social studies* availability and need for various types of accommodation, the problems of living alone, safety in the home and at work

*communications* survey of household equipment and its cost.

**3.** Kitchen design centres often use computers to design kitchens to customers' specifications. A visit to one of these could be useful. Large furniture shops also have areas fitted out as kitchens or bedrooms and would be worth a visit.

Chapter 5

# Decorating a Flat

These days many people do their own home decorating and maintenance, which involves skills such as plastering, paper hanging and painting. Many people are also employed to do this sort of work, both in new houses and for people who do not want to do it themselves.

This chapter covers the basic calculations needed in order to decorate your own or someone else's home. When planning to decorate it is always important to try to estimate the amounts of different materials you will need.

## 1. Perimeter and Area

In order to estimate the materials needed to decorate a room you will need to be able to work out two things.

(a) *the perimeter* – distance around the edge

(b) *the area* – a measure of the amount of space on a surface

### 1.1 Units of Area

In the last chapter you will have worked with both metric and imperial units of length.

Whatever unit is used to measure length, area is measured in *square units,* as shown below.

| *Unit of length* | *Unit of area* |
|---|---|
| in | square in or $in^2$ |
| m | $m^2$ |

There are also special measures used for land areas, such as

metric : 1 hectare = $10\,000\,m^2$
imperial : 1 acre = $4840\,yd^2$

In measuring up a room for decoration you are most likely to need $m^2$, $yd^2$ or $ft^2$.

## 1.2 Areas of Standard Shapes

Most areas which you will need to calculate for decorating will be made up of certain basic shapes.

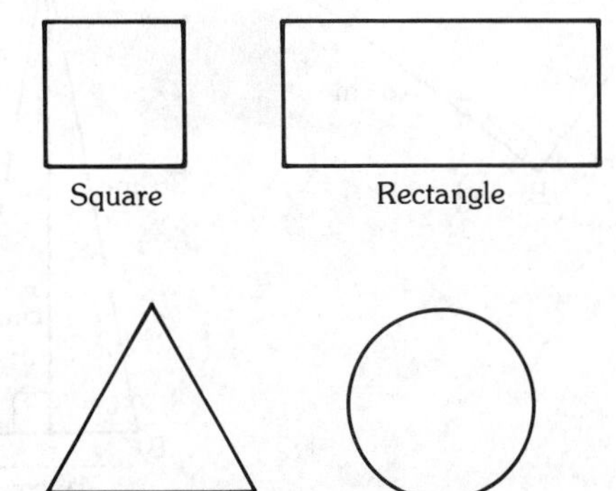


There are standard formulae, which can be used to find the areas of each of these shapes. These are

*square or rectangle*
area = length × width = $lw$
*triangle*
area = $\frac{1}{2}$ base × height = $\frac{1}{2}bh$
*circle*
area = $\pi$ × radius × radius = $\pi r^2$,
where $\pi$ is about 3.14

---

Work out the areas of the following basic shapes. Make sure that you give the correct units in your answers.

**1.** a square of side 8 m

**2.** a rectangle 8 in by 5 in

**3.** a rectangle 4 ft 6 in by 8 ft 6 in (answer in $ft^2$)

**4.** a triangle with base 10 cm and height 8 cm

**5.** a triangle with base 7 ft and height 4 ft

**6.** a circle with radius 7 cm (answer to the nearest $cm^2$)

**7.** a circle with radius 10.5 in (answer to the nearest 10 $in^2$)

**8.** a semicircle with radius 4 ft 6 in (answer in $ft^2$ to the nearest $ft^2$).

---

## 1.3 Perimeter

For most shapes the perimeter is easy to find – you just have to measure the lengths of each of the edges and add them up.

For the circle it is a bit more difficult and you have to use a standard formula.

the perimeter of a circle (called the *circumference*)
= $\pi$ × twice the radius
= $\pi$ × diameter = $\pi d$

---

Find the perimeter of each of the shapes given in questions 1–3 and 6–8 of section 1.2 above.

---

## 1.4 Right Angles and Pythagoras

An angle of exactly 90° is normally called a *right angle* and a triangle with one angle equal to 90° is a *right-angled triangle*. There is a simple result which connects the length of the 3 sides of a right-angled triangle which mathematicians have known and used for thousands of years – this is usually known as Pythagoras' Theorem.

In a right-angled triangle the side opposite the 90° angle is called the hypotenuse and Pythagoras' Theorem states that

the square on the hypotenuse equals the sum of the squares on the other two sides

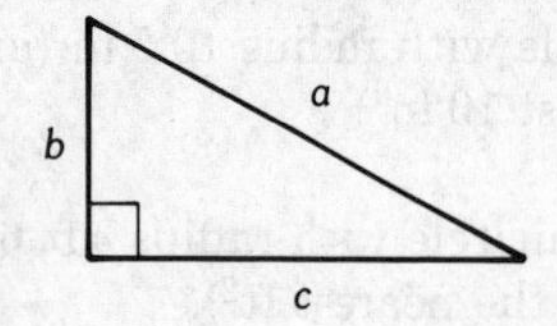


In the diagram above, Pythagoras' Theorem gives

$$a^2 = b^2 + c^2$$

($a^2$ means $a \times a$ of course)

So if $b = 6$ cm and $c = 10$ cm

$$a^2 = 6^2 + 10^2$$
$$= 36 + 100$$

then $a^2 = 136$

To find $a$ you need to find the square root, so

$$a = \sqrt{136} = 11.7 \quad \text{to 1 decimal place}$$

to do this on a calculator you need the $x^2$ and $\sqrt{\ }$ keys

Sometimes Pythagoras' Theorem needs some rearranging, for instance using the diagram above, find $b$ if $a = 12$ cm and $c = 7$ cm.

$$12^2 = b^2 + 7^2$$

giving $b^2 = 12^2 - 7^2$

$$= 144 - 49 = 95$$

so $b = \sqrt{95} = 9.7$ to 1 decimal place

**1.** Using the diagram above fill in this table.

| $a$ | $b$ | $c$ |
|---|---|---|
| i)...... | 3 cm | 4 cm |
| ii)...... | 5 cm | 12 cm |
| 10 cm | 6 cm | iii)...... |
| 12 cm | iv)...... | 5 cm |
| v)...... | 11 cm | 4 cm |

**2.** Find the value of $x$ for each of the following diagrams.

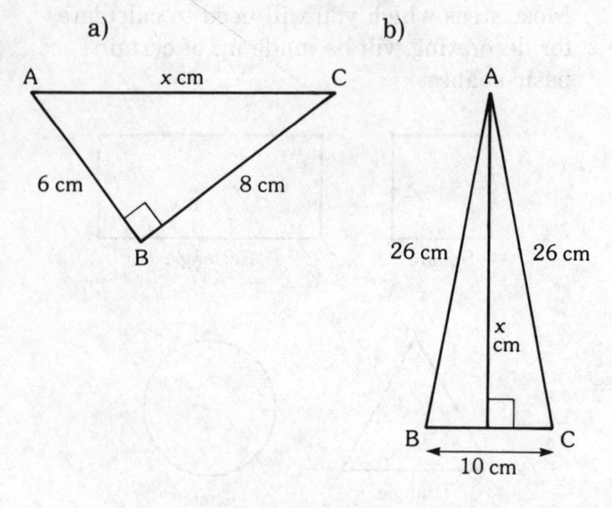


**3.** Find the area of the triangle ABC in each of the cases in question **2.**

**4.** A rectangle has adjacent sides of length 10 m and 6 m. Find the distance from one corner to the opposite corner (the length of a diagonal).

## 1.5 Composite Shapes

The floor plan of a room is rarely one of the standard shapes but is usually made up of a mixture of two or more of them. You therefore need to be able to work out the areas and perimeters of more complicated shapes by splitting them up into standard shapes, for instance

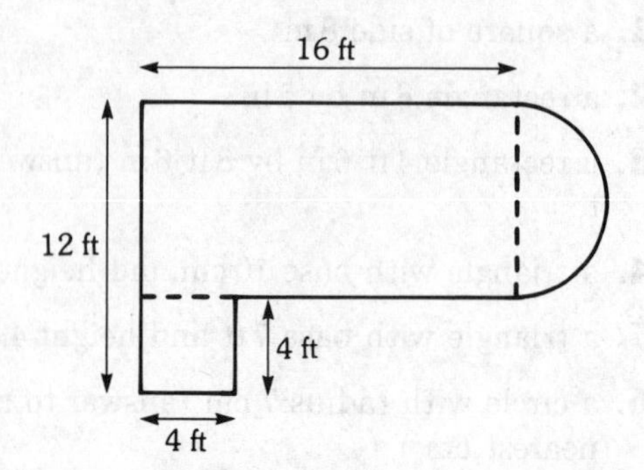

This problem can be dealt with by splitting the area up into 3 sections as shown by the dotted lines.

$$\begin{aligned} \text{area} &= \text{large rectangle} + \text{small square} + \text{semicircle} \\ &= 16 \times 8 + 4 \times 4 + \tfrac{1}{2}\pi \times 4^2 \\ &= 128 + 16 + 25 = 169\ \text{ft}^2 \quad \text{to the nearest ft}^2 \end{aligned}$$

$$\begin{aligned} \text{perimeter} &= 16 + 12 + 4 + 4 + 12 + \tfrac{1}{2}\pi \times 8 \\ &= 48 + 13 = 61\ \text{ft} \quad \text{to the nearest ft} \end{aligned}$$

---

Work out the areas and perimeters of the following floor plans.

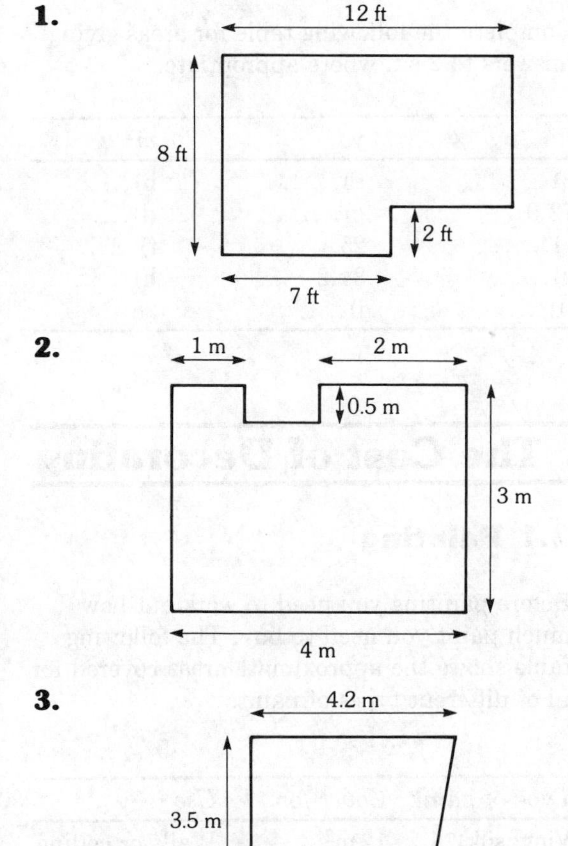


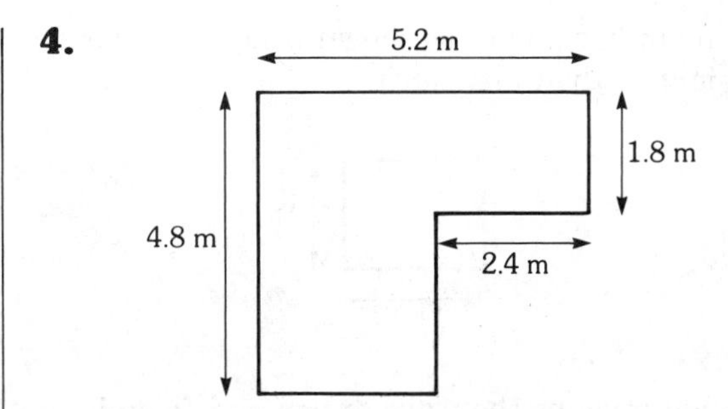


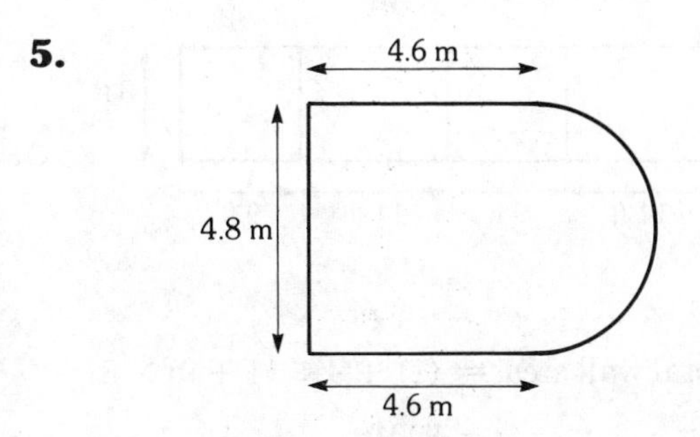


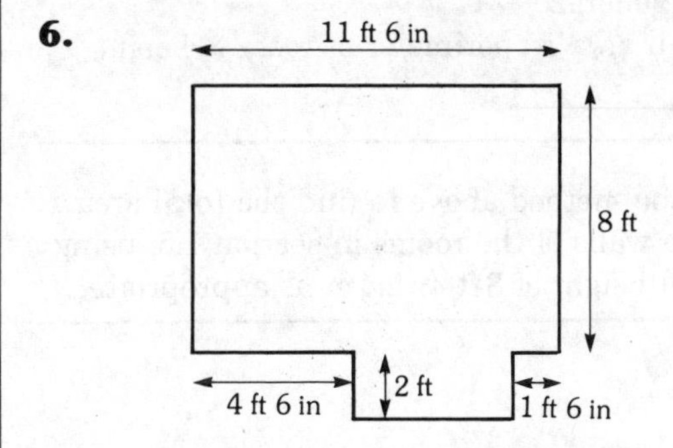


---

## 1.6 Area of the Walls of a Room

When decorating a room it may be necessary to work out the area of the walls of the room. This can be done either by working out the area of each wall individually and then adding the areas, or by using the following method.

A room has a plan as shown below and the height of the walls is 8 ft.

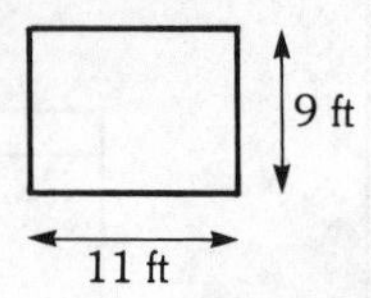


If you imagine the walls drawn end to end they would form a long strip as shown.

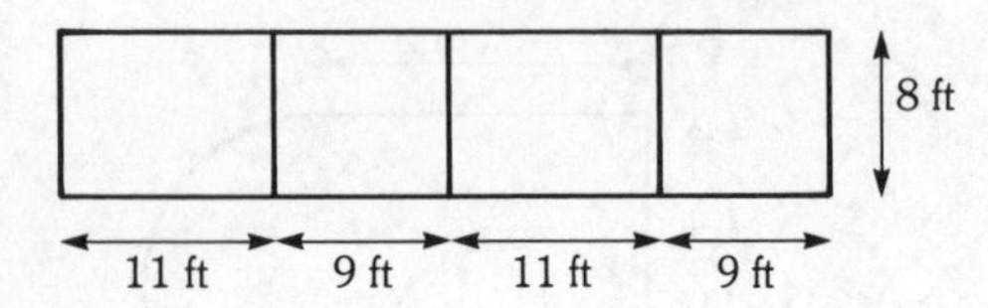


Then

$$\text{total wall area} = (11 + 9 + 11 + 9) \times 8 = 320\,\text{ft}^2$$

in general:
wall area = perimeter of room × height

---

Use the method above to find the total area of the walls of the rooms in section 1.5, using a wall height of 8 ft or 2.4 m as appropriate.

---

### 1.7 Unit Conversion

Sometimes an object is measured in one unit, and then the measurement needs to be converted into another unit. In the case of lengths this is dealt with in chapter 4 but you may also need to convert areas from one unit to another, such as $\text{ft}^2$ to $\text{yd}^2$, or $\text{yd}^2$ to $\text{m}^2$. For instance,

A room is measured as 12 ft × 9 ft.
The floor area is therefore 108 $\text{ft}^2$.
The floor area can also be written as 4 yd × 3 yd.
Then the area becomes 12 $\text{yd}^2$.

Notice that the area in $\text{ft}^2$ is the area in $\text{yd}^2 \times 9$, i.e.

$$1\,\text{yd}^2 = 9\,\text{ft}^2$$

This is part of the general result that the unit conversion factor for area is the *square* of the unit conversion factor for length.

$$1\,\text{ft} = 12\,\text{in} \Rightarrow 1\,\text{ft}^2 = 144\,\text{in}^2$$

$$1\,\text{m} = 100\,\text{cm} \Rightarrow 1\,\text{m}^2 = 10\,000\,\text{cm}^2$$

Now 1 yd is approximately equal to 0.91 m, so

1 $\text{yd}^2$ is about 0.84 $\text{m}^2$
1 $\text{ft}^2$ is about 0.093 $\text{m}^2$

---

Complete the following table for areas giving answers to 2 s.f. where appropriate.

| *ft²* | *yd²* | *m²* |
|---|---|---|
| 81 | a)...... | b)...... |
| 72.9 | c)...... | d)...... |
| e)...... | 25 | f)...... |
| g)...... | 34.2 | h)...... |
| i)...... | j)...... | 28 |

## 2. The Cost of Decorating

### 2.1 Painting

Before painting you need to work out how much paint you need to buy. The following table shows the approximate areas covered for 1 l of different types of paint.

| *Type of paint* | *Cover for 1 l* | *Use* |
|---|---|---|
| Vinyl silk or matt | 12 $\text{m}^2$ | Walls or ceiling |
| Undercoat | 11 $\text{m}^2$ | Woodwork |
| Gloss | 17 $\text{m}^2$ | Woodwork |

1. Using the wall areas calculated in section 1.6 and taking off 4 m$^2$ in each case for doors and windows, find out how many litres of paint would be needed to put *two coats* of vinyl paint on the walls of each of the rooms in section 1.5.

2. Paint can be bought in 1 l and 2.5 l cans. By getting costs from a local shop find the cost of painting the walls of each of these rooms.

3. For each of the rooms above use the area of the ceiling (the same as that of the floor already calculated) to find out how many litres would be needed for 2 coats on the ceiling, and the cost of this paint.

4. A decorating magazine lists the following figures for gloss and undercoat.

   1 door needs about 0.15 l
   1 average window needs about 0.15 l
   11 m of skirting board needs about 0.15 l

   Assuming that the rooms have 1 door, 1 average window and skirting boards find out how much undercoat and how much gloss paint would be needed for each room in section 1.5.
   Undercoat and gloss are sold in 1 l tins. How much would these paints cost for each of the rooms?

5. Using the figures worked out above find the total cost of paint for redecorating each of the rooms giving your answer to the nearest £1.

## 2.2 Papering

If you are going to put wallpaper on the walls you need to know how many rolls to buy. This will obviously depend on the size of the room. A decorating magazine gives the following figures for a room of normal height.

| *Perimeter of room (ft)* | 40 | 44 | 48 | 52 | 56 | 60 | 64 | 68 | 72 | 76 |
|---|---|---|---|---|---|---|---|---|---|---|
| *Number of rolls needed* | 5 | 6 | 6 | 7 | 8 | 8 | 9 | 9 | 10 | 10 |

1. Use the table above and the perimeters already found to calculate the number of rolls of wallpaper needed for each of the rooms in section 1.5.

2. Choose a wallpaper you like in a local shop and find the cost of papering each room.

## 2.3 Carpets

If you are redecorating a room you may want to fit a new carpet as well. The cost of carpeting will depend on the floor area of the room. Carpet is usually priced per yd$^2$.

1. Using the areas already calculated and converting them to yd$^2$ if necessary, find the cost of carpeting each of the rooms in section 1.5 assuming that the carpet will cost £10.95 per yd$^2$. Give your answers to the nearest £10.

2. Look around local shops and find out the costs per yd$^2$ of the cheapest carpets suitable for a living room. Update your answers to question **1** in view of these new costs.

# Decorating Harish's Flat

Having been in his flat (see chapter 4, page 34) for a while, Harish has decided that he would like to decorate his bedsitting room and hall. He wants to paper the walls of the bedsitting room but to paint the walls in the hall. For convenience he decides to use the same colour for the woodwork throughout.

**1.** The two ceilings are to be painted using the same colour of vinyl matt paint. From the plan work out the total area of the two ceilings in $m^2$.

**2.** If 1 l covers 12 $m^2$ and the paint can be bought in 1 l and 2.5 l cans, work out how much paint would be needed for 2 coats on the ceilings. Find the cost of this paint from a local shop.

**3.** If the walls are 2.2 m high, work out the total area of the two long walls of the hall and the amount of paint which would have to be bought for two coats on these walls. Find the cost of the paint.

**4.** Work out the perimeter of the bedsitting room in ft and by using the table given in this chapter (page 41) find how many rolls of wallpaper would be needed to paper it. Choose a suitable paper in a local shop and cost it.

**5.** In total there are 4 doors and 1 average sized window to be painted with 1 coat of undercoat and 1 coat of gloss. How much of each type of paint would be needed for these?

**6.** Find the total length of skirting board for the room and hall and work out how much paint would be needed (see page 40) if 1 coat of undercoat and 1 of gloss are to be used.

**7.** From questions **5** and **6** work out how much undercoat and how much gloss would have to be bought. Cost these in your local shop.

**8.** Using your figures from questions **2, 3, 4** and **7** above, find the total cost of decorating the flat to the nearest £1.

**9.** Having decorated the flat Harish decides that the floor coverings are a bit shabby and that he needs some cheap carpet. The total floor area will be the same as the ceiling area in question **1** above. Use this figure to find the cost of carpeting the bedsitting room and hall using the same carpet if it costs £6.95 per $yd^2$.

**10.** Finally, each of the three rooms needs new curtains. Each window is 1.2 m high and a local chain store is advertising curtains at the following prices per pair.

| *Window height* (m) | *1.0* | *1.2* | *1.4* |
|---|---|---|---|
| *Window width* (m) | | | |
| 0.8 m | £4.95 | £5.95 | £6.95 |
| 1.0 m | £6.95 | £7.95 | £8.95 |
| 1.2 m | £8.95 | £9.95 | £10.95 |
| 1.4 m | £10.95 | £11.95 | £12.95 |
| 1.6 m | £12.95 | £13.95 | £14.95 |

Using this table find the total cost of curtains for the three rooms.

# Teacher's Notes

**1.** This topic links in well with a variety of other skills, both practical and classroom-based. It not only includes several basic mathematical techniques but can be easily used as the mathematical part of a general vocational project.

**2.** Extensions to other subject areas could include

*science* materials used in decorating – paints, pastes, etc.

*social sciences* local housing conditions, substandard housing and slums

*career development* jobs available in the construction and decorating business

*communications* colour and decor; the problems of living alone.

**3.** On vocational and pre-vocational courses this topic could be tied in with caring, construction, retail and distribution options. The topic is, of course, of value in its own right as part of preparations for life. College or school art departments could also be brought in for discussions on colour schemes, fabric design, etc.

Chapter 6

# Running a Car

With bus services getting fewer, particularly for those living outside big towns, many people find that owning a car is almost a necessity. Running a car is quite a costly business and this project looks at a variety of problems associated with motoring, particularly the cost of being a car owner.

Many people also have to drive as part of their job and this project may therefore be of use to you if you are one of those people.

## 1. Useful Units

As in many other aspects of life, motoring brings you into contact with problems involving both metric and imperial units. It is therefore important to understand the units used and to be able to work with them.

### 1.1 Distance, Speed and Time

In the UK, distance between places is almost always measured in miles, but in many other countries the kilometre (km) is used. It is important to know the connection between them.

$$1 \text{ mile} = 1.609 \text{ km}$$
$$\text{or} \quad 1 \text{ km} \simeq \tfrac{5}{8} \text{ mile}$$

**1.** Using the information given complete the following table.

| *miles* | 0 | 10 | 20 | 30 | 40 | 50 | 60 | 70 | 80 |
|---|---|---|---|---|---|---|---|---|---|
| *km* | 0 | | | | | | | | |

Now draw a conversion graph to show the relationship between miles and km. Use your graph to find the equivalents in miles of

a) 60 km  b) 100 km  c) 72 km.

Units of time are of course the same wherever you are, i.e. the hour, minute and second.

Speed is therefore usually measured in miles per hour (m.p.h.) or km per hour (km/h).

**2.** Use your conversion graph to find the equivalents in km/h of the following speeds given in m.p.h.
a) 30 m.p.h. b) 40 m.p.h.
c) 50 m.p.h. d) 70 m.p.h.
These speeds are the standard speed limits on British roads – what are the equivalent speed limits in, say, France?

Speed, distance and time are connected by the formula

$$\text{average speed} = \frac{\text{distance}}{\text{time}}$$

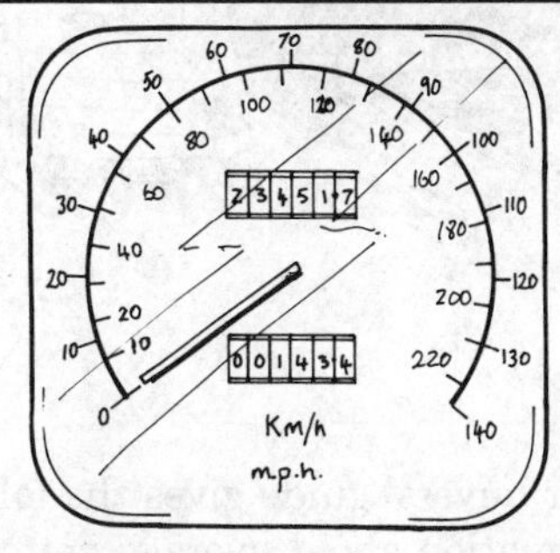


**3.** Use the formula given to complete the following table.

| *Average speed* | *Distance* | *Time* |
|---|---|---|
| 60 m.p.h. | 120 miles | a)...... |
| 50 m.p.h. | 140 miles | b)...... |
| 70 m.p.h. | 290 miles | c)...... |
| d)...... | 360 miles | 6 h |
| e)...... | 150 miles | $2\frac{1}{2}$ h |
| f)...... | 40 miles | 40 min |
| 60 m.p.h. | g)...... | $2\frac{1}{2}$ h |
| 40 m.p.h. | h)...... | $4\frac{1}{4}$ h |
| 30 m.p.h. | i)...... | 40 min |

**4.** To go from my house to central London I would have to travel 50 miles on the motorway and 24 miles on other roads. If my average speed on the motorway is likely to be 60 m.p.h. and on other roads 30 m.p.h., how long should my journey take?

## 1.2 Petrol Consumption

Petrol is priced and sold in one of two units, gallons (g), and, increasingly in the UK, litres (l).

The connection between them is given below.

$$1\text{ g} = 4.546\text{ l} \quad \text{or} \quad 1\text{ l} \simeq 1\tfrac{3}{4}\text{ pt}$$

**1.** Using the connection between gallons and litres given, complete the following table and draw a g/l conversion graph.

| *g* | 0 | 2 | 4 | 6 | 8 | 10 | 12 | 15 |
|---|---|---|---|---|---|---|---|---|
| *l* | 0 | | | | | | | |

Use your graph to find the equivalents in gallons of the following.
a) 20 l b) 30 l c) 45 l

**2.** Petrol prices are usually quoted in £ per gallon or p per litre. Work out the p per l prices of the following.
a) £1.80 per g b) £1.92 per g
c) £1.94 per g d) £2 per g

Petrol consumption figures are usually quoted on advertisements for new cars. These figures are given in miles per gallon (m.p.g.). They can be worked out by dividing the number of miles travelled by the number of gallons used.

**3.** Find the average petrol consumption of a car in the following situations.
a) 200 miles of regular motoring on 6 gallons of petrol
b) long journey of 300 miles on 7 g
c) 250 miles of motorway driving on 7 g

**4.** By looking at new car advertisements or at car test results you will see that the petrol consumption of a car depends on the speed at which it is being driven, and how much stopping and starting is involved, for instance

| *Speed* (m.p.h.) | *30* | *56* | *75* |
|---|---|---|---|
| Renault 5 (m.p.g.) | 48.7 | 68.9 | 50.4 |
| Citroen BX (m.p.g.) | 36.7 | 50.4 | 37.7 |

Using the same piece of graph paper draw a graph of petrol consumption against speed for each of these cars.

Look at some newspapers or manufacturers' brochures to get the figures for 3 more cars (preferably British) and add them to your graph.

**5.** A car buyers' guide gives the following information about average petrol consumption.

| *Model* | | *Consumption (m.p.g.)* |
|---|---|---|
| Metro | 1.0 l | 43 |
| | 1.3 l | 39 |
| Fiesta | 1.1 l | 41 |
| | 1.3 l | 36 |
| Nova | 1.1 l | 44 |
| | 1.2 l | 42 |
| | 1.3 l | 41 |
| Uno | 900 cc | 47 |
| | 1.1 l | 41 |

Put this information on to a graph of consumption against engine size, marking each make of car with a separate symbol. Can you see trends on your graph? Which is the most economical make of 'small' car? Find some more figures for other models and see how they compare.

# 2. The Cost of a Car

Before you buy a car it is a good idea to work out whether or not you can afford it. As well as looking at the cost of buying a car you must also look at the general running costs, including the cost of petrol. Being a car owner is not cheap.

## 2.1 Buying a Car

The cost of a car can be anything between £50 and £90 000 depending on what you want in terms of luxury, reliability, looks, speed and so on. Whatever the price, you are likely to pay for your car in one of three ways.

(a) by paying cash

(b) by taking out a loan, sometimes from a firm which is associated with the car dealer

(c) by a series of regular payments to the dealer ('easy terms')

Method (b) is probably the most common and also the most expensive in the long run, because both the dealer and the firm lending you the money must make a profit on the deal. For either method (b) or (c) the normal system is that you pay a deposit (sometimes your old car is taken as the deposit) and regular monthly payments.

The table below shows figures taken from advertisements in local papers.

| | *Price* | *Deposit* | *Monthly payments* |
|---|---|---|---|
| Colt 1200 | £5265 | 20% | 24 at £193.05 |
| Nova 1.2 | £4670 | £500 | 48 at £130.33 |
| Renault | £1995 | 25% | 36 at £60.26 |
| Cortina | £3358 | £599 | 36 at £126.08 |
| Nissan | £4240 | £848 | 48 at £107.41 |

Extend the table above by working out in each case

a) the total easy terms price, i.e.

deposit + total of instalments

b) the extra paid on easy terms as a % of the basic price, i.e.

$$\frac{\text{extra on easy terms}}{\text{basic price}} \times 100$$

c) the extra cost on easy terms as a % of the balance left after the deposit has been paid, i.e.

$$\frac{\text{extra on easy terms}}{\text{basic price} - \text{deposit}} \times 100$$

If you wanted to buy a car and did not have enough money to pay cash then obviously you would have to get a loan or pay on easy terms.

If you could pay cash then you need to work out whether it is worth investing your money and still paying by easy terms. This is quite a complicated calculation based on the answer to c) above.

## 2.2 Petrol Costs

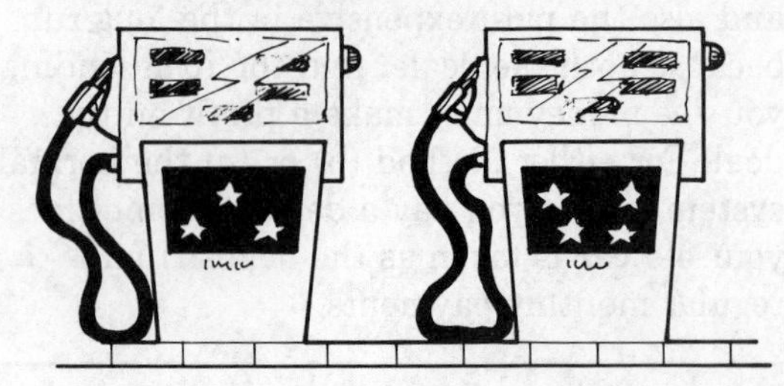

For most drivers, one of the most expensive parts of owning a car is paying for the petrol. How much you have to pay out will depend on how many miles you travel, the petrol consumption of your car and, of course, the price of petrol.

**1.** Assuming that petrol costs £1.75 per gallon, complete the following table giving the yearly expenditure on petrol to the nearest £.

| *Average yearly mileage* | *Average consumption* | *Average cost* |
|---|---|---|
| 8000 miles | 40 m.p.g. | a)...... |
| 10 000 miles | 36 m.p.g. | b)...... |
| 10 000 miles | 28 m.p.g. | c)...... |
| 14 000 miles | 32 m.p.g. | d)...... |
| 11 000 miles | 24 m.p.g. | e)...... |

Now add 4 more lines to this table from information given by people you know. Ask them what their average yearly mileage and average petrol consumption are and complete the table.

**2.** Petrol costs must of course be compared with the cost of other means of travel. If a car does about 32 m.p.g. on average and petrol costs £1.75 per gallon, find the petrol costs for each of the following journeys.
a) 8 miles b) 30 miles c) 80 miles
d) 300 miles
Find some places which are these distances away from your town and find out how much it would cost to get there
i) by bus ii) by train (if it is possible).

## 2.3 Other Costs

Apart from the petrol cost there are various other items which make up the cost of running a car. These include

| | |
|---|---|
| insurance | – depends on age of driver, type of car, etc. |
| motor tax | – a fixed amount, £100 per year in 1987 |
| depreciation | – measures the loss in value of the car |
| repair costs | – depends on how much you do yourself. |

**Insurance**
The following table shows approximate insurance costs for an 18-year-old driver wishing to insure a car for the first time. It has been assumed that he or she will be the only driver and that he or she lives in a medium sized town.

| | *Age of car* | |
|---|---|---|
| | *2 years* | *5 years* |
| Allegro 1100 | £295 (£175) | £278 (£165) |
| Escort 1100 | £245 (£147) | £234 (£138) |
| Mini GT | £332 (£196) | £313 (£185) |
| Capri 1600 | £395 (£234) | £373 (£221) |

| | *Age of car* | |
|---|---|---|
| | *7 years* | *10 years* |
| Allegro 1100 | £248 (£147) | £234 (£138) |
| Escort 1100 | £208 (£123) | £195 (£116) |
| Mini GT | £278 (£165) | £263 (£155) |
| Capri 1600 | £332 (£196) | £313 (£185) |

The main figures given are for a 'fully comprehensive policy with a £50 voluntary excess' and those in brackets are for a 'third party, fire and theft' policy.

**1.** Using the table on the opposite page find the insurance costs for the following cars.
a) a 5-year-old Allegro 1100, third party fire and theft only
b) a 2-year-old Mini GT, fully comprehensive
c) a 7-year-old Escort 1100, fully comprehensive
d) a 10-year-old Capri 1600, third party, fire and theft

**Depreciation**

A new car loses a great deal of value in the first year but an older car tends to lose less of its value. A motoring organisation gives the following figures as a guide to depreciation costs.

| | *Age of car* New | *Age of car* 1 year old | *Age of car* 2 years old |
|---|---|---|---|
| Percentage of beginning of year value lost in a year | 20% | 18% | 15% |

| | *Age of car* 3 years old | *Age of car* More than 3 years old |
|---|---|---|
| Percentage of beginning of year value lost in a year | 15% | 12% |

**2.** Using these figures work out approximate depreciation costs for a year (to 2 significant figures) for the following cars.

| | *Price* |
|---|---|
| a) a 5-year-old Allegro 1100 | £900 |
| b) a 2-year-old Mini GT | £2200 |
| c) a 7-year-old Escort 1100 | £700 |
| d) a 10-year-old Capri 1600 | £600 |
| e) a 7-year-old Renault 12 | £800 |
| f) a 1-year-old Renault 9 | £3800 |

**Repair Costs**

Repair costs depend on the type and age of your car, how much motoring you do and how much work you can do on your car yourself. A recent survey gave the following approximate figures for repair costs per mile of driving.

| *Engine size* (cc) | Up to 1000 | 1000–1499 |
|---|---|---|
| *Cost per mile* (p) | 5.3 | 5.6 |

| *Engine size* (cc) | 1500–1999 | 2000–2999 |
|---|---|---|
| *Cost per mile* (p) | 6.3 | 8.7 |

These figures can be reduced by about 0.6 p per mile if the owner carries out his/her own regular servicing.

**3.** Using the figures given earlier in this section for insurance, motor tax, depreciation and repair costs, work out the approximate cost of a year's motoring for an 18-year-old buying the cars in parts a), b), c) and d) of question **2** above. Assume a mileage of 10 000 miles per year. Also find the average weekly cost of driving each of these cars (excluding petrol).

# Dave's Travel Costs

Dave is 18 years old and lives in a small town. He has just got a job at a larger town 6 miles away and as he has passed his driving test he is considering buying a car. He wishes to work out how much more it would cost him to run a car for a year rather than using public transport.

**1.** The fare on the bus from Dave's house to his new job would be £6.30 for a weekly ticket. Assuming that he works 5 days a week for 48 weeks, how much would he have to pay out in a year using the bus?

**2.** Dave sometimes uses his parents' car in the evening but he estimates that he pays out about £3 a week in bus fares or petrol costs for social activities. How much would this be in a 52-week year?

**3.** Dave has seen a 5-year-old Escort for sale at a local garage at a reasonable price. He estimates that it should do 35 m.p.g. if looked after well. If petrol costs £1.75 a gallon, how much would he have to spend on petrol getting to work and back in a week? (Answer to the nearest 10 p.)

**4.** What is the total distance Dave would be travelling in a year getting to and from work?

**5.** With a car Dave could get around a bit easier and he estimates that he would do about 50 miles a week on top of his journeys to work. How many miles would that be in a 52-week year?

**6.** Including weekends and holidays Dave estimates that he will be driving about 8000 miles a year. Using the petrol consumption and cost figures given in question **3** above work out the petrol cost for this mileage to the nearest £10.

**7.** Motor tax will cost Dave £100 and he has been quoted £136 for a third party, fire and theft insurance policy. He likes working with cars and hopes that he can do most of the regular servicing himself. He estimates that he will have to allow about £100 a year for repairs. What would be the total cost of these three items in a year?

**8.** The Escort is being offered at £1400 but Dave has made an arrangement with the dealer whereby he can put down a deposit of £500 and pay off the rest in 24 instalments of £46.20. How much will he be paying in total for the car?

**9.** After a year the car will be worth about 12% less than the £1400 it is worth now. Work out its value at the end of the year and estimate the depreciation by taking this value away from the total cost in question **8**.

**10.** Using the figures you have worked out in questions **6, 7** and **9** above, estimate, to the nearest £10, the cost of a year's motoring for Dave.

**11.** By adding together Dave's present costs in questions **1** and **2** and rounding them off to the nearest £10, estimate the extra cost to Dave of owning a car for a year.

How much would this be per week to the nearest £1?

**12.** Someone who lives near Dave has asked him for a lift to work and has offered to pay half the petrol cost. Using the mileage in question **4** work out how much this would be worth to Dave in a year to the nearest £10.

**13.** Owning a car will cost Dave quite a bit. Is it worth it?

---

# Teacher's Notes

**1.** The subject of buying a second-hand car is one which many college and 5th and 6th year school students will see as relevant. A member of staff or contact who has some expertise in this field would be a useful asset and could be asked to give a talk.

**2.** Other subject areas could be brought in as follows.

*science* the basic principles of a car engine – fuels and combustion

*social sciences* the environmental effects of the car, the future of public transport, the problems of non-drivers

*career development* opportunities in the motor trade such as in selling and car repair

*information technology* car registration records, police records, computer aided design

*communications* the need for cars, the future of the motor industry, forms relating to car ownership and use

**3.** This topic could be part of much wider projects in a variety of vocational areas such as transport of people and goods in general, mobility for handicapped people, automotive engineering, mass production systems, etc.

**4.** A visit to a car assembly plant would be of interest as would a visit to a local car auction or a survey of second-hand car prices.

Chapter 7

# Wages, Salaries and Benefits

Whether they have a job or not, everyone needs money to live on. People get money either as a wage or salary or some form of state benefit. This project is designed to help you to get used to working out how much you should be getting. The work here will obviously help you if you are hoping for a job dealing with wages, salaries or state benefits.

## 1. Wages and Salaries

If you are working you are likely to be paid in one of four ways.

- at an hourly rate with different rates for overtime
- on a piecework system which could include a bonus scheme
- on a commission basis (probably on top of some other method)
- by a standard weekly wage or yearly salary for the job

### 1.1 Payment by the Hour

If you are paid by the hour it is fairly easy to work out how much money you should be getting, because

> money earned = hourly rate × number of hours worked

For instance, if you work 38 hours at £2.25 per hour, you earn 38 × £2.25 = £85.50.

**1.** Complete the following table.

| *Hours worked* | *Hourly rate* | *Amount earned* |
|---|---|---|
| 36 | £3.00 | a)...... |
| 38 | £3.50 | b)...... |
| 38 | £3.75 | c)...... |
| 40 | £2.82 | d)...... |
| $37\frac{1}{2}$ | £4 | e)...... |
| $38\frac{1}{2}$ | £2.80 | f)...... |
| 40 | g)...... | £120 |
| 40 | h)...... | £78 |
| 36 | i)...... | £90 |
| 38 | j)...... | £118.56 |

If you work more than your basic hours you usually get paid at a better hourly rate for those extra hours. This will normally be calculated on the basis of 'time-and-a-half' or 'time-and-a-quarter' or, if you are lucky, 'double time'. This means that, for example, at 'time-and-a-quarter' the overtime hourly rate is $1\frac{1}{4} \times$ normal hourly rate.

**2.** A man is paid £120 for a basic 40-hour week and overtime is paid at time-and-a-half. Work out

a) his basic hourly rate

b) his overtime hourly rate

c) the amount of overtime pay he gets for 3 hours overtime

d) the amount of overtime pay he gets for $5\frac{1}{2}$ hours overtime

e) his total wage for a week in which he works 44 hours

f) the number of hours he has worked if his total wage for the week is £147.

**3.** The union has negotiated a new deal under which the man gets £126 for a 36-hour week but overtime is only paid at time-and-a-quarter. Answer question 2 above for these new rates, giving answers to the nearest penny.

## 1.2 Piecework

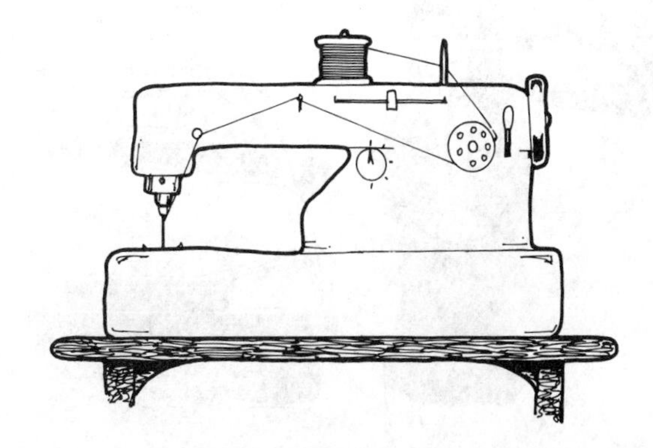

Under a piecework system you will be paid a certain amount for every item produced in a certain time plus a bonus for items above a certain number. For example, the basic rate for machining a zip into a skirt is 12 p and a bonus of 2 p is paid for every zip over 100.

If a woman machines 120 zips she receives

$100 \times 12\,p = £12$

$+\ 20 \times 14\,p = £2.80$ for the extra zips over 100

i.e. a total of £14.80

Work out the total amount earned producing 120 items in the following cases.

**1.** 7 p each for the first 100 and a bonus of 3 p each above that

**2.** 11 p each for the first 80 and a bonus of 4 p each above that

**3.** 9 p each for the first 110 and a bonus of 3 p each above that

**4.** 15 p each for the first 90 and 19 p each above that

## 1.3 Payment on Commission

For this work you will need to be able to work out percentages. If this is a problem then you should turn to part B of this book and work through the Percentage sections of chapters 3 and 4.

If your job is to sell things then you may be paid on a commission basis, with or without a basic wage or salary for the job.

A car saleswoman gets 2% commission on every car she sells. If she sells a car for £3500 her commission will be 2% of £3500, i.e.

$$\frac{2}{100} \times 3500 = £70$$

---

**1.** The car saleswoman above sells 3 cars in a week costing £2500, £4000 and £3700. In addition to her commission she gets a basic wage of £20 per week. Calculate her total commission and total wage for this week.

**2.** At another showroom a salesman gets £40 a week basic plus commission of $1\frac{1}{2}$%. Find his total commission and total wage in a week in which he sells 4 cars costing £2600, £5800, £4000 and £1800.

**3.** A door-to-door salesman is paid only by commission at a rate of 12% of total sales. In a week his daily sales figures are £60, £210, £180, £340 and £110. Work out how much he earns in that week.

---

## 1.4 Weekly Wages and Yearly Salaries

Most people are paid by the week or by the month although their actual salary may be stated as a yearly one. To find out the monthly amount you have to divide the yearly salary by 12.

---

Work out the monthly amounts to the nearest £ paid to people whose yearly salaries are

**1.** £5400 **2.** £6000 **3.** £8600 **4.** £9600

**5.** £10 800 **6.** £7200 **7.** £11 000.

---

# 2. State Benefits

Whether you are working or not you may be eligible for some form of state benefit. The major types are

Family Income Supplement (FIS)
Unemployment Benefit
Supplementary Benefit
State Pensions

## 2.1 Family Income Supplement (FIS)

FIS is a benefit paid to people with children who are working but have a low income. The amount payable depends on a person's income and a detailed form has to be filled in in order to claim it. The table below gives some idea of who could claim in 1986.

> Family Income Supplement is payable to working people who have children and who are single, married or living together.
>
> The working partner must be working at least 30 hours a week or, if he or she is a single parent, 24 hours a week.
>
> The person claiming must have at least one child still at school.
>
> With one child the family income must be less than £97.50 per week, with two children less than £109 per week, with three children less than £120.50 etc.

---

Use the table above to work out whether or not the following people are entitled to FIS. If the answer is no state why you have given that answer.

**1.** a couple with 2 children where only the woman works and earns £90 for a 40-hour week. The children are aged 5 and 8

**2.** a single parent with a child aged 12, earning £42 for a 20-hour week

**3.** a single person with no children earning £80 for a 36-hour week

**4.** a couple who are both unemployed and have 2 children aged 2 and 6

**5.** a couple with 5 children of school age where only the man works and earns £120 for a 40-hour week

Check on the rates now – have they changed much?

---

## 2.2 Supplementary Benefits and other Forms of Income

Supplementary Benefit is the name given to a whole series of state benefits which can be claimed by people in different circumstances. The table below gives examples of some of the weekly amounts payable in 1986.

*Basic rates*

| Married couple | Single person (18+) | Single person (16–17) |
|---|---|---|
| £47.85 | £23.60 | £18.20 |

*Extra allowances*

| Blindness | Heating | Special diets |
|---|---|---|
| £1.25 | £4.40 | £3.70 |

Rent is paid in full and for people with a mortgage, the interest will be paid.

*Children (dependent)*

| 18 or over | 16–17 | 11–15 | Under 11 | |
|---|---|---|---|---|
| £23.60 | £18.20 | £15.10 | £10.10 | each child |

Other income is subtracted from benefits before they are paid.

These rates will have changed since 1986 so find out what the new ones are.

---

**1.** Using the 1986 rates find the total amount payable in Supplementary Benefits to the following.

a) a married couple with 2 children aged 10 and 12 paying £24.50 rent

b) a single person over 18 paying rent of £18.70 and claiming heating benefit

c) a married couple with dependent children aged 14, 16 and 19, one of whom is on a special diet. They pay £28.40 rent and need extra heating. They have an income of £14.80 a week

d) a married couple paying £22.50 interest on their mortgage with dependent children aged 15, 15, and 20. Their income is £16.40

e) a single blind person over 18 living alone, paying £19.50 rent and with no other income

**2.** Men over 65 and women over 60 can claim a state pension. The rates for the state pension change from year to year. Find out what they are now for a single person and for a married couple.

**3.** People who have worked and have then become unemployed can normally claim unemployment benefit rather than supplementary benefit. In 1986 the weekly rate was £30.45. What is it now?

# 3. Gross and Net Pay

If you are working you get paid a 'gross' amount but at the end of the week or month you do not take home the gross amount. The amount you get after various deductions is called your 'net' pay. The two main deductions are income tax and national insurance - these are amounts which your employer pays directly to the government.

## 3.1 Income Tax

The rates at which you pay tax are fixed by the government and are published each year in the PAYE coding guide. The 1986/7 coding guide includes the following information.

| Personal allowances (income which is not taxed) | single person | £2335 |
|---|---|---|
| | married man | £3655 |
| | wife's earned income | £2335 |

It is also possible to claim tax relief on items such as expenses connected with your job, superannuation schemes and for special personal circumstances.

*Tax rates*

| *Taxable income* | *Rate* |
|---|---|
| £1–£17 200 | 29% basic rate |
| next £3000, to £20 200 | 40% |
| next £5200, to £25 400 | 45% |
| and higher rates for higher values of taxable income | |

In calculating the amount of tax to be paid you need to use

taxable income = gross income − allowances

tax paid = tax rate × taxable income

Calculate the taxable income and the amount of tax per year to be paid in the following cases.

| | *Status* | *Income (per year)* | *Extra allowances* |
|---|---|---|---|
| **1.** | single woman | £7000 | £350 |
| **2.** | single man | £6000 | none |
| **3.** | married man | £12 000 | £800 |
| **4.** | married man | £15 000 | working wife + £300 |
| **5.** | single man | £22 000 | £500 |
| **6.** | single woman | £11 000 | none |

## 3.2 National Insurance and Pension Funds

Your National Insurance contributions pay for things like sick pay and unemployment benefit and both you and your employer pay a share. Your share is a percentage of your gross pay.

As well as getting a state pension many people now receive a pension from their firm which they and their employers pay for as a percentage of their earnings. In some cases this pension contribution is called superannuation.

---

At a time when National Insurance contributions were fixed at 9%, Jones Crisps Ltd ran a pension fund for their workers for which the workers paid 5% of their wage or salary before deductions.

Calculate the separate amounts paid for National Insurance and pension fund *per month* in each of the following cases.

**1.** Jim earning £600 a month

**2.** Sarah earning £540 a month

**3.** John earning £8000 a year

**4.** Dave earning £11 000 a year

**5.** Jean earning £10 400 a year

Find out what the present rate of National Insurance contributions is and work out the above again using that figure.

---

## 3.3 Pay Slips

If you are working you should receive with your money every week or month a 'pay slip' or 'pay advice' which gives details of your gross and net incomes and the various deductions. An example is shown below.

| **Name** P.R. BOWNESS | Feb. 1985 |
|---|---|
| TAXABLE ALLOWANCES £ | DEDUCTIONS £<br>Income Tax 207.30<br>Nat. Ins. 72.23 |
| Total gross pay 1008.25<br>Superannuation 60.50<br>Taxable pay 947.75 | NON-TAXABLE ALLOWANCES |
| TOTALS FOR THIS TAX YEAR<br>Taxable pay 10 425.28 Tax paid 2253.30<br>Superannuation 665.47<br>National Insurance 781.50 | NET PAY |

*Note*: taxable income = taxable pay (above) – personal allowance

---

Answer the following questions.

**1.** What was the yearly gross pay for P. R. Bowness? (Assume February rates)

**2.** What was superannuation as a % of gross pay?

**3.** If P. R. Bowness had personal tax allowances of £2795 per year, on how much did he actually pay tax in February 1985?

**4.** What was the rate of tax?

**5.** What should the net pay have been?

# Family Incomes

Dave left school last year and he works at Jones Crisps Ltd loading and unloading vans. He is paid £3 an hour for a 36-hour week and any overtime is paid at time-and-a-quarter. His sister Liz works in a carpet shop where she earns a basic wage of £60 per week plus a commission of 2% of all her sales.

**1.** Work out Dave's gross pay for a week in which he works 39 hours.

**2.** The same week Liz sells £1800 worth of carpets. What is her gross pay?

**3.** Dave and Liz can each claim the basic personal allowances of £2335 against tax. How much is this per week?

**4.** Dave has to pay 9% of his gross pay for National Insurance and 5% to a pension fund. What would these come to between them this week?

**5.** Take the weekly personal allowance and pension contribution away from Dave's gross income to find his taxable income.

**6.** Work out how much tax Dave will have to pay at a rate of 29% of this taxable income.

**7.** Taking National Insurance, pension and tax away from Dave's gross pay find his net pay.

**8.** Liz's pension fund contributions are $5\frac{1}{2}$% of her gross pay. Work out her net pay in the same way, if she also has to pay 9% of her gross pay for National Insurance.

**9.** Next month Liz will be marrying Stuart who works in a bank and earns £6800 per year, paying 6% into a pension fund (superannuation). Assuming that the tax rate is 29%, the National Insurance rate is 9% and that Stuart can claim the basic single person's allowance of £2335, write out a monthly pay slip for Stuart based on the one on page 57, leaving out the totals for this tax year.

**10.** If Liz and Stuart were already married and Stuart was claiming the married person's allowance of £3655, how much extra would his net pay be than it is now? Would Liz still have to pay tax?

# Teacher's Notes

**1.** Most students in their final year at college or school will be looking forward to obtaining income from some source in the near future if they are not doing so already. Some may be hoping to work in a salaries or benefits office and this topic will act as a lead into future employment.

**2.** Extensions to other subject areas could include

*social sciences* benefits and how they are paid, the problems of low income families, the social effects of unemployment

*career development* wage rates in different industries, part-time employment, cooperatives and other structures

*information technology* wages and salaries packages, employment record keeping, preparation of a data file on benefits

*communcations* applying for jobs and benefits, form filling and interpretation of information.

**3.** This work could be linked in with vocational elements in courses by looking at salaries and wages paid in various types of employment. A course with a caring element could usefully expand the work on eligibility for benefits, pensions and the poverty level in the UK.

**4.** Most towns have some sort of welfare rights advisory service which may be able to supply an interesting speaker. The Department of Health and Social Security can supply leaflets on rates of benefits, and they also publish a special guide to benefits for young people.

Chapter 8

# A Roof over your Head

Everyone needs somewhere to live. At present you may be living with your parents but sometime you will probably want to move out into a flat, a bedsitter or even a home of your own.

In this chapter you will be carrying out calculations related to the costs of renting and buying houses, choosing where to live and moving in. Some of the work you will be doing will also be used in later chapters.

## 1. Renting or Buying

### 1.1 Paying Rent

When they first move out on their own, away from parents, most people rent a bedsitter or flat on a weekly or monthly basis. Many people continue to pay rent for a flat or a house for the rest of their lives.

Rents don't usually stay fixed but may increase every few years.

**1.** Calculate the rent paid in a year for flats or houses where the weekly rent is
a) £18 b) £22.50 c) £16.60
d) £20.80

**2.** A local council decides to increase all its rents by 8%. Find the new weekly rents if the old rents are
a) £16.50 b) £21 c) £22.40
d) £18.10

**3.** Find the percentage increases for each of the following rent rises.
a) £24 to £27 b) £18 to £20.70
c) £25 increased by £2
d) £20 increased by £2.50

**4.** Three students decide to share the cost of a flat equally. How much will each have to pay in a year if the monthly rent is
a) £120 b) £110.60 c) £98.50?

### 1.2 Buying a House

In most cases people who want to buy their own house have to borrow money. The lenders are usually banks, building societies or local authorities and, whoever you borrow from, the loan is called a mortgage.

Mortgages are usually very long term loans – 20, 25 or 30 years – and are paid back on a monthly basis. Banks, building societies and local authorities do not lend you money for nothing. They charge you interest on the loan.

At this stage you should work through the sections on simple interest and compound interest in part B, chapter 4.

A mortgage is a large loan for a long time and therefore a lot of interest is charged. If you buy a house using a mortgage you end up paying a lot more than the price you were actually charged for the house.

Mortgages are of two basic types.

repayment mortgage – you pay the building society on a monthly basis

endowment mortgage – you pay premiums for an insurance policy which covers the cost of the mortgage

There are also other schemes which are a mixture of these but whatever the scheme you have to pay a monthly amount which depends on how much you borrowed and the present rate of interest.

---

The following table produced by a building society shows the monthly repayments payable on a loan of £10 000 on two different types of mortgage.

*Repayment mortgage (12.75%)*

| | *Monthly repayment* | *Net monthly repayment* |
|---|---|---|
| 25 years | £111.81 | £84.32 |
| 20 years | £116.85 | £90.80 |
| 15 years | £127.29 | £102.92 |

*Endowment mortgage (13.25%)*

| | *Monthly repayment* | *Net monthly repayment* |
|---|---|---|
| 25 years | £110.42 | £77.29 |
| 20 years | £110.42 | £77.29 |
| 15 years | £110.42 | £77.29 |

(with an endowment mortgage you also make payments to a life assurance company.)

Net repayment means the amount payable after tax relief has been taken off.

**1.** Find the net monthly repayments for the following loans.
   a) £10 000 repayment mortgage over 20 years
   b) £20 000 endowment mortgage over 15 years
   c) £15 000 repayment mortgage over 25 years
   d) £25 000 endowment mortgage over 20 years

**2.** Find the total repayment and total net repayment over the whole period for the loans in question 1.

**3.** Interest rates change with time. Find out what the rates are now and make up a table similar to the one above for the present rates.

**4.** Find out the costs of a few houses in your area and the loans needed to buy assuming that you borrow 90% of the cost. Use your table from question **3** to work out the net monthly repayments for these houses.

## 1.3 The Rates Bill

Your local council gets its money in two major ways – from central government and from local people and businesses as rates.

If you are living in a house of any type that house will have a 'rateable value' which depends on the number of rooms, area of land, the surroundings and facilities such as central heating. This rateable value is set by the local rating officer. Each year the local council decides on the rate in the £ which it is going to set and sends out rate bills to house owners. These bills are calculated using the formula

> amount you pay per year = rateable value × rate in the £

For example, a council sets a rate of 92 p in the £ and a house has a rateable value of £260.

amount to pay = 260 × 0.92 = £239.20

---

**1.** In Puddington the local rate is 95 p in the £. Find the yearly amounts to be paid by the owners of houses with rateable values of
a) £320 b) £210 c) £264 d) £218

**2.** Four people living in different towns all have houses with a rateable value of £260. Find the rate in the £ for each of these areas if the amounts to be paid are
a) £228.80 b) £244.40 c) £257.40
d) £270.40.

**3.** A leaflet produced by the Puddington Borough Council states that the average Puddington householder lives in a house with a rateable value of £190 and will have to pay £287.85 in rates this year. Included in this amount will be

£143.26 for education
£27.17 for roads
£24.40 for social services
£22.23 for the police
£7.12 for environmental health

Find
a) the rate in the £ for Puddington this year
b) the rate in the £ for the police services
c) the amount to be paid for a house with a rateable value of £205
d) the education bill as a % of the total.

**4.** Find out what the rate in the £ is in your area this year. Use this to calculate the amount to be paid for houses with rateable values of
a) £184 b) £212
c) the same as your house.

---

If you rent a house or flat you may be paying rates as part of your rent, or you may be billed separately.

# 2. Saving up

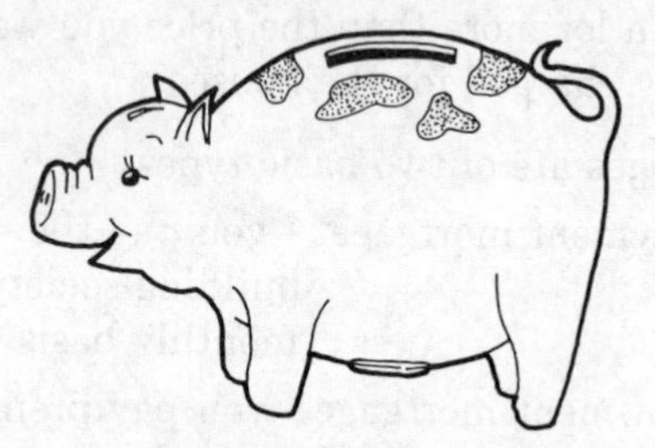

## 2.1 How Fast can you Save?

If you are going to buy a house you are unlikely to be able to get a mortgage to cover the full cost. Generally you will have to find about 10% of the cost from your own pocket – this is called a 90% mortgage. The first thing to decide is whether you will be able to save enough money to do this.

---

**1.** If your building society has offered you a 90% mortgage how much extra will you have to find to buy a house priced at
a) £25 000 b) £22 900 c) £18 400
d) £49 900?

**2.** To buy a house priced at £20 000 you would have to produce £2000 to add to a 90% mortgage. How long will it take to save this amount at a rate of
a) £50 a month b) £20 a week
c) £80 a month?

**3.** House prices don't stay the same from year to year – they increase with inflation. Assuming that house prices are increasing at 5% per year find the cost in 1 year's time of houses presently priced at
a) £25 000  b) £18 000  c) £16 600
d) £22 500.

**4.** a) A basic 'young couple's house' costs about £18 000 at present in some parts of the country. Assuming that this price will increase by 5% per year make up a table showing the likely costs for the next 5 years.
b) To buy such a house on a mortgage you would need to find 10% of the cost. Add another line to your table showing the 10%s over the next 5 years. Draw a graph to show these figures.
c) Assuming that a couple can save £20 a week draw another graph on the same axes showing the amount they could save over the next 5 years. How long will it take to save up the 10%?

**5.** By looking in an estate agents or the local paper pick an average priced 3-bedroomed semi-detached house in your area. Using this figure, work through question **4** again using an approximation to the present rate of inflation.

---

## 2.2 Investing your Savings

Most people who are trying to save up for something don't hide their money under the mattress each week, they invest it – in shares, a building society, a bank etc. When money is invested it gains interest. This interest can be calculated using the principle of compound interest. In the work you have already done on compound interest you will have worked out the interest year by year (method 1 in part B chapter 4). There is, however, another way of working which makes it easier to deal with long-term investment (method 2 in part B chapter 4). For example, £200 is invested at 8% p.a. After 1 year the amount becomes

$$200 + \frac{8}{100} \times 200$$

$$= 200\left(1 + \frac{8}{100}\right)$$

$$= 200\,(1 + 0.08)$$

$$= 200 \times 1.08$$

$$= £216$$

After 2 years the amount becomes

$$216 + \frac{8}{100} \times 216$$

$$= 216\left(1 + \frac{8}{100}\right)$$

$$= 216 \times 1.08$$

$$= £233.28$$

Notice that for each year you multiply by 1.08.

After 3 years the amount becomes

$£233.28 \times 1.08 = £251.94$ to the nearest p

You can now build up the following table.

| | |
|---|---|
| After 1 year | $£200 \times 1.08$ |
| After 2 years | $£200 \times 1.08 \times 1.08$ |
| | $= £200 \times 1.08^2$ |
| After 3 years | $£200 \times 1.08 \times 1.08 \times 1.08$ |
| | $= £200 \times 1.08^3$ |

---

For this exercise you will need a calculator. If you are not sure how to use it work through part B, chapter 5 first.

Find the total amount saved to the nearest p in the following cases.

**1.** £200 invested at 8% p.a. for 5 years

**2.** £500 invested at 10% p.a. for 5 years

**3.** £1000 invested at 11% p.a. for 3 years

**4.** £800 invested for 4 years at $8\frac{1}{2}$% p.a.

## 2.3 Regular Savings

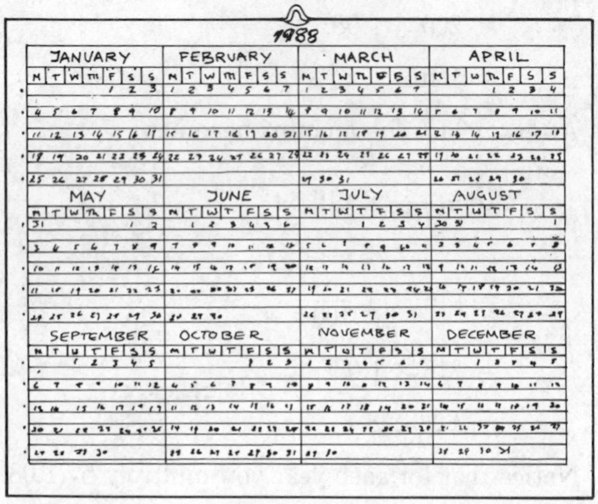

The work in section 2.2 applies if you are investing an amount of money and leaving it in a bank or building society for a number of years. The situation when you join a regular savings plan is more complicated to deal with. The graph opposite shows the total amounts in a monthly savings account up to 5 years for every £10 per month invested. The two sets of points are based on interest rates of 8% and 12% p.a.

You can see that different interest rates only begin to have a noticeable effect after about 2 years.

If you saved £10 per month for 4 years at 8% p.a. you would end up with about £570.

If you could save £30 per month you would have $3 \times 570 = £1710$.

Use the graph opposite to answer the following questions.

**1.** A couple can save £20 per month when the interest rate is 12% p.a.
How much will they have saved after $4\frac{1}{2}$ years?

**2.** After 5 years how much will a person have saved if she has invested £25 a month at an interest rate of 8% p.a.?

**3.** Estimate the total amounts saved over 5 years for every £10 a month invested if the interest rate is a) 10% b) 11% c) 13%

**4.** A couple can afford to save £25 per month. How long will it take them to save £1200 if the interest rate is 12% p.a.?

**5.** Copy the graph opposite on to graph paper and draw on your paper a line showing how much has actually been invested at a rate of £10 per month for up to 5 years. What is the total interest earned over 5 years at a rate of 12% p.a.?

**6.** A couple want to save up for the deposit on a house. They think that they will need about £2000 and they want to try to save this within 3 years. If the interest rate is 12% p.a. how much will they have to invest per month in order to do this?

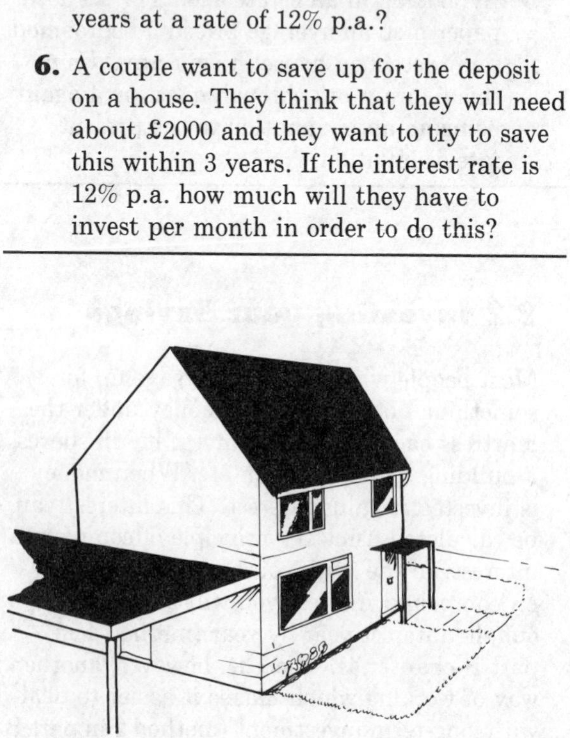

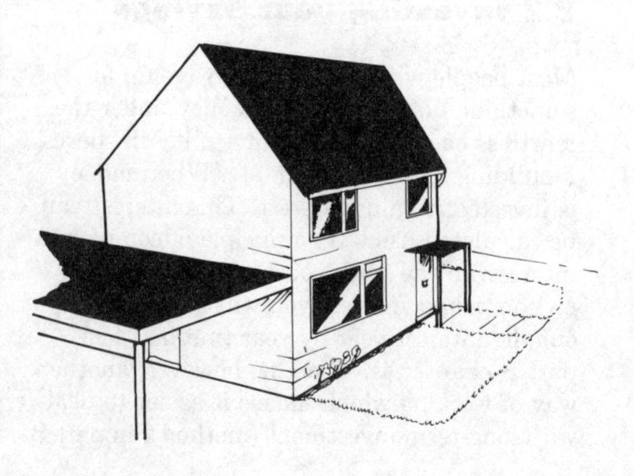

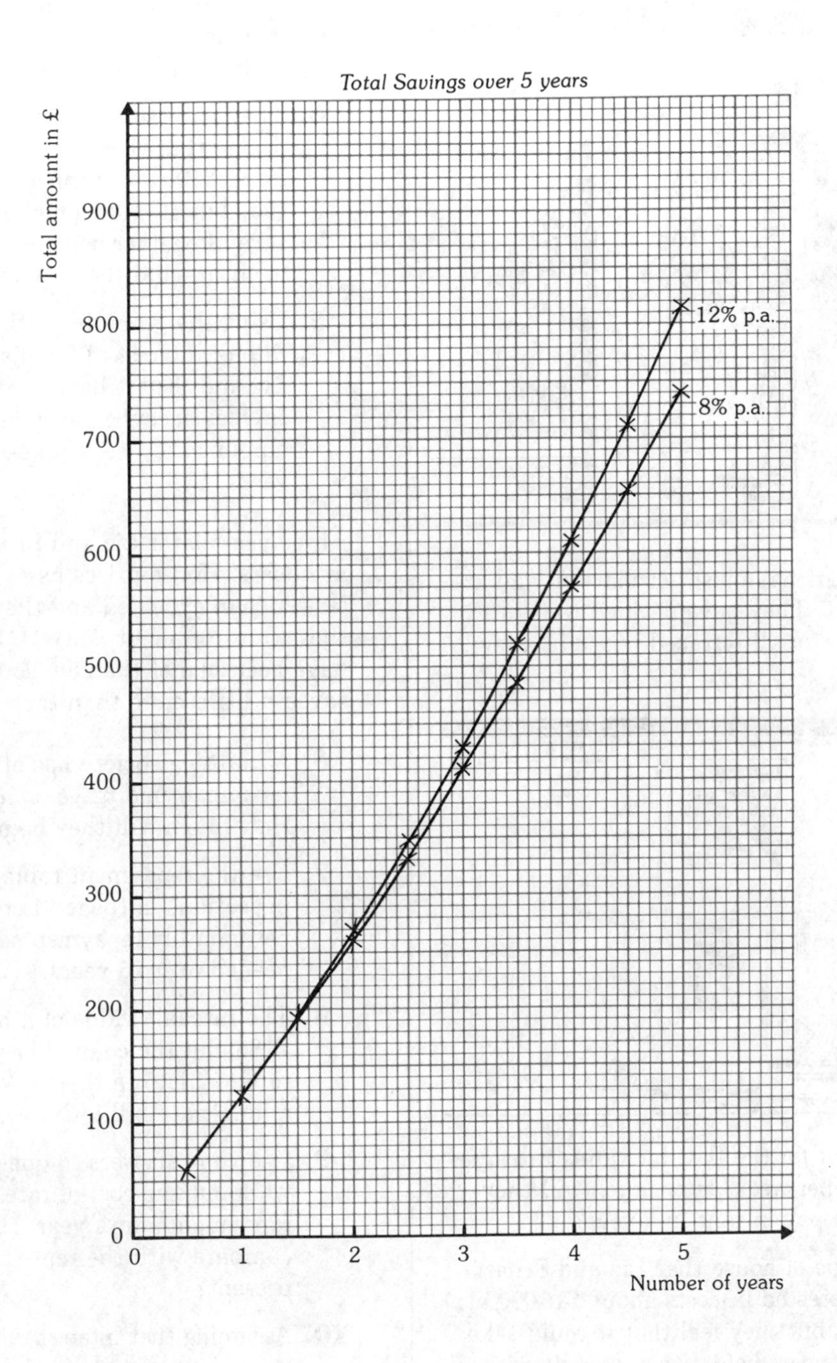
Total Savings over 5 years
Total amount in £
900
800
700
600
500
400
300
200
100
0
0
1
2
3
4
5
Number of years
12% p.a.
8% p.a.

# From Renting to Buying

Liz and Stuart have just got married and are renting a flat. They have decided to try to save to buy a small terraced house in a few years' time.

**1.** The rent for the flat, including rates, is £32.50 per week. How much will they have to pay in rent in a year?

**2.** The type of house that Liz and Stuart are interested in costs about £16 000 at present but they feel that it could take up to 3 years before they can afford one. Assuming that house prices are rising by 5% per year, how much would you expect such as a house to cost in 3 year's time?

**3.** Liz and Stuart expect to be able to get a 90% mortgage so they will have to provide about 10% of the price. How much is the 10% likely to be in 3 years' time?

**4.** They have some furniture already but decide that they need to try to save an extra £1000 to furnish the house, when they buy it. How much do they need to save altogether over the 3 years for the furniture and their share of the house?

**5.** Using the graph in section 2.3 with a rate of interest of 8% p.a. how much will Liz and Stuart have to put into a regular savings account per month to save the amount they need (question **4**) in 3 years?

Three years later Liz and Stuart are looking at houses. House prices have risen slightly slower than expected and they can get what they want for about £18 000. Interest rates have been higher than 8% so that they have saved a little more than they expected.

**6.** A building society has offered a 90% repayment mortgage on an £18 000 house. How much will they be borrowing?

**7.** Use the repayment table in the exercise in section 1.2 (page 61) to work out the net monthly repayments on the mortgage needed over 25 years.

**8.** The rateable value of a house they like is £185 and the council has recently set a rate of £1.16 in the £. What will the annual rate bill be?

**9.** Use your answers to questions **7** and **8** to work out the cost of rates and mortgage repayments for a year. How does this compare with the rent they are paying at present?

**10.** Assuming that interest rates, and therefore the mortgage repayment, stay the same for 25 years how much will they have paid to the building society in total in that time?

# Teacher's Notes

**1.** This project would obviously fit in well as part of any business orientated course and could be linked with general work on banks and building societies. The numeracy work is not always easy and a calculator is really a necessity for this chapter.

**2.** Links with other subject areas could include

*social studies* survey of house prices, housing problems, availability of rented accommodation etc., council housing, rights and responsibilities of landlords and tenants

*career development* a visit to a building society could lead to discussions on job prospects. There could also be a link with local housebuilding firms

*information technology* estate agents now use computers and would probably be happy to demonstrate their system. Programs on mortgage repayments are widely available

*communications* getting information on loans, interest rates etc., dealing with estate agents, interpreting literature.

**3.** The ideas suggested above show how this chapter could be linked into courses in business studies, construction, and caring skills.

**4.** This chapter could be extended to include work similar to that in chapters 4 and 5 to carry Liz and Stuart through into their house.

Chapter 9

# Planning a Holiday

Most people like to get away from home from time to time, perhaps for a weekend, a week, or a longer holiday. In this chapter you will be looking at the basic things you have to do as part of planning a holiday, such as checking times and distances, planning routes, costing the holiday, etc.

The work in this chapter does not only concern holidays but will come in useful if you are planning any sort of trip, for work or for pleasure.

## 1. Time and Timetables

### 1.1 The 24-hour Clock

Most timetables use the 24-hour clock system – instead of stopping at 12 noon and starting again at 0, time is measured from 0 to 24 hours. You have probably seen a 24-hour clock at home or in a public place.

---

**1.** Turn the following 24-hour clock times into 12-hour clock times remembering to put a.m. or p.m. where appropriate.

a) 1035 b) 0720 c) 1315 d) 1750
e) 2230 f) 1200 g) 2359 h) 0000

**2.** Turn the following 12-hour clock times into 24-hour times.

a) 7.30 a.m. b) 7.30 p.m.
c) 12.15 p.m. d) 12.15 a.m.
e) 10.00 p.m. f) 1.15 p.m.

---

## 1.2 Time Differences

If you are planning a trip or some other event it is often useful to know how long it will take. This may involve finding journey times from start and finishing times, or finding the finishing time, given the starting time and the time to be taken. For example,

(a) A journey starts at 8.50 a.m. and finishes at 10.25 a.m. How long does it take?

The easiest way to work it out, usually, is to split the time up.

8.50 to 9.00 is 10 min

9.00 to 10.25 is 1 h 25 min

total time is 1 h 35 min

You can check this answer by using the subtraction method, i.e.

$$\begin{array}{r} \text{h min} \\ 10.25 \\ -\ 8.50 \\ \hline \\ \hline \end{array}$$

but remember that there are only 60 minutes in an hour.

(b) Driving to London usually takes me $2\frac{1}{4}$ h. If I start at 9.55 a.m. what time should I get there?

2 h on from 9.55 gives 11.55 a.m.

15 min on from 11.55 a.m. gives

12.10 p.m. – the arrival time.

In this case you could use the addition method all in one go, i.e.

$$\begin{array}{r} \text{h min} \\ 9.55 \\ +2.15 \\ \hline \\ \hline \end{array}$$

again remember that 1 h = 60 min.

Copy and complete the following table.

| *Starting time* | *Finishing time* | *Journey time* |
|---|---|---|
| 10.00 a.m. | 3.00 p.m. | a)...... |
| 9.20 a.m. | 4.30 p.m. | b)...... |
| 0950 | 1345 | c)...... |
| 1615 | 0730 | d)...... |
| 1240 | 1630 | e)...... |
| 5.50 p.m. | f)...... | $2\frac{1}{2}$ h |
| 11.30 p.m. | g)...... | 2 h 50 min |
| 0550 | h)...... | 4 h 20 min |
| i)...... | 3.15 p.m. | 2 h 40 min |
| j)...... | 0920 | 1 h 45 min |

## 1.3 Using a Timetable

If you understand the 24-hour clock system and have successfully completed the questions in section 1.2 then you should, with a little care, be able to deal with a bus or train timetable. The important things to know are your start and finishing places and the day of the week – from then on it is easy.

# NORTHAMPTON - LONDON

## YOUR DAILY NATIONAL EXPRESS SERVICES FROM 20 JUNE

| *Service No.* | *455* | *455* | *455** | *455** | *455* |
|---|---|---|---|---|---|
| NORTHAMPTON, Greyfriars Bus Station | 0600 | 0645 | 0900 | 1100 | 1300 |
| Golders Green, London Transport Bus Station | 0705 | 0800 | 1015 | 1215 | 1415 |
| Marylebone, Town Hall | 0717 | 0817 | 1032 | 1232 | 1432 |
| LONDON, Victoria Coach Station | 0730 | 0830 | 1045 | 1245 | 1445 |

| *Service No.* | *455* | *455* | *455** | *455* |
|---|---|---|---|---|
| NORTHAMPTON, Greyfriars Bus Station | 1500 | 1700 | 1900 | 2100 |
| Golders Green, London Transport Bus Station | 1615 | 1815 | 2015 | 2215 |
| Marylebone, Town Hall | 1632 | 1832 | 2032 | 2232 |
| LONDON, Victoria Coach Station | 1645 | 1845 | 2045 | 2245 |

| *Service No.* | *451** | *455* | *455* | *455* | *455** |
|---|---|---|---|---|---|
| LONDON, Victoria Coach Station | 0800 | 1000 | 1200 | 1400 | 1545 |
| Marylebone, Gloucester Place nr. Dorset Close | 0810 | 1010 | 1210 | 1410 | 1555 |
| Golders Green, London Transport Bus Station | 0825 | 1025 | 1225 | 1425 | 1610 |
| NORTHAMPTON, Greyfriars Bus Station | 0945 | 1145 | 1345 | 1545 | 1730 |

| *Service No.* | *451* | *455* | *455* | *451* | *455* |
|---|---|---|---|---|---|
| LONDON, Victoria Coach Station | 1645 | 1800 | 2000 | 2130 | 2300 |
| Marylebone, Gloucester Place nr. Dorset Close | 1655 | 1810 | 2010 | .... | 2310 |
| Golders Green, London Transport Bus Station | 1710 | 1825 | 2025 | 2155 | 2325 |
| NORTHAMPTON, Greyfriars Bus Station | 1830 | 2000 | 2130 | 2300 | 0115 |

*Not Sundays

## CORBY-MILTON KEYNES BUS SERVICE

### Sundays

| *Service No.* | *249* | *254* | *254* | *297* | *255* | *250* | *254* | *297* | *265* | *254* | *249* | *254* | *297* |
|---|---|---|---|---|---|---|---|---|---|---|---|---|---|
| CORBY, Bus Station | 0800 | 1018 | 1118 | .... | 1218 | 1310 | 1318 | .... | 1350 | 1418 | 1510 | 1518 | .... |
| Danesholme, Minden Close | .... | 1032 | 1132 | .... | 1232 | .... | 1332 | .... | .... | 1432 | .... | 1532 | .... |
| Hallwood Road/Neale Avenue | .... | 1042 | 1142 | .... | 1246 | .... | 1342 | .... | .... | 1442 | .... | 1542 | .... |
| KETTERING, Bus Station *arr.* | 0820 | 1055 | 1155 | .... | 1255 | 1330 | 1355 | .... | 1420 | 1455 | 1530 | 1555 | .... |
| Kettering, Bus Station *dep.* | 0820 | 1100 | .... | 1200 | 1300 | .... | → | 1400 | .... | 1500 | 1530 | → | 1600 |
| Northampton, Bus Station | .... | 1152 | .... | .... | 1352 | .... | .... | .... | .... | 1552 | .... | .... | .... |
| Wellingborough, Church Street | 0838 | .... | .... | 1230 | .... | .... | .... | 1430 | .... | .... | 1548 | .... | 1630 |
| Milton Keynes | .... | .... | .... | .... | .... | .... | .... | .... | .... | .... | .... | .... | .... |

## CORBY-MILTON KEYNES BUS SERVICE

### Sundays

| *Service No.* | *255* | *250* | *254* | *297* | *254* | *254* | *297* | *265* | *255* | *254* | *297* | *254* | *254* |
|---|---|---|---|---|---|---|---|---|---|---|---|---|---|
| CORBY, Bus Station | 1618 | 1710 | 1718 | .... | 1818 | 1918 | .... | 1940 | 2018 | 2118 | .... | 2218 | 2323 |
| Danesholme, Minden Close | 1632 | .... | 1732 | .... | 1832 | 1932 | .... | .... | 2032 | 2132 | .... | 2232 | 2335 |
| Hallwood Road/Neale Avenue | 1646 | .... | 1742 | .... | 1842 | 1942 | .... | .... | 2046 | 2142 | .... | .... | .... |
| KETTERING, Bus Station *arr.* | 1655 | 1730 | 1755 | .... | 1855 | 1955 | .... | 2010 | 2055 | 2155 | .... | 2250 | 2353 |
| Kettering, Bus Station *dep.* | 1700 | .... | → | 1800 | 1900 | → | 2000 | .... | 2100 | → | 2200 | 2255 | .... |
| Northampton, Bus Station | 1752 | .... | .... | .... | 1952 | .... | .... | .... | 2152 | .... | .... | 2338 | .... |
| Wellingborough, Church Street | .... | .... | .... | 1830 | .... | .... | 2030 | .... | .... | .... | 2230 | .... | .... |
| Milton Keynes | .... | .... | .... | .... | .... | .... | .... | .... | .... | .... | .... | .... | .... |

**1.** Use the coach timetable opposite to find

a) the normal journey time from Northampton to Victoria coach station

b) the normal journey time from Golders Green to Northampton

c) the shortest journey time from London Victoria to Northampton

d) the latest coach I could catch from Northampton to be at London Victoria by 1200.

**2.** Use the bus timetable opposite to answer the following questions.

a) If I wish to go by bus from Corby to Wellingborough, what is the earliest time I can get to Wellingborough?

b) How long does the 1018 from Corby take to get to Northampton?

c) If I wish to get to Wellingborough by 2.00 p.m. what time is the latest bus I can catch from Corby?

d) How long should I have to wait at Kettering for the connection to Wellingborough if I catch the 1710 from Corby?

e) What is the normal journey time from Corby to Kettering?

f) If I live in Wellingborough what time is the last bus I can catch home from Corby?

# 2. Routes and Distances

## 2.1 Planning your Route

If you are planning your own holiday you will need to work out, using maps and timetables, how to get to your destination, whether you are going by car or by public transport.

**1.** Look at the simplified map of major roads in the Yorkshire/Lancashire area, below.

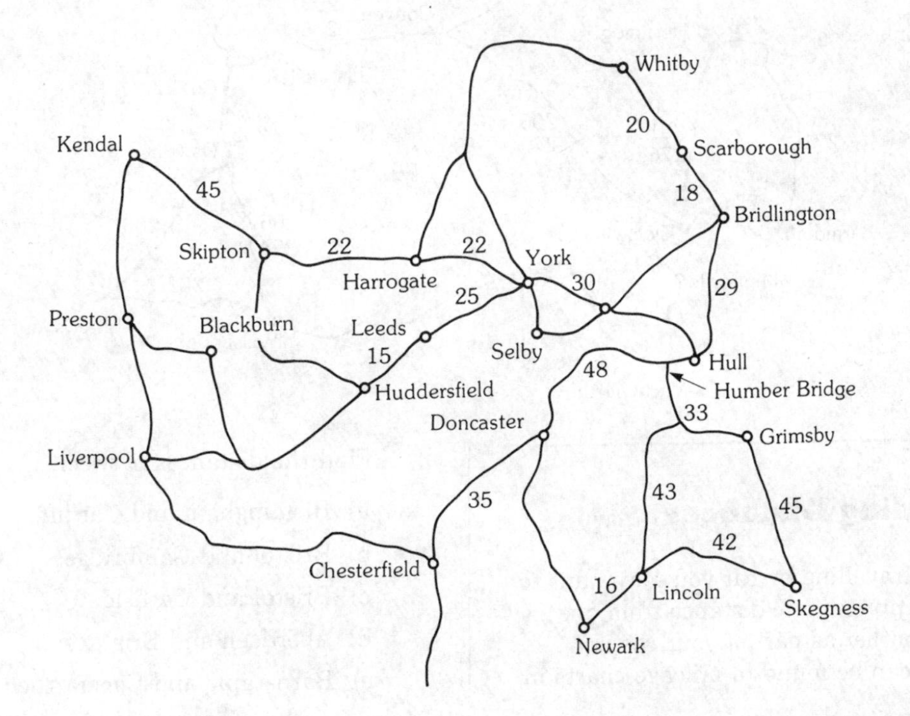

Describe the towns you would pass through or near to, going from Hull to

a) Whitby b) Newark c) Liverpool

d) Kendal.

Look at a map book with more detail and see if there are any better routes, for example using motorways.

**2.** The map below shows some major roads in North and Mid Wales. The numbers by the roads show the bus routes along these roads.

Which bus routes would be used on the following bus journeys and where would the changes be made

a) Chester to Welshpool
b) Chester to Machynlleth
c) Corwen to Welshpool
d) Barmouth to Welshpool?

map books, by adding up distances on maps or by a combination of both.

**1.** The table below shows a section of a mileage chart for distances between towns in the UK.

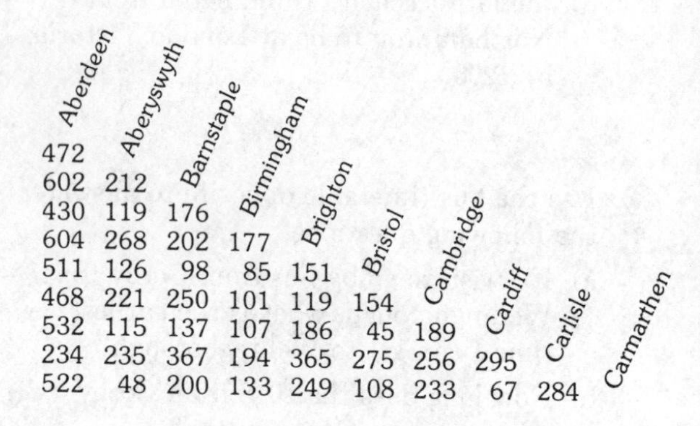

| Aberdeen | Aberyswyth | Barnstaple | Birmingham | Brighton | Bristol | Cambridge | Cardiff | Carlisle | Carmarthen |
|---|---|---|---|---|---|---|---|---|---|
| 472 | | | | | | | | | |
| 602 | 212 | | | | | | | | |
| 430 | 119 | 176 | | | | | | | |
| 604 | 268 | 202 | 177 | | | | | | |
| 511 | 126 | 98 | 85 | 151 | | | | | |
| 468 | 221 | 250 | 101 | 119 | 154 | | | | |
| 532 | 115 | 137 | 107 | 186 | 45 | 189 | | | |
| 234 | 235 | 367 | 194 | 365 | 275 | 256 | 295 | | |
| 522 | 48 | 200 | 133 | 249 | 108 | 233 | 67 | 284 | |

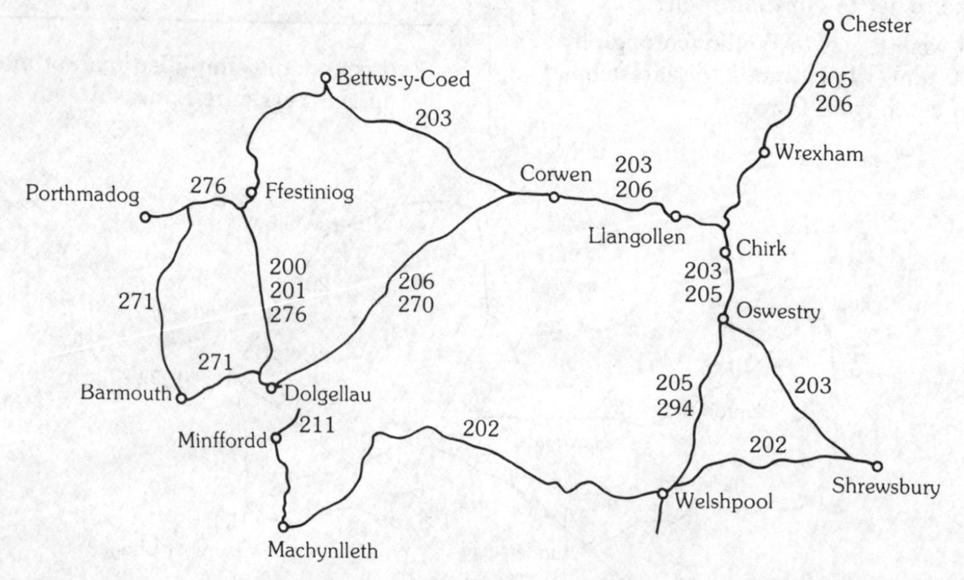


## 2.2 Finding Distances

If you are travelling by car you may want to know the approximate distance from one place to another as part of your costing. Distances can be found in mileage charts in

Find the distances between

a) Birmingham and Cardiff

b) Bristol and Cambridge

c) Bristol and Carlisle

d) Aberdeen and Brighton

e) Barnstaple and Carmarthen.

**2.** The numbers on the map in section 2.1 question 1 give the distances in miles between the towns marked. Use this map to find the distances in miles from Hull to
a) Whitby b) Newark c) Skipton
d) Chesterfield.

**3.** Using the mileage chart will give you the distances from one major town to another. To find distances to places which are near the towns on the chart, look at the chart and a map of Britain. Use this system to find the distances between
a) Bristol and Eastbourne
b) Barnstaple and Swansea
c) Birmingham and Weston-Super-Mare
d) Brighton and Ely.

## 2.3 Estimating Distances

Often it is not necessary to know the exact distance between places, an estimate to the nearest mile or $\frac{1}{4}$ mile (or even 10 or 50 miles) will usually do. The accuracy needed depends on what you need the information for, for example whether you are walking or travelling by plane.

Usually, to estimate the length of a journey you haven't done before you need a known distance to compare it with, for example from Northampton to London is about 70 miles. From the map Northampton to Portsmouth looks about $1\frac{1}{2}$ times as far. You could therefore estimate that the distance is about 100 miles.

**1.** Look at a map of Britain. Remembering that London is about 70 miles from Northampton, estimate the distance from
a) London to Bristol
b) Birmingham to Leeds
c) Liverpool to Glasgow
d) Edinburgh to Inverness
e) Sheffield to Hull.

You should be able to check your estimates against real distances using a mileage chart or by adding up distances on the map.

**2.** You should also be able to estimate shorter distances in your area. Estimate the distances in appropriate units between
a) your home and the nearest library
b) the library and the Post Office
c) your home and the nearest post box
d) your school or college and the nearest railway station.

Check these estimates using a map or some other means.

# 3. Costing a Holiday

## 3.1 Travel by Public Transport

When travelling by train or bus costing is a fairly easy business as long as you watch out for any special offers which can be used. Some people can also get cheaper fares by having student or senior citizen railcards or some form of 'rover' ticket.

A coach company leaflet lists the following fares from Northampton (1986 prices).

| To | Single £ | Period return £ | Peak period return £ |
|---|---|---|---|
| Aberdeen | 24.50 | 32.75 | 37.25 |
| Blackpool | 11.00 | 14.50 | 16.50 |
| Bournemouth | 11.50 | 15.50 | 17.00 |
| Exeter | 13.50 | 18.00 | 20.50 |

Period returns are *not* available when the outward journey is on a Friday (any time of year) or Saturday (July and August only). On these occasions you have to pay the peak rate.

**1.** a) How much would you save buying a period return to Aberdeen rather than 2 singles?
b) Can a period return ticket be used on
i) Saturday 15th June
ii) Friday 4th August?

**2.** Reductions of $\frac{1}{3}$ of all fares are allowed for
children aged 5 and under 17
(under 5s free)
men and women over 60
students and youth trainees.

What would be the cost for
a) a 70-year-old woman travelling with a period return to Bournemouth
b) a couple with a 4-year-old travelling with peak period returns to Exeter
c) a family of 2 adults and 2 children aged 8 and 11 travelling with singles to Blackpool
d) a man and his student daughter travelling with period returns to Blackpool?

**3.** Pick 3 holiday areas in the UK and find out the fares from your town to each of them by coach and by train. Make a note of any restrictions on these tickets.

**4.** If you are working on this exercise you are likely to be under 16 or a student or trainee. Find out about reductions which are available to you on different means of travel.

## 3.2 Using a Car

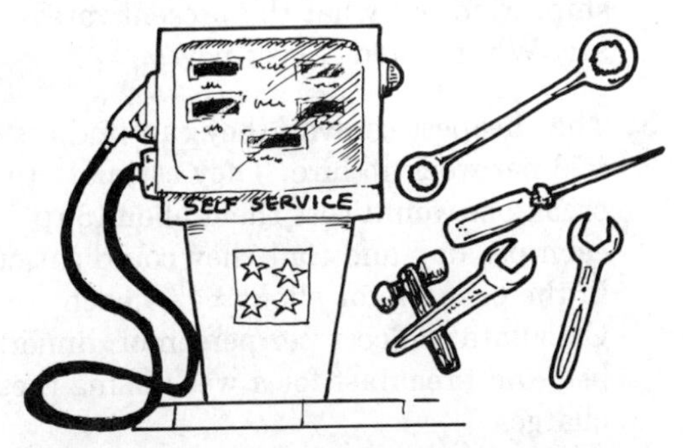


When costing travel by car you need to know how far you are going to travel, the average number of miles per gallon for the car and the cost of petrol per gallon. Any figures you work out will, of course, be approximate and will depend on how much of the journey is in towns, on motorways, etc.

**1.** Assuming that a car will do 40 m.p.g. on average, work out the cost of petrol per mile to 1 decimal place if the price of petrol is
a) £1.85 per gallon b) £1.65 per gallon
c) £1.55 per gallon.

**2.** Round up your figures for question **1** b) above to the next whole number of pence (to allow for other costs) and use this value to work out the cost, to the nearest £, of journeys of
a) 120 miles b) 200 miles
c) 500 miles d) 360 miles.

**3.** When driving fast on motorways the petrol consumption goes up to 34 m.p.g. Using a petrol price of £1.75 per gallon repeat the calculations for the costs of the journeys in question **2** assuming that most of the journeys are on motorways.

**4.** Draw a graph to show the cost of petrol for a journey of 100 miles for average petrol consumptions from 30 m.p.g. to 50 m.p.g. assuming a petrol price of £1.50 per gallon.

**5.** In fact the total cost per mile for an average 'old' car is nearer 15 p. Using this figure find the real costs of journeys of
a) 120 miles b) 200 miles
c) 500 miles d) 360 miles.

**6.** If you haven't got a car you may decide to hire one for a holiday. The following section is taken from the price list of a car hire firm.

| | *1 day* | *Weekend (2 days)* | *1 week* |
|---|---|---|---|
| | £ | £ | £ |
| Renault 5 | 11 | 24 | 66 |
| Vauxhall Astra | 13 | 28 | 75 |
| Ford Sierra | 15 | 32 | 100 |

Insurance costs £2 per day or £10 per week. Mileage over 200 miles per day is charged at 5 p per mile. A deposit of £20 is required.

Using the figures in the table find the cost of hiring a car, including the deposit, in the following cases.
a) a Renault 5 for a weekend travelling 200 miles
b) an Astra for a day covering 120 miles
c) a Sierra for a week covering 500 miles
d) a Renault 5 for 9 days covering 1700 miles
e) a Sierra for 2 weeks to cover 1200 miles
f) an Astra for a day to drive 300 miles

Are there situations where it could be cheaper to hire a car than to use your own?

# A Holiday for Dave and Stuart

Dave and his friend Stuart have decided to go for a week's holiday in North Devon. For accommodation there are four possibilities.

(a) a guest house (they cannot afford a hotel)

(b) youth hostels

(c) hiring a caravan

(d) camping using a small tent belonging to Stuart's family.

Dave has an old car (see chapter 6, page 50) but he is not sure that he trusts it to get that far so they decide to investigate the cost of coach travel as well.

**1.** The cheapest place they can find is a farm which charges £75.00 per person per week for dinner, bed and breakfast, or £8.00 per night bed and breakfast. If they estimate that a meal out will cost them a minimum of £4.00 each, how much would they save each by booking dinner, bed and breakfast for 7 nights?

**2.** Charges at suitable youth hostels would be as follows.

| | |
|---|---|
| Bed | £2.80 |
| Breakfast | £1.40 |
| Evening meal | £1.80 |

plus £3.00 membership fee

Calculate the cost per person for 7 nights' dinner, bed and breakfast plus membership. Find out what the present rates are. What would it cost now?

**3.** The cheapest caravan they can find costs £56 per week to hire. They estimate that breakfast would cost them about 50 p each per day and that they could eat out in the evening for about £4.00 each. Calculate the cost per person of dinner, bed and breakfast for a week using these charges.

**4.** The charge for 2 persons, a tent and a car at a campsite in the area is £3.50 per night. Assuming the same food costs as in question 3, work out the cost per week for each of them if they camped.

**5.** Dave and Stuart feel that lunch should be a bar snack or sandwiches which they estimate will cost about £3 each per day. They will also need about £4 each per day for general spending money. Add these amounts on to the food and accommodation costs to find the total holiday costs each without travel, for each of the four types of accommodation.

**6.** Dave and Stuart both live in Kettering in Northamptonshire and the area they have picked for their holiday is around Ilfracombe in North Devon. Work out the distance from Kettering to Ilfracombe by car. Assuming that they will drive about 100 miles while staying in the area, find the total mileage for the whole holiday.

**7.** Using the total mileage cost of Dave's car worked out in chapter 6, page 50, question 6, estimate the cost of taking the car on holiday.

**8.** The single fare by coach from Kettering to Ilfracombe is £6.40. By adding on £12 each for trips and fares while in the area work out the total travel cost per person if they use public transport.

**9.** Work out the total holiday costs for the following options.

a) guest house and travel by public transport
b) youth hostelling and public transport
c) caravanning and using the car
d) camping and using the car.

**10.** Apart from cost, list some of the advantages and disadvantages of the different types of accommodation.

# Teacher's Notes

**1.** Most people have some kind of holiday and young people with a limited income obviously need to think about costs. The work on times and timetables is also an important part of general life skills which everyone needs to be able to deal with.

**2.** Extensions to other subject areas could include

*social studies* public and private transport, effects of road and motorway building, effects of the tourist trade on the UK economy

*science* photography

*personal development* the planning of a residential element to a course, working with others

*communications* advantages and disadvantages of different types of holiday, description of an enjoyable holiday, survey of holiday habits, obtaining information.

**3.** Students on vocational courses in construction or engineering could look at the construction of caravans and holiday homes. Caring courses could investigate holidays for elderly or handicapped people, while students on any type of retail course might look at the advertising and marketing of holidays in the UK.

**4.** Most tourist information centres in the UK can supply vast amounts of free or low-cost literature which students could obtain for themselves. A local coach operator may also be willing to discuss his programme with students.

Chapter 10

# A Holiday Abroad

This chapter leads on from the previous one, using the work done there and extending the ideas to include the extra problems you will meet when going abroad. As most countries you are likely to visit use metric measurements, the earlier work you have done on these will also be useful.

As travel and tourism become more important to the economy of the country, more jobs are becoming available in travel agents, information centres, etc. The work in this chapter could be useful if you are thinking of applying for this sort of job.

## 1. Planning a Holiday

### 1.1 Getting the Dates Right

If you are planning your own holiday completely, obviously you can go for as long a time as you please. Package holidays, on the other hand, last for a fixed time such as 8, 11 or 15 days. Remember that 15 days means 14 nights away from home not 15 whole days actually at your resort, so that if you leave on 11th June for an 8-day holiday you will be back home on the 18th June.

**1.** Find the return dates for the following holidays.

a) 8 days leaving on 29th June
b) 15 days leaving on 18th July
c) 11 days leaving on 28th August
d) 8 days leaving on 25th February 1984
e) 15 days leaving on 18th February 1982
f) 2 weeks Saturday to Saturday leaving on 20th July

**2.** Complete the following table for holidays in France.

| *Outward Channel crossing* | *Inward Channel crossing* | *Full days in France* |
|---|---|---|
| 20th July | 6th August | a)...... |
| 18th June | 13th July | b)...... |
| 28th August | c)...... | 14 |
| d)...... | 6th July | 16 |

### 1.2 Time Differences

If you are travelling from the UK to another country in Europe you are almost certain to have to adjust your watch. Most countries are

1 or 2 hours ahead of our time. For each particular country you would have to check up what to do, but in general, South-East Europe is 2 hours ahead of us and the rest of Europe 1 hour ahead. So for example, when it is 10.00 p.m. in the UK, it is 11.00 p.m. in Germany.

Some countries do not have different systems for summer and winter, which complicates matters even further as the changes you have to make depend on the time of year.

---

**1.** Greece is 2 hours ahead of the UK. If the flight to Greece takes $3\frac{1}{2}$ hours find the arrival times in Greek time for the following UK departure times.
a) 10.30 a.m. b) 9.00 p.m. c) 1800

**2.** On the last day of your holiday your watch is working in Greek time. If the flight home takes $3\frac{1}{4}$ hours, find the arrival times in the UK, in UK time, for the following departure times from Greece.
a) 1600 b) 5.30 a.m. c) 9.30 p.m.

**3.** France is 1 hour ahead of the UK. If you do not wish to drive in France any later than 6.00 p.m. French time, how many hours of driving will you have in France for the following ferry times from the UK (in UK time)?
a) 10.30 a.m. (ferry takes $3\frac{1}{2}$ hours)
b) 1200 (ferry takes $1\frac{1}{2}$ hours)
c) 3.00 p.m. (ferry takes $1\frac{3}{4}$ hours)
Because of some of the very fast Channel crossings available today, and the time difference, it is now possible to arrive back in England before you left France!

---

## 1.3 Currency Changes

One big difference in a holiday abroad is getting used to the money compared with our own (sterling).

If you go to France you will be using the franc, in Romania the leu, in Portugal the escudo.

It is important to understand this problem both ways round, i.e. how many francs should you get for £1 and what is 1 franc worth?

The banks in the UK will be able to give you the rate for £1 sterling and the rest you can work out for yourself, for example

£1 = 11.8 French francs (FF)

1 FF = $1 \div 11.8 \simeq 9$ p

5 FF = $5 \times 9$ p = 45 p

---

In this exercise give answers to the nearest p.

**1.** If £1 = 15.4 krone (Kr) in Denmark, find the sterling equivalents of
a) 1 Kr b) 5 Kr c) 20 Kr d) 100 Kr.

**2.** If £1 = 62 escudo (Esc) in Portugal, find the sterling equivalents of
a) $2\frac{1}{2}$ Esc b) 20 Esc c) 100 Esc.

**3.** If £1 = 41 dinar (Din) in Yugoslavia, find the sterling equivalents of
a) 20 Din b) 500 Din c) 50 para (1 Din = 100 para).

**4.** By looking in a local bank or in a newspaper, find the present-day sterling equivalents of the following.
a) a 20-forint note in Hungary
b) a 500-franc note in Belgium
c) a 50-groschen coin in Austria
d) a 10-deutschmark coin in West Germany

---

### 1.4 Choosing the Right Weather

If you are able to take your holidays at any time of the year then you can choose the best time to go to get the sort of temperature you prefer, for example in Greece the temperature in July or August can be as high as 32 °C (90 °F) – a bit hot you might think!

On the same sheet of graph paper draw graphs to show the maximum and minimum average daily temperatures in °C in different places using the following tables.

| | | *Jan* | *Feb* | *Mar* | *Apr* | *May* | *Jun* |
|---|---|---|---|---|---|---|---|
| Innsbruck (Austria) | Max°C | 2 | 4 | 11 | 15 | 21 | 23 |
| | Min°C | −6 | −4 | 0 | 3 | 8 | 11 |

| | | *Jul* | *Aug* | *Sep* | *Oct* | *Nov* | *Dec* |
|---|---|---|---|---|---|---|---|
| Innsbruck (Austria) | Max°C | 25 | 24 | 20 | 15 | 7 | 3 |
| | Min°C | 12 | 12 | 9 | 4 | 1 | −4 |

| | | *Jan* | *Feb* | *Mar* | *Apr* | *May* | *Jun* |
|---|---|---|---|---|---|---|---|
| Athens (Greece) | Max°C | 12 | 13 | 16 | 19 | 25 | 29 |
| | Min°C | 6 | 6 | 8 | 11 | 16 | 19 |

| | | *Jul* | *Aug* | *Sep* | *Oct* | *Nov* | *Dec* |
|---|---|---|---|---|---|---|---|
| Athens (Greece) | Max°C | 32 | 32 | 28 | 23 | 18 | 14 |
| | Min°C | 22 | 22 | 19 | 16 | 11 | 8 |

| | | *Jan* | *Feb* | *Mar* | *Apr* | *May* | *Jun* |
|---|---|---|---|---|---|---|---|
| Bergen (Norway) | Max°C | 4 | 4 | 6 | 9 | 13 | 16 |
| | Min°C | −1 | −1 | 0 | 3 | 6 | 9 |

| | | *Jul* | *Aug* | *Sep* | *Oct* | *Nov* | *Dec* |
|---|---|---|---|---|---|---|---|
| Bergen (Norway) | Max°C | 18 | 17 | 19 | 10 | 6 | 4 |
| | Min°C | 11 | 11 | 8 | 5 | 2 | 1 |

If you cannot think in °C then put a °F scale on your graph as well so that you get a better idea of the temperature.

Put a line across your graph to show the outdoor temperature where you are today.

## 2. Costing the Holiday

### 2.1 Going by Car

If you are going on holiday by car there are three major types of expenditure, apart from living costs and spending money, which you have to consider.

(a) petrol and other costs for the car

(b) ferry costs (or tunnel costs when it opens)

(c) insurance

**a) The Cost of Driving**

A motoring holiday abroad is likely to involve quite a high mileage so it is better to take into account the whole cost of using the car, i.e. to use a figure of about 18 p per mile for an average car.

The chart below shows approximate distances in miles from Calais to some major centres in Europe.

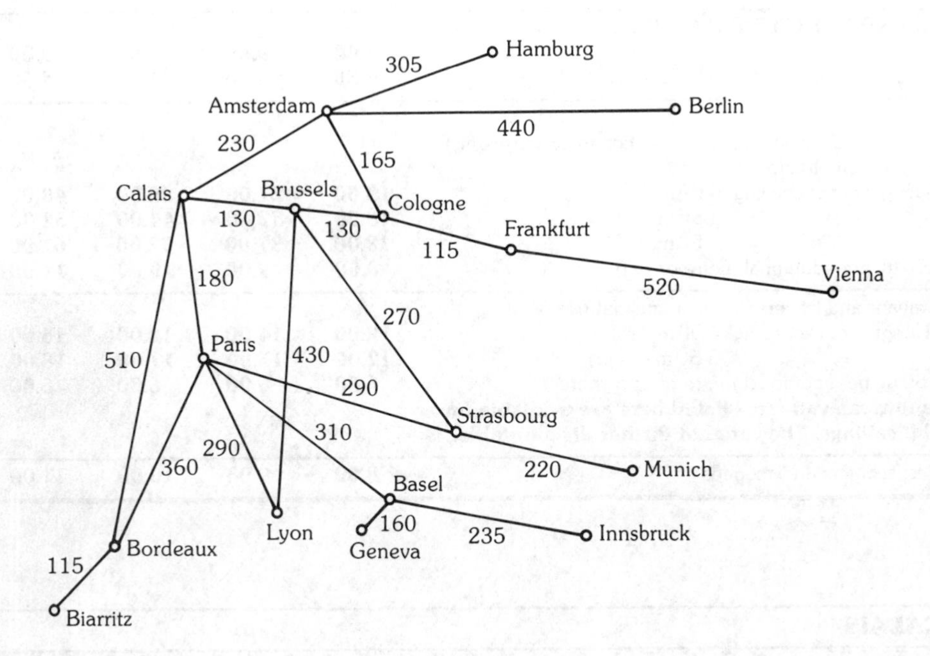


1. Use the chart above to find the distances from Calais to
   a) Biarritz b) Lyon via Paris
   c) Munich via Brussels d) Vienna
   e) Innsbruck.
2. Assuming that your home is 120 miles from Dover and that you will do about 300 miles of driving while based at your centre, estimate the total mileage for holidays based at each of the centres in question 1. (Remember to come home afterwards!)
3. Using a figure of 18 p per mile, find the approximate cost to the nearest £10 of using the car for holidays based at each of the centres in question 1.
4. Of this 18 p per mile about 2 p is for basic costs, such as tax and insurance, 6 p for petrol (slightly higher in places such as France) and the other 10 p for repairs and depreciation. If you hired a car for the holiday you would only be paying the 6 p per mile for petrol, plus hire charges. Get some information on hire charges and work out the car costs of a holiday based in Innsbruck using a hired car.

**b) Crossing the Channel**

The cost of crossing the Channel depends on route, time of year, time of day, size of car and, of course, the number of people going on holiday. For a particular holiday you would have to study the brochures carefully to work out the cost.

The following are extracts from an old time-table of one of the ferry companies based in the UK.

# TARIFF DOVER-CALAIS

**All fares are for single journeys unless otherwise stated** *Port taxes included*

| **Travelling with a Vehicle** | Tariff E £ | Tariff D £ | Tariff C £ | Tariff B £ | Tariff A £ |
|---|---|---|---|---|---|
| **DRIVERS AND VEHICLE PASSENGERS** | | | | | |
| Adults | **9.00** | **9.00** | **9.00** | **9.00** | **9.00** |
| Children (4 and under 14 years. Under 4 free) | **4.50** | **4.50** | **4.50** | **4.50** | **4.50** |
| **Vehicles** | | | | | |
| Cars, motor caravans, minibuses, vans (non-commercial use only) and motorcycle combinations. | | | | | |
| Overall length not exceeding 4.00 m | **18.00** | **27.00** | **36.00** | **43.00** | **49.00** |
| 4.50 m | **18.00** | **32.00** | **44.00** | **52.00** | **58.00** |
| 5.50 m | **18.00** | **37.00** | **52.00** | **61.00** | **67.00** |
| over 5.50 m, per additional metre or part thereof | **8.00** | **9.00** | **10.00** | **11.00** | **12.00** |
| Towed caravans and trailers (non-commercial use only). | | | | | |
| Overall length not exceeding 4.00 m | **12.00** | **14.00** | **14.00** | **15.00** | **15.00** |
| 5.50 m | **12.00** | **17.00** | **17.00** | **18.00** | **18.00** |
| over 5.50 m, per additional metre or part thereof | **6.00** | **6.00** | **6.00** | **6.00** | **6.00** |
| ★★★ **The low caravan fares listed here are available on ALL sailings. They are not further discountable.** | | | | | |
| Motorcycles, scooters and mopeds | **8.00** | **9.00** | **10.00** | **11.00** | **12.00** |

## DOVER-CALAIS

| AUG | W 1 | T 2 | F 3 | S 4 | S 5 | M 6 | T 7 | W 8 | T 9 | F 10 | S 11 | S 12 | M 13 | T 14 | W 15 | T 16 | F 17 | S 18 | S 19 | M 20 | T 21 | W 22 | T 23 | F 24 | S 25 | S 26 | M 27 | T 28 | W 29 | T 30 | F 31 |
|---|---|---|---|---|---|---|---|---|---|---|---|---|---|---|---|---|---|---|---|---|---|---|---|---|---|---|---|---|---|---|---|
| **0200** | D | D | C | C | D | D | D | D | D | C | C | D | D | D | D | D | C | C | D | D | D | D | D | D | D | D | D | D | D | D | D |
| **0400** | D | D | D | D | D | D | D | D | D | D | D | D | D | D | D | D | D | D | D | D | D | D | D | D | D | D | D | D | D | D | D |
| **0600** | C | C | C | C | C | C | C | C | C | C | C | C | C | C | C | C | C | C | C | C | C | C | C | C | C | C | C | C | C | C | C |
| **0730** | C | C | C | C | C | C | C | C | C | C | C | C | C | C | C | C | C | C | C | C | C | C | C | C | C | C | C | C | C | C | C |
| **0900** | B | A | A | A | B | B | B | B | A | A | A | B | B | B | B | A | A | A | B | B | B | B | B | B | B | B | B | B | B | B | B |
| **1030** | B | A | A | A | B | B | B | B | A | A | A | B | B | B | B | A | A | A | B | B | B | B | B | B | B | B | B | B | B | B | B |
| **1200** | B | A | A | A | B | B | B | B | A | A | A | B | B | B | B | A | A | A | B | B | B | B | B | B | B | B | B | B | B | B | B |
| **1330** | B | B | A | A | B | B | B | B | B | A | A | B | B | B | B | B | A | A | B | B | B | B | B | B | B | B | B | B | B | B | B |
| **1500** | B | B | A | A | B | B | B | B | B | A | A | B | B | B | B | B | A | A | B | B | B | B | B | B | B | B | B | B | B | B | B |
| **1630** | B | B | B | B | B | B | B | B | B | B | B | B | B | B | B | B | B | B | B | B | B | B | B | B | B | B | B | B | B | B | B |
| **1800** | B | B | B | B | B | B | B | B | B | B | B | B | B | B | B | B | B | B | B | B | B | B | B | B | B | B | B | B | B | B | B |
| **1930** | C | B | B | B | C | C | C | C | B | B | B | C | C | C | C | B | B | B | C | C | C | C | C | C | C | C | C | C | C | C | C |
| **2100** | C | B | B | B | C | C | C | C | B | B | B | C | C | C | C | B | B | B | C | C | C | C | C | C | C | C | C | C | C | C | C |
| **2230** | C | C | C | C | C | C | C | C | C | C | C | C | C | C | C | C | C | C | C | C | C | C | C | C | C | C | C | C | C | C | C |
| **2359** | D | C | C | C | D | D | D | D | C | C | C | D | D | D | D | C | C | C | D | D | D | D | D | D | D | D | D | D | D | D | D |

## CALAIS-DOVER

| AUG | W 1 | T 2 | F 3 | S 4 | S 5 | M 6 | T 7 | W 8 | T 9 | F 10 | S 11 | S 12 | M 13 | T 14 | W 15 | T 16 | F 17 | S 18 | S 19 | M 20 | T 21 | W 22 | T 23 | F 24 | S 25 | S 26 | M 27 | T 28 | W 29 | T 30 | F 31 |
|---|---|---|---|---|---|---|---|---|---|---|---|---|---|---|---|---|---|---|---|---|---|---|---|---|---|---|---|---|---|---|---|
| **0015** | D | D | D | C | C | D | D | D | D | D | C | C | D | D | D | D | D | C | C | D | D | D | D | D | C | C | C | C | D | D | D |
| **0200** | D | D | D | C | C | D | D | D | D | D | C | C | D | D | D | D | D | C | C | D | D | D | D | D | C | C | C | C | D | D | D |
| **0400** | D | D | D | D | D | D | D | D | D | D | D | D | D | D | D | D | D | D | D | D | D | D | D | D | D | D | D | D | D | D | D |
| **0600** | D | D | D | D | D | D | D | D | D | D | D | D | D | D | D | D | D | D | D | D | D | D | D | D | D | D | D | D | D | D | D |
| **0730** | C | C | C | C | C | C | C | C | C | C | C | C | C | C | C | C | C | C | C | C | C | C | C | C | C | C | C | C | C | C | C |
| **0915** | C | C | C | C | C | C | C | C | C | C | C | C | C | C | C | C | C | C | C | C | C | C | C | C | C | C | C | C | C | C | C |
| **1045** | B | B | B | B | B | B | B | B | B | B | B | B | B | B | B | B | B | B | B | B | B | B | B | B | B | B | B | B | B | B | B |
| **1215** | B | B | B | B | B | B | B | B | B | B | B | B | B | B | B | B | B | B | B | B | B | B | B | B | B | B | B | B | B | B | B |
| **1345** | B | B | B | A | A | B | B | B | B | B | A | A | B | B | B | B | B | A | A | B | B | B | B | B | A | A | A | B | B | B | B |
| **1515** | B | B | B | A | A | B | B | B | B | B | A | A | B | B | B | B | B | A | A | B | B | B | B | B | A | A | A | B | B | B | B |
| **1645** | B | B | B | A | A | B | B | B | B | B | A | A | B | B | B | B | B | A | A | B | B | B | B | B | A | A | A | B | B | B | B |
| **1815** | B | B | B | A | A | B | B | B | B | B | A | A | B | B | B | B | B | A | A | B | B | B | B | B | A | A | A | B | B | B | B |
| **1945** | B | B | B | A | A | B | B | B | B | B | A | A | B | B | B | B | B | A | A | B | B | B | B | B | A | A | A | B | B | B | B |
| **2115** | C | C | B | B | B | C | C | C | C | B | B | B | C | C | C | C | B | B | B | C | C | C | C | B | B | B | B | C | C | C | C |
| **2245** | C | C | B | B | B | C | C | C | C | B | B | B | C | C | C | C | B | B | B | C | C | C | C | B | B | B | B | C | C | C | C |

**PASSENGERS AND CARS SHOULD BE AT THE DOCKS AT LEAST 30 MIN BEFORE SAILING TIME**

**1.** A family intends leaving home at 5.00 a.m. to get a ferry from Dover. What is the earliest ferry they can get if their estimated journey time from home to Dover is
a) $1\frac{1}{2}$ hr b) 4 h c) 5 h?

**2.** The family consists of 2 adults and 2 children aged 5 and 8. They will be travelling on Monday 6th August. Find the total cost of the ferry from Dover to Calais for the 4 people.

**3.** Their car is 4.40 m long. Find the cost for the car in each of the cases in question **1**.

**4.** Use your answers to questions 2 and 3 to find the total cost of the Dover–Calais crossing for the family, who live a 4 h drive from Dover.

**5.** The family intend to return from Calais to Dover on the 1345 boat on Saturday 18th August. What will be their total ferry costs for both journeys?

**6.** Assuming that this family lives 120 miles from Dover and is taking a holiday based at Innsbruck, use your work from section 2.1 to find the total cost of car and ferry for them.

### c) Insuring Yourself

When going abroad for a holiday most people take out some sort of personal insurance. Many people going by car also take out insurance on their car. This is in addition to the 'green card' driver's insurance which is required to make your normal insurance valid abroad.

The table below shows the rates for a system of holiday and vehicle insurance.

## Holiday Insurance Premiums

Free for children under 4 years, reduced rate for children over 4 and under 14 years.

| | ADULTS | CHILDREN |
|---|---|---|
| **Up to 3 days** | **£4.00** | **£2.70** |
| **Up to 6 days** | **£5.50** | **£3.70** |
| **Up to 9 days** | **£6.40** | **£4.30** |
| **Up to 17 days** | **£8.40** | **£5.60** |
| **Up to 24 days** | **£9.90** | **£6.60** |
| **Up to 31 days** | **£11.60** | **£7.70** |
| **Cost per week extra** | **£1.60** | **£1.10** |

The Winter Sports exclusions of Sections 1, 2 and 3 can be reinstated for twice the above premiums.

### Vehicle Breakdown Extension

| | |
|---|---|
| **Up to 3 days** | **£13.80** |
| **Up to 6 days** | **£16.20** |
| **Up to 9 days** | **£18.20** |
| **Up to 17 days** | **£21.30** |
| **Up to 24 days** | **£23.30** |
| **Up to 31 days** | **£26.30** |
| **Costs per week extra** | **£3.00** |

**The cost for trailers and caravans is £8.00 for any period.**

**N.B.1** The premium payable is calculated on the number of days required, including your outward voyage sailing date and your inward voyage sailing date. The period of cover does however extend to include the direct journeys in the United Kingdom between your home and the port, subject to a maximum of 36 hours for each journey.

**1.** What would be the cost of holiday insurance for 2 adults and 1 child for a week?

**2.** How much would vehicle insurance cost for 6 weeks?

**3.** How much would the insurance be per person for an 8-day skiing holiday?

**4.** Work out the total cost of holiday and vehicle insurance for the family in the exercise of section 2.1(b).

**5.** Add your answer to question 4 to the total ferry and car costs to find the cost of travel and insurance for the family in section 2.1(b).

**6.** To find a total holiday cost you would have to add on the cost of food, accommodation and spending money. Get some information on campsite fees in Austria and estimate the total cost of a camping holiday for this family for the dates given.

## 2.2 Package Holidays

If you go on a package holiday you don't have the bother (or fun?) of organising everything yourself. All you have to do is to book your holiday, turn up at the airport and find the money to pay for it. The main problem is making sure that you understand the brochure and know exactly what you are getting.

The table below shows a section of a brochure produced by a tour operator.

Use the table to work out the following.

**1.** the basic cost without supplements of a 14-night holiday for 1 person at the Maricel starting Saturday 8th June

**2.** the flight supplement for the holiday in question 1 if travelling from Luton

**3.** the complete cost including flight supplement and insurance of a 7-night holiday for 1 person at the De Mar Sol leaving Heathrow on Thursday 19th September

**4.** the complete cost of a 7-night holiday for 2 adults at the Maricel, leaving Manchester on Tuesday 30th July at 0855

### FLIGHTS TO PALMA
### Prices per person in £ including airport charges

| Hotel | Playa Marina Sol (b) (Half board) | | | | Maricel (Full board) | | | | De Mar Sol (Half board) | | | |
|---|---|---|---|---|---|---|---|---|---|---|---|---|
| Holiday number | Q2075 | | | | Y2006 | | | | Y2011 | | | |
| No. of nights in hotel | 7 | | 14 | | 7 | | 14 | | 7 | | 14 | |
| Insurance per person | £8.50 | | £9.90 | | £8.50 | | £9.90 | | £8.50 | | £9.90 | |
| | Adult | Child | Adult | Child | Adult | Child | Adult | Child | Adult | Child | Adult | Child |
| 23 Mar-1 Apr | — | — | — | — | **192** | 164 | **300** | 255 | **224** | 191 | **361** | 307 |
| 2-8 Apr | **164** | 156 | **234** | 211 | **209** | 178 | **326** | 278 | **248** | 211 | **405** | 345 |
| 9-30 Apr | **161** | 137 | **236** | 189 | **218** | 186 | **348** | 296 | **268** | 228 | **458** | 390 |
| 1-19 May | **168** | 143 | **252** | 202 | **229** | 195 | **368** | 313 | **302** | 257 | **514** | 437 |
| 20-27 May | **175** | 167 | **263** | 237 | **235** | 200 | **374** | 318 | **309** | 263 | **520** | 442 |
| 28 May-6 Jun | **172** | 155 | **260** | 221 | **234** | 199 | **373** | 318 | **308** | 262 | **519** | 442 |
| 7-20 Jun | **177** | 160 | **268** | 228 | **239** | 204 | **378** | 322 | **313** | 267 | **527** | 448 |
| 21-27 Jun | **184** | 166 | **303** | 258 | **244** | 208 | **372** | 317 | **324** | 276 | **567** | 482 |
| 28 Jun-4 Jul | **214** | 193 | **327** | 278 | **244** | 208 | **372** | 317 | **361** | 307 | **608** | 517 |
| 5-18 Jul | **224** | 209 | **340** | 306 | **247** | 210 | **374** | 318 | **374** | 318 | **634** | 539 |
| 19 Jul-4 Aug | **236** | 220 | **353** | 318 | **252** | 215 | **380** | 323 | **386** | 329 | **646** | 550 |
| 5-11 Aug | **235** | 219 | **350** | 315 | **250** | 213 | **378** | 322 | **384** | 327 | **644** | 548 |
| 12-25 Aug | **231** | 215 | **346** | 312 | **248** | 211 | **375** | 319 | **382** | 325 | **639** | 544 |
| 26 Aug-8 Sep | **227** | 212 | **342** | 308 | **245** | 209 | **373** | 318 | **368** | 313 | **610** | 519 |
| 9-29 Sep | **221** | 195 | **326** | 278 | **242** | 206 | **367** | 312 | **358** | 305 | **587** | 499 |
| 30 Sep-6 Oct | **176** | 155 | **255** | 217 | **227** | 193 | **347** | 295 | **308** | 262 | **508** | 432 |
| 7-25 Oct | **170** | 150 | **250** | 213 | **223** | 190 | **343** | 292 | **304** | 259 | **504** | 429 |

**Flight Supplements from UK airports**

| No of nights | Depart day/time | Arr. back day/time | First dep. | Last dep. | Add |
|---|---|---|---|---|---|
| **FROM GATWICK** Flight time 2hr 5 (B. Airtours 737/Tristar) | | | | | |
| 7 or 14 | Tue 2200 | Wed 0350 | 7 May | 22 Oct | **£0** |
| 7 or 14 | Tue 1715 | Tue 1505 | 7 May | 22 Oct | **£13** |
| 7 or 14 | Sat 0835 | Sat 1425 | 30 Mar† | 19 Oct | **£23** |
| **No Sat flight supplement on deps to 6/5 (1550, return 2140)** | | | | | |
| **FROM HEATHROW** Flight time 2hr 5 (Iberia 727) | | | | | |
| 7 or 14 | Thu 1715 | Thu 1615 | 4 Apr | 24 Oct | **£29** |
| 7 or 14 | Sat 1715 | Sat 1615 | 6 Apr | 19 Oct | **£39** |
| **(Flight supplements in April-Thur deps £14; Sat deps £24)** | | | | | |
| **FROM LUTON** Flight time 2hr 30 (Dan Air/Monarch 1-11) | | | | | |
| 7 or 14 | Sat 1540 | Sat 2125 | 30 Mar† | 20 Apr | **£7** |
| 7 or 14 | Sun 2145 | Mon 0320 | 5 May | 20 Oct | **£7** |
| **FROM MANCHESTER** Flight time 2hr 30 (BA 757/Airtours 737, Tristar) | | | | | |
| 7 or 14 | Tue 2300 | Wed 0515 | 7 May | 22 Oct | **£15** |
| 7 or 14 | Tue 0855 | Tue 2320 | 7 May | 22 Oct | **£26** |
| 7 or 14 | Sat 1600 | Sat 2225 | 30 Mar† | 27 Apr | **£13** |
| 7 or 14 | Sat 1600 | Sat 2225 | 4 May | 19 Oct | **£37** |
| **FROM BIRMINGHAM** Flt time 2hr 20 (Airtours 737/Aviaco Airbus) | | | | | |
| 7 or 14 | Sat 1500 | Sat 2050 | 30 Mar† | 27 Apr | **£15** |
| 7 or 14 | Sat 1810 | Sat 1710 | 4 May | 19 Oct | **£38** |
| **FROM BRISTOL** Flt time 2hr 10 (Aviaco DC9/727) | | | | | |
| 7 or 14 | Sat 1800 | Sat 1715 | 30 Mar† | 27 Apr | **£12** |
| 7 or 14 | Sat 1115 | Sat 1030 | 4 May | 19 Oct | **£35** |
| **FROM EAST MIDLANDS** Flight time 2hr 20 (Aviaco DC9/BA 1-11) | | | | | |
| 7 or 14 | Sat 2005 | Sat 1920 | 30 Mar† | 27 Apr | **£17** |
| 7 or 14 | Sat 0700 | Sat 1245 | 4 May | 19 Oct | **£39** |
| **FROM GLASGOW** Flt time 2hr 55 (B. Airways 757/Aviaco 727) | | | | | |
| 7 or 14 | Sat 0830 | Sat 1510 | 30 Mar† | 20 Apr | **£33** |
| 7 or 14 | Sat 1620 | Sat 1535 | 4 May | 19 Oct | **£45** |

NOTE: † First 14 night departure is one week earlier than dates shown. All last 14 night departures are one week earlier than dates shown.
*NB: B'ham, Glasgow, E. Midlands - no 14nt deps on 20 & 27/4.

**5.** Work out the total cost for a family of 2 adults and 2 children taking the holiday in question **4**.

**6.** Get a brochure from a local travel agent, choose a holiday and work out the total cost for your family.

---

# A Package Holiday Abroad

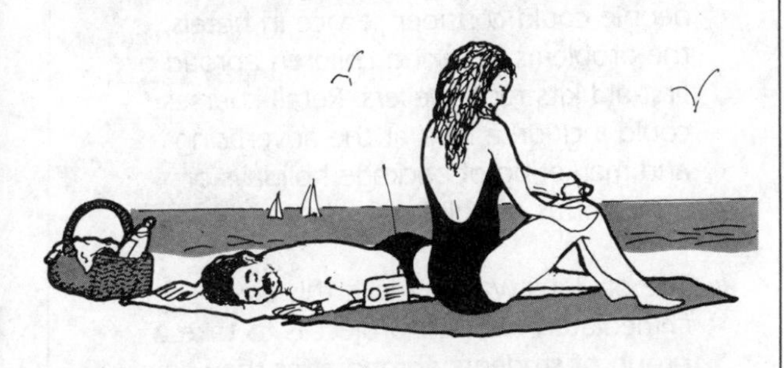

It is winter and Colin and Cheryl are wondering whether they could afford a 2-week holiday in Spain in the summer. They decide to cost it out.

Cheryl can take her holiday at any time but Colin has been told that he can only take 2 weeks holiday in the summer and that it must be in July.

**1.** Look at a brochure which covers Spain and choose a holiday. Write down the details, for example travel company, place, hotel, dates, cost per person, times of flights.

**2.** Colin decides to return to work on his first full day back home, unless that is at a weekend. On which day would he be back at work?

**3.** They must get to the airport at least 1 hour before the flight time. Allowing 3 hours to get to the airport at what time should they leave home?

**4.** If they estimate that it will take 1 hour to clear customs and $2\frac{1}{2}$ hours to get home from the airport at what time will they expect to get home?

**5.** How much will the holiday cost for both of them, including any extra charges mentioned in the brochure? (State what these extras are.)

**6.** They estimate that they will need about £50 each for spending money for a full-board holiday, £100 each for half-board or £150 each if they have chosen bed and breakfast only. How much spending money will they need between them for the holiday you have chosen?

**7.** If the exchange rate is 193.82 pesetas to the £ how many pesetas will they get for the spending money in question **6**?

**8.** If travel to and from the airport will cost about £11 each what should be the total cost of the holiday for both of them including spending money?

**9.** To book the holiday they must pay a £10 deposit each and the remainder must be paid at least one month before the day of departure. What is the last day on which they can pay the balance?

**10.** If they book the holiday on 1st February how long have they got to save up the money to pay the balance? How much should they put away each week to do this?

# Teacher's Notes

**1.** This chapter is really about using leisure time and may not be easy to fit in with anything except a more general project on holidays.

**2.** Extensions to other subjects could include
*personal development* a group activity to plan a holiday
*career development* job prospects in the travel and tourist industry
*communications* reading and interpreting brochures and insurance documents, filling in forms, finding out about resorts, use of *Holiday Which*, etc.
*information technology* computerisation of booking systems and on-line information links
*social studies* difference in lifestyle and standard of living in European countries, effects of tourism and increasing air traffic

**3.** Engineering students would have a useful project connected with this work in looking at car spares kits for holidays abroad. Students on courses involving services to people could consider service in hotels, the problems of taking children abroad or first-aid kits for travellers. Retail courses could include a look at the advertising and marketing of package holidays or other forms of holiday.

**4.** The obvious way to make this an immediately relevant project is to take a group of students abroad after they have planned the route, worked out costs, etc.

Chapter 11

# Shapes and Patterns

In chapter 5 page 37 you worked with the three standard shapes which you will meet most frequently – the rectangle, the triangle and the circle. In this chapter the work will be extended to take in some other shapes which you are likely to see around you, in books, in nature and in man-made objects.

Standard shapes are often used in patterns on textiles, wallpaper, woodwork and so on, and this chapter also provides an opportunity for you to design your own patterns.

## 1. Standard Shapes

### 1.1 Quadrilaterals

Quadrilateral is the name given to any figure with 4 straight sides. Squares and rectangles are special sorts of quadrilaterals but there are others.

One of the simplest of these figures is the *kite*. The kite has two pairs of equal sides as marked in the diagram. Its area can be calculated by dividing it into two equal triangles as shown by the broken line.

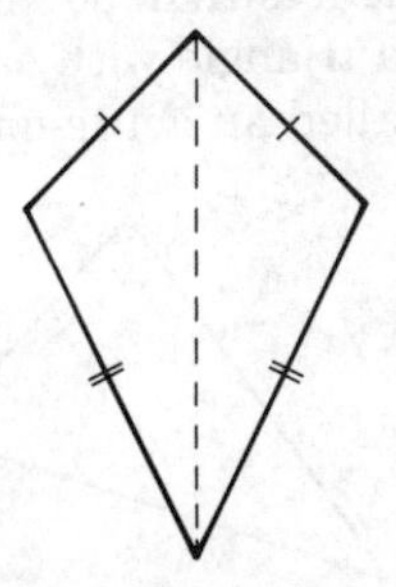

*NOTE* Lines which join opposite corners of a quadrilateral like the broken line shown are called *diagonals*.

Another standard figure is the *rhombus*. This is a bit like a kite but *all* the sides are equal – it is sometimes called a diamond. Again a diagonal divides it into 2 equal triangles.

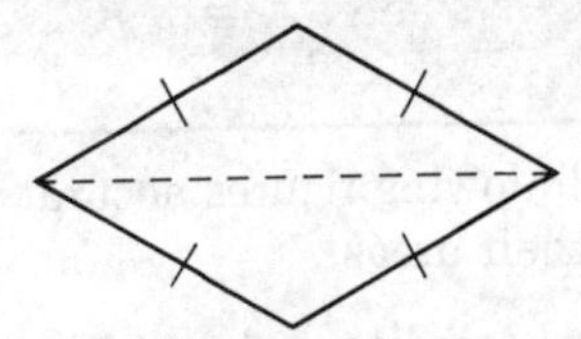

The *trapezium* is a four-sided figure which generally has no equal sides but which has two sides which are parallel (shown by arrows on a diagram).

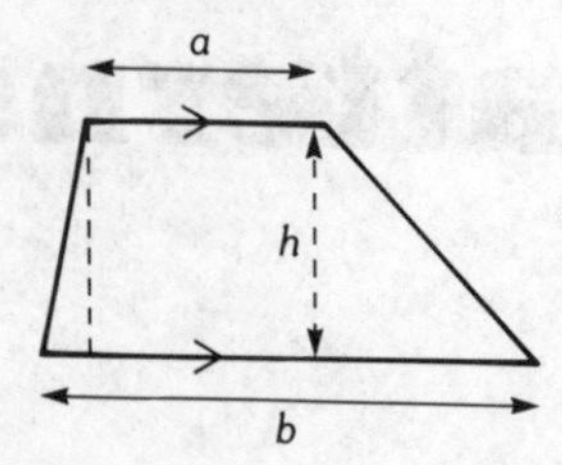


*Parallel lines* are lines which are always the same distance apart (like railway lines).

The area of a trapezium can be found by dividing it into a rectangle and two triangles as shown, or by using the formula

area of trapezium = average width × height

or $A = \left(\frac{a+b}{2}\right) \times h$

The general name for a quadrilateral which has two pairs of parallel sides is a *parallelogram*. The square, rectangle and rhombus are in fact special cases of the parallelogram. Again a diagonal divides the parallelogram into two equal triangles.

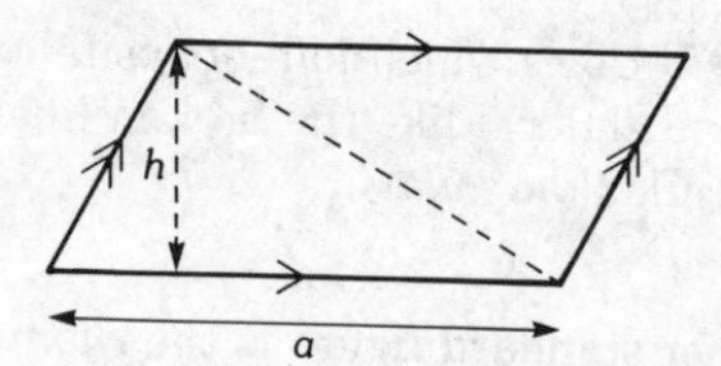


Since the area of a triangle is $\frac{1}{2}$ base × height,

area of parallelogram = base × height
or $A = a \times h$

Draw the following figures accurately and calculate their areas.

**1.** a kite whose diagonals are 6 cm and 4 cm long

**2.** a rhombus whose diagonals are 6 cm and 4 cm long

**3.** a trapezium whose parallel sides are 7.5 cm and 4.5 cm long with height 3.7 cm

**4.** a parallelogram with base 6.2 cm and height 3.2 cm

In most of these questions you will find that you have drawn a different picture from other people in your group but that your answers for the areas are the same – for which one does this not happen, and why?

## 1.2 Triangles

Triangles come in all shapes and sizes and there are certain words used to describe special types of triangle.

A triangle with two equal sides is called an *isosceles* triangle.

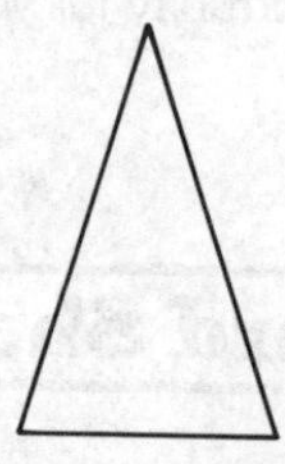

A triangle with all 3 sides equal is called an *equilateral* triangle.

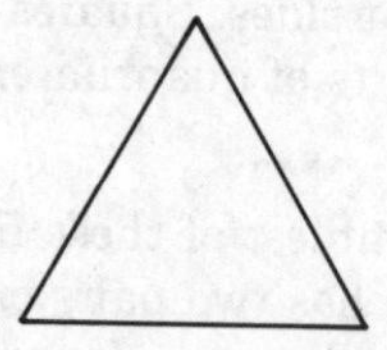

Since an angle less than 90° is called an acute angle, a triangle with each angle less than 90° is called an *acute-angled triangle.*

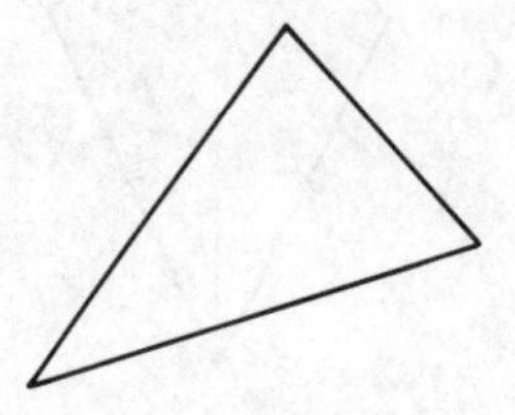

Angles between 90° and 180° are obtuse angles so that a triangle with one angle more than 90° is called an *obtuse-angled triangle.*

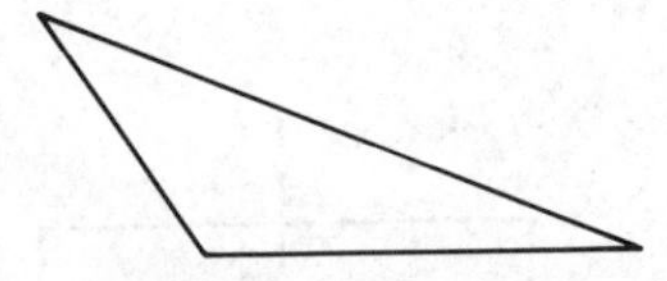

An angle of 90° is called a right angle so that a triangle with one angle equal to 90° is called a *right-angled triangle.*

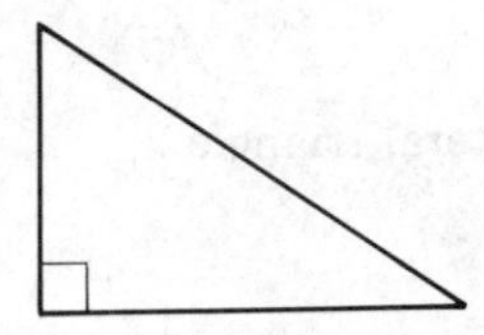

Whatever the shape of the triangle the formula for the area used in chapter 5 still applies.

$$\text{area} = \tfrac{1}{2}\,\text{base} \times \text{height}$$

In the following questions you will need to draw triangles from the given information, take measurements from your diagram and carry out some calculations. For this you will need a ruler, a pair of compasses, a pencil and a protractor.

---

For the following triangles

a) state which of the names in this section could be applied to each one
b) measure each angle
c) find the sum of the angles for each triangle
d) measure the height
e) find the area of the triangle.

**1.** sides 4 cm, 5 cm, 8 cm

**2.** sides 4.5 cm, 6 cm, 7.5 cm

**3.** sides 6 cm, 4 cm, 6 cm

**4.** sides 5 cm, 6 cm, 7 cm

---

## 1.3 Polygons

Strictly the word polygon means any figure with straight sides but it is usually used for figures with more than four sides. Each type of polygon has a special name, for example a *pentagon* has 5 sides;

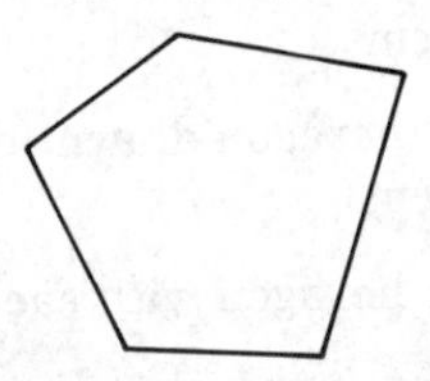

a *hexagon* has 6 sides;

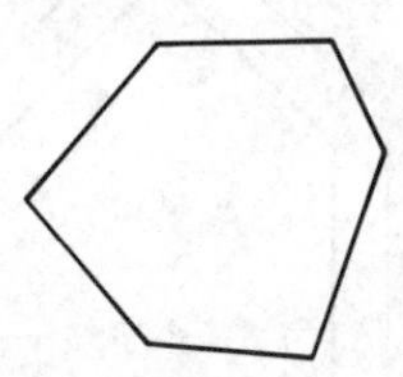

an *octagon* has 8 sides.

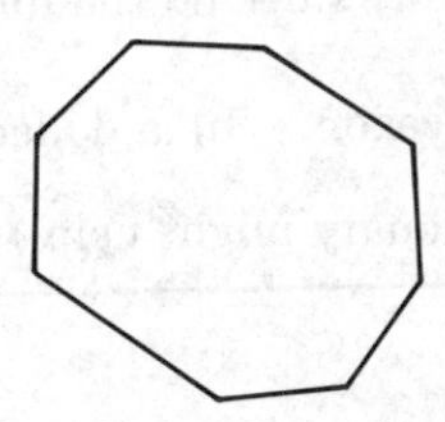

You may see some of these shapes around you if you look at coins, building plans, wallpaper patterns etc.

Most polygons you will see around you have all sides equal – they are then called *regular polygons*. The area of a regular polygon can be found by dividing it up into equal triangles. The best way to draw a regular polygon is to draw a circle and then divide the angle at the centre of the circle up into however many equal parts you need.

Draw the following figures, measure their angles, find the sum of the angles for each figure and calculate its area.

**1.** a regular pentagon drawn in a circle of radius 4 cm

**2.** a regular octagon drawn in a circle of radius 4 cm

**3.** a regular hexagon drawn in a circle of radius 5 cm

**4.** a regular hexagon with each side 5 cm long

**5.** the pentagon shown in the diagram below

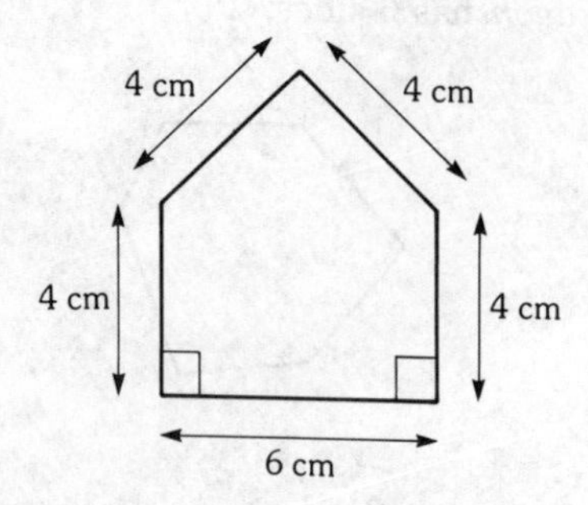


**6.** How many sides do the following figures have?
a) a decagon  b) a dodecagon

(A dictionary might help.)

# 2. Working with Shapes

## 2.1 Symmetry

If you look around you at the furniture in a room, at the faces of other people or at posters on the wall you are likely to see examples of what is called *line symmetry*.

An object has line symmetry if a line can be drawn on it which divides it into two exactly equal parts, for example a kite or a capital letter T. Some shapes may have more than one line of symmetry, for example

a rectangle has 2,

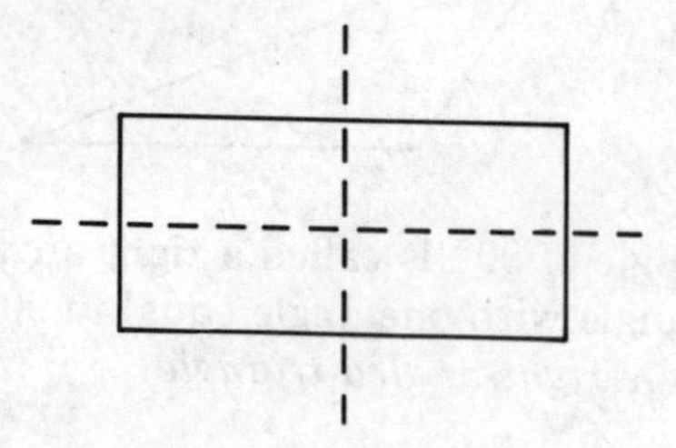

an equilateral triangle 3,

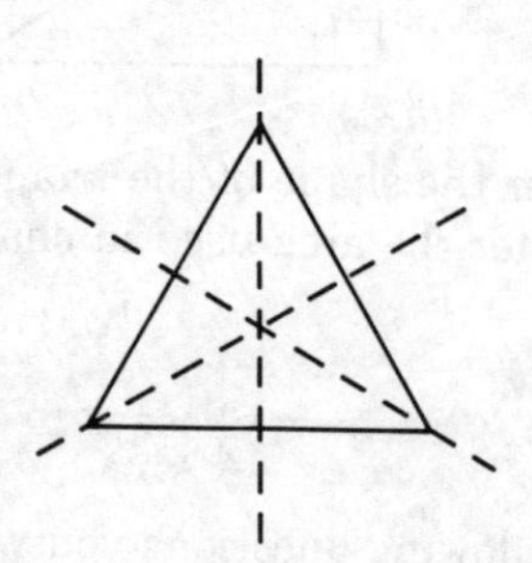

and so on.

There is also another type of symmetry which is called *rotational symmetry*. An object has rotational symmetry if it can be turned about a central point through an angle less than 360° without changing the picture, for example, a square turned through 90° about its centre looks exactly the same.

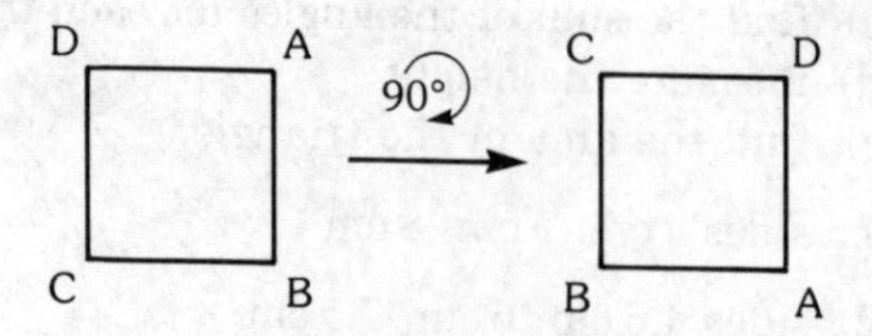


The *order of rotational symmetry* is the number of times an object can be turned before it gets back to its original position, for example, a square has rotational symmetry of order 4.

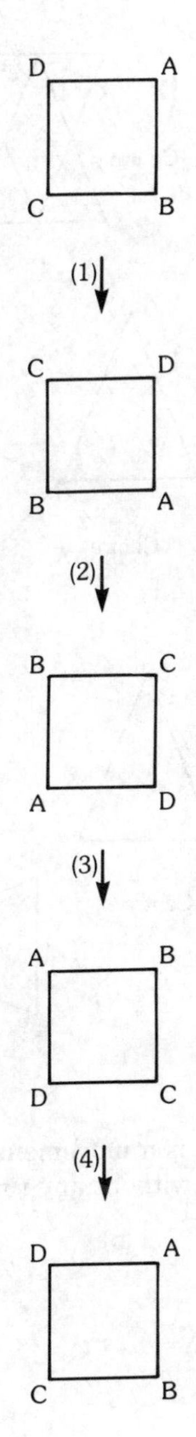


Some shapes, such as a square, have both line and rotational symmetry. Other standard shapes have one or the other, or sometimes neither.

**1.** Complete the following table.

| *Shape* | *Number of lines of symmetry* | *Order of rotational symmetry* |
|---|---|---|
| Square | | |
| Rectangle | | |
| Parallelogram | | |
| Kite | | |
| Rhombus | | |
| Isosceles triangle | | |
| Equilateral triangle | | |
| Regular hexagon | | |

**2.** If a figure is symmetric you can use this fact to find out about its lengths and angles. Draw the following figures including their diagonals and mark on your diagrams any lengths or angles which are equal.

a) a parallelogram b) a rhombus
c) a kite

## 2.2 Simple Transformations

If you look around at patterns for textiles, wallpapers, etc. you will find that they are often made up of a few basic shapes used over and over again in different ways.

One way of doing this is to carry out certain basic transformations on simple shapes. In this section you will be looking at the effect of certain transformations on an isosceles triangle.

**Reflection**

In a reflection an object is repeated on the other side of a mirror line so that the object and image together have line symmetry, for example

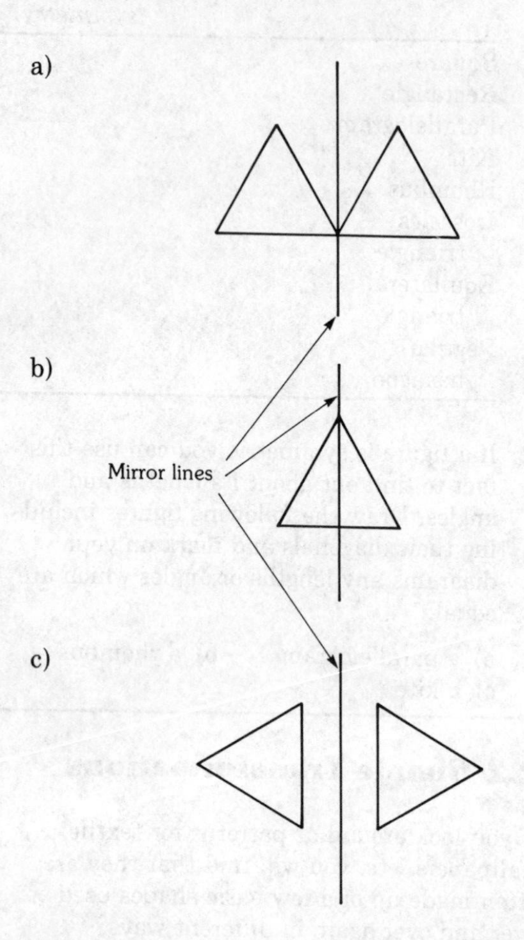


Notice that in b) the object and image are the same.

**Rotation**

Here the object is turned about a fixed point (the centre) through any angle, for example

a)

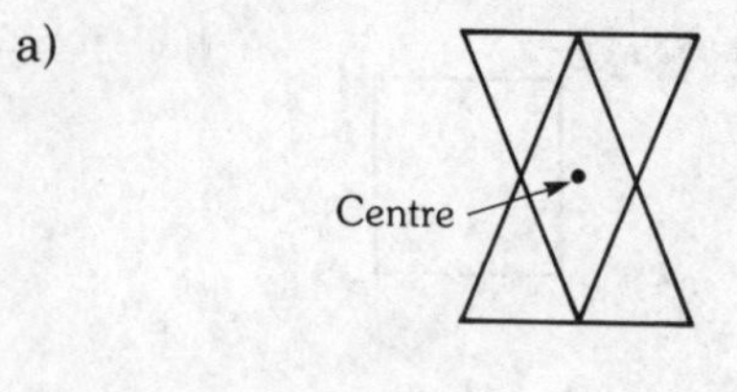


b)

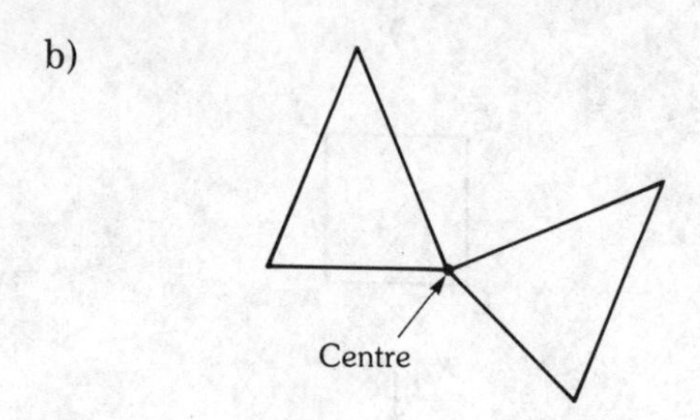


c)

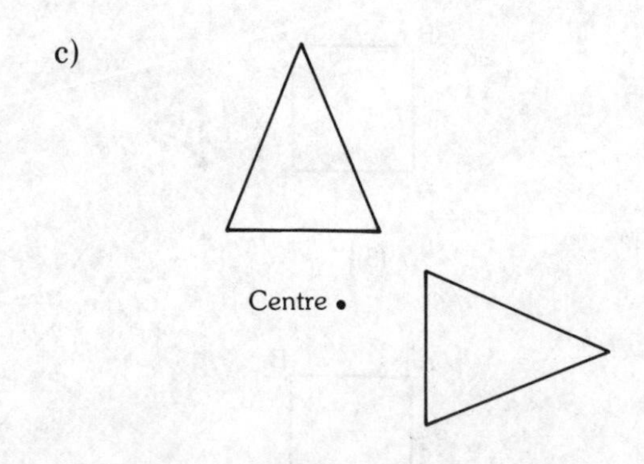


**Translation**

A translation is a movement of an object in any direction without any turning, for example

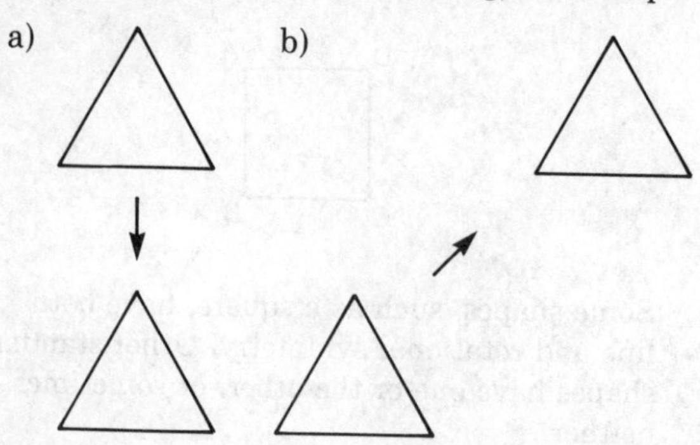

These 3 transformations are called *isometries* because they do not change the shape or size of the object, only its position.

A further isometry which is useful in design is a combination of a reflection and a translation. It is called a *glide reflection*. It is made up of a reflection and a translation parallel to the mirror line, for example

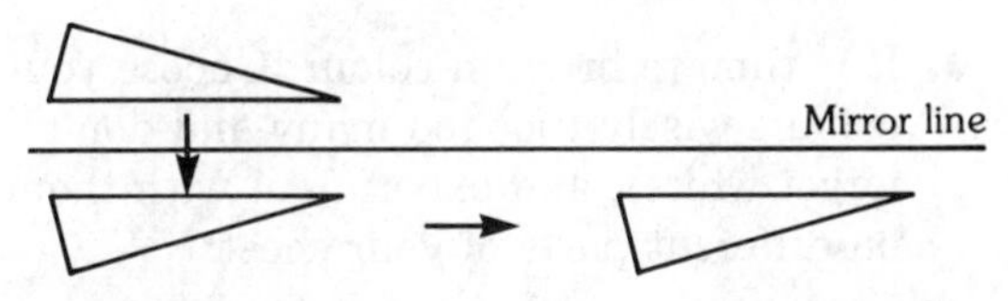


**Stretch**
A stretch is what it says – the object is stretched in one or more directions, for example

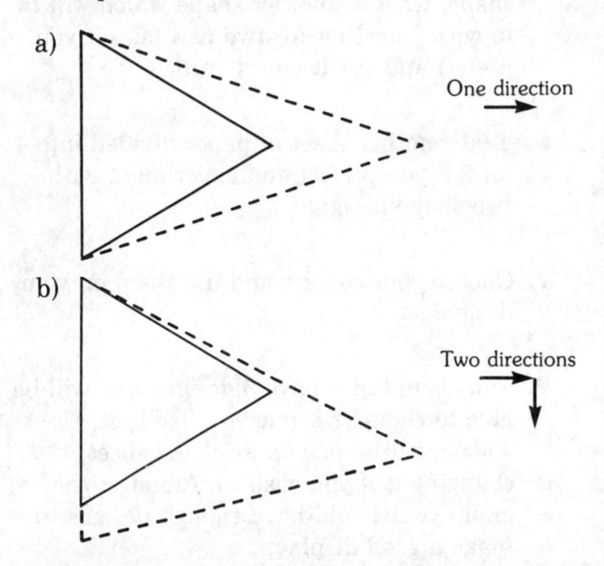


Here, of course, you will generally end up with an image which is a different shape from the object.

**Enlargement**
When a picture is enlarged it is made bigger in all directions in the same ratio, for example

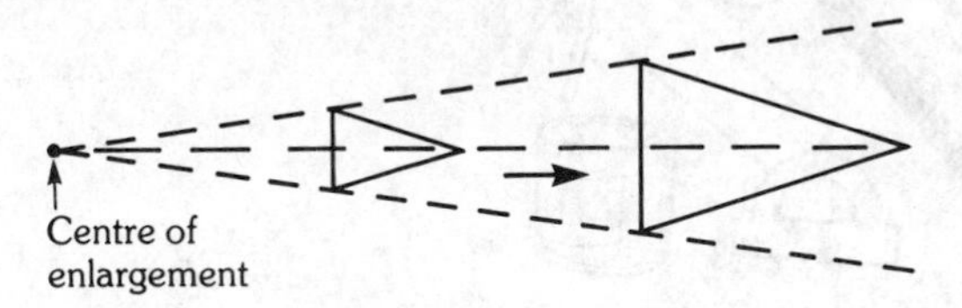


Here the image is the same shape as the object but a different size – the shapes are said to be *similar*.

The fraction $\dfrac{\text{length of line in image}}{\text{corresponding object length}}$ is

called the *scale factor*.

Note that the word *enlargement* is also used when objects are made smaller so that the scale factor could be less than 1 or even negative.

A negative scale factor means that the lines are drawn in the opposite direction from the original, so that the centre of enlargement is between the object and the image.

---

Draw on card and cut out an isosceles triangle ABC with AB = 4 cm, BC = 6 cm and CA = 6 cm. Use this basic shape as the object for the following transformations drawing the object and image on your paper.

**1.** a reflection with AB as the mirror line

**2.** a reflection with a line through B parallel to AC as the mirror line

**3.** a rotation of 90° clockwise about C

**4.** a translation of 3 cm in the direction AB and 2 cm perpendicular to this direction

**5.** a reflection in a line through B perpendicular to AB followed by a translation of 5 cm parallel to the mirror line

**6.** a stretch in the direction AB so that AB becomes twice as long

**7.** an enlargement with C as the centre and with scale factor 2

**8.** an enlargement with the mid-point of AB as the centre and scale factor −2.

---

# Making up a Design

Sarah is on an art and design course at a college and her main interest is in designing fabrics. Unfortunately she is not a great artist so she decides that she will have to stick to working with simple shapes which she can draw easily. For this work you will need some card, drawing instruments and some sheets of A2 paper.

**1.** Sarah's first task is to choose a basic shape to work with. Choose one shape from those used in this chapter, draw it on card and cut it out. It is best not to make its overall length or width more than 8 cm.

**2.** Divide your paper into 4 roughly equal parts by drawing a cross on it and draw your basic shape near the middle of each part.

**3.** Now it's up to you – for the time being pretend you are Sarah. Use each section of your paper to experiment on making designs using your basic shape and some of the transformations mentioned in the chapter. You may want to use one transformation per section, a mixture, or any other combination.

**4.** It is time to bring in colour. Choose your colours wisely (not too many and don't forget white), and experiment with them on different parts of your sheet.

**5.** Sarah has now produced some quite good designs and is feeling more confident so she decides to try to use more than one shape. Choose another shape which will fit in with your first (or two new ones if you prefer) and cut it out of card.

**6.** Take another sheet of paper divided into 4 (or 2 if you prefer) and experiment with two-shape designs.

**7.** Choose your colours and use them on your designs.

**8.** You should now have 8 designs and will be able to choose the one you like best. Use your favourite one on a full A2 sheet, changing it if you wish to. Amongst the group you should have enough designs to make a good display.

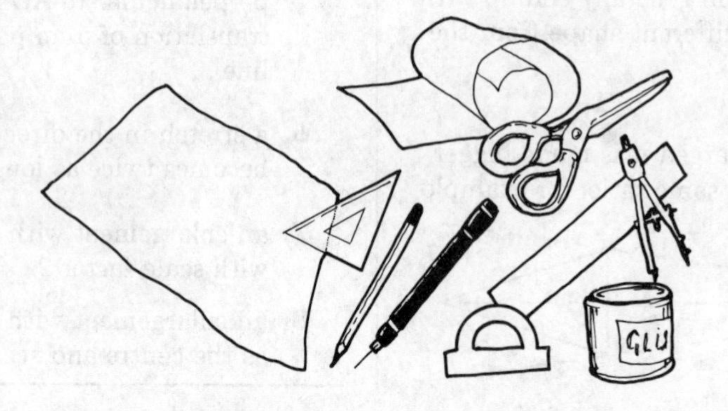

# Teacher's Notes

**1.** Very few students are totally useless at mathematics; most people can achieve something. Hopefully this project will appeal to some students who find the numerical parts of mathematics difficult. It will show them a new side of the subject.

**2.** Possible extensions to other subject areas are

*career development* careers in design and display, shop display techniques

*information technology* use of computer graphics for design work, production of a design or logo

*communications* survey of fabric and paper designs, street patterns

*practical skills* using designs to produce something such as a tile, a poster etc.

**3.** The most obvious vocational link is in the field of design and display work but there are also possible links with construction work, such as layout of paved areas etc. A group following a caring vocational option may wish to look at the use of pattern and colour by young children.

Chapter 12

# Energy in the Home

No matter where they live in the world, people use energy in their homes for cooking, heating and lighting. In this country, most people cook by gas or electricity. Heating could be by coal, oil, gas, electricity or a mixture of these. Where you live may determine which system you use.

Bringing energy to your home provides jobs for thousands of people. Many people are employed making equipment such as electric cookers, gas fires, light bulbs and washing machines. Others spend their time reading meters, helping to make up bills and advising customers on energy costs.

This project is designed to help you to work out the costs of using energy in your own home and to help you to think about the use of energy in factories, shops and offices.

## 1. How Much do you Use?

The electricity and gas supplies to most homes are fed through a meter so that you and the Gas and Electricity Boards can tell how much you have used.

### 1.1 Reading a Meter

The dials on a gas meter usually look like the ones opposite. You only need to read the bottom set of 4 dials, which in the diagram opposite give a reading of 7519 read from left to right. The readings you get from the meter are in hundreds of cubic feet.

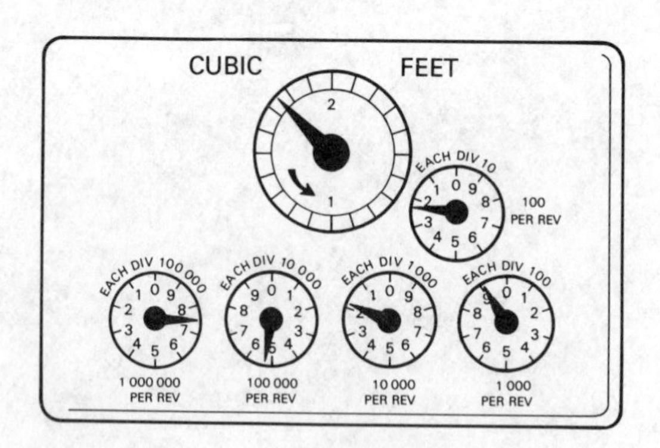


Notice that if the pointer is between two numbers you take the smaller of the two except when the pointer is between 9 and 0. In this case take 9 as the reading.

Now read the following meters.

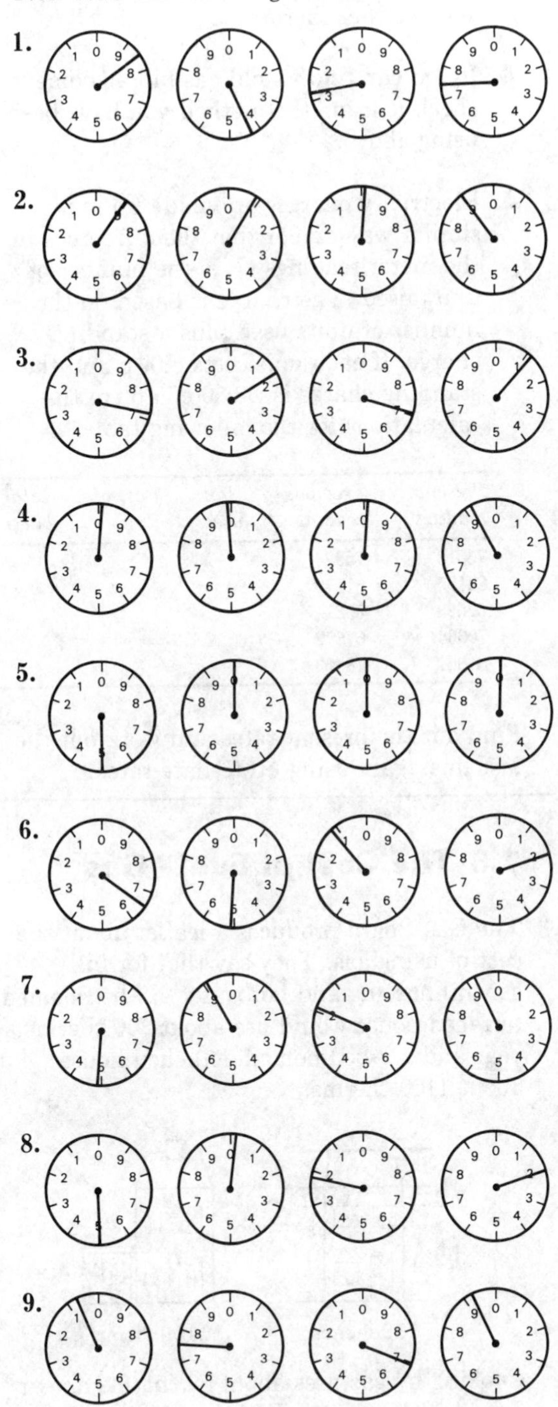


Some gas meters look like the one below.

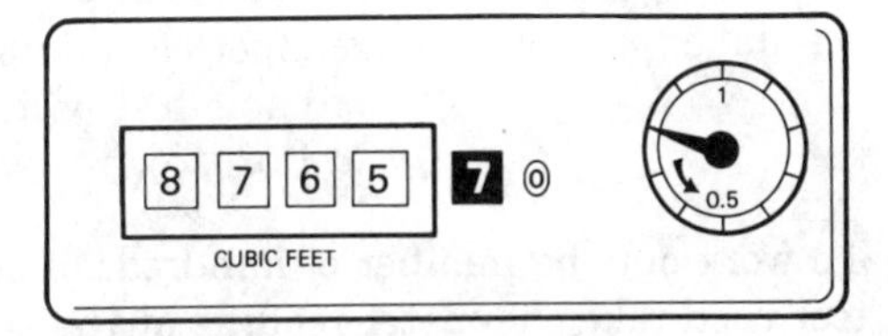


They are called 'digital' and are much easier to read.

Electricity meters come in various shapes and sizes. Your local electricity showroom can provide you with a leaflet about digital meters and Economy 7 meters. The standard electricity meter below is similar to the gas meter shown on the previous page. Read it just like a gas meter.

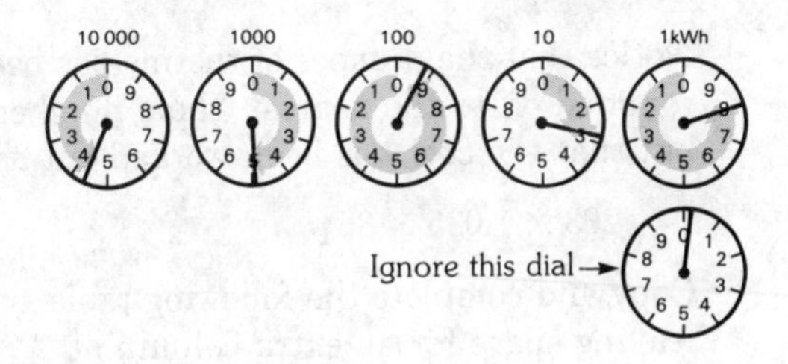


The meter shown reads 44928.

Electricity is measured in units called kilowatt hours (kW h).

## 1.2 Paying for Fuel

Some people, particularly those living in private flats or bed-sitters, may pay for their gas and electricity by using a coin-in-the-slot meter. However, the normal system is the credit meter system where you are sent a bill at the end of every quarter (3 months) which is based on the amount of fuel you have used in that quarter.

Gas charges contain a fixed amount called a *standing charge* and then an amount based on the number of therms of gas used. For natural gas

> number of therms = number of hundreds of cubic feet used × 1.035

To work out the number of hundreds of cubic feet used take the meter reading at the beginning of the quarter away from the meter reading at the end of the quarter.

---

**1.** Follow through this example and try the questions after it.

| *Reading at end* | *Reading at beginning* | *Cubic feet used (100s)* | *Therms used* | *Cost at 35 p per therm* |
|---|---|---|---|---|
| 3712 | 3509 | 203 | 210 | £73.54 |

Notice that the number of therms has been rounded off to the nearest whole number but that the cost has been worked out as

$$203 \times 1.035 \times 35\,\text{p}$$

Copy and complete the following table, leaving space for an extra column at the end.

| *Reading at end* | *Reading at beginning* | *Cubic feet used (100s)* | *Therms used* | *Cost at 35 p per therm* |
|---|---|---|---|---|
| 2798 | 2512 | | | |
| 3112 | 2975 | | | |
| 4889 | 4772 | | | |
| 3002 | 2775 | | | |
| 0972 | 0638 | | | |
| 5827 | 5598 | | | |

**2.** The standing charge will probably have increased by now but in 1986 it was £8.60. Add £8.60 to each of the costs in the questions above to get the total bill in each case, which you can put in a column headed *Total bill*. In example **1** this would be £82.14. Calculate also the answers to the top and bottom sets of figures in the table using the present standing charge and cost per therm.

**3.** If you can find an old gas bill at home check it using the method you have been using above.

**4.** Electricity charges are made up in a similar way except that the difference in the meter readings gives the number of *units* used. The charge is based on the number of units used plus a standing charge. If each unit costs 5.09 p and the standing charge is £8.50 complete the calculations for the following table.

| *Present reading* | *Previous reading* | *Units used* | *Cost of units* | *Total bill* |
|---|---|---|---|---|
| 76513 | 75924 | | | |
| 84155 | 81179 | | | |
| 62623 | 59871 | | | |
| 70064 | 67872 | | | |
| 81021 | 78863 | | | |

Find out the present rates and work out the last line again using up-to-date rates.

---

## 1.3 The Cost of Fuel - Gas

The Gas Board produces a leaflet about the cost of using gas. They say that for full central heating and hot water a 2-bedroomed terraced house would use about 750 therms a year and a 2-bedroomed detached house about 1100 therms.

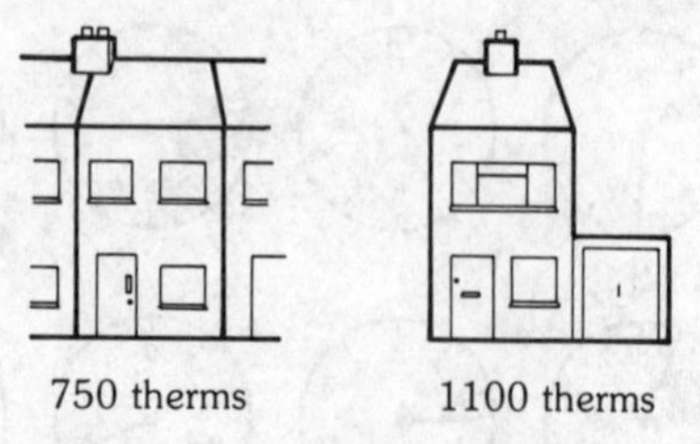


Cooking by gas uses about 50 therms a year for a couple who are both working or 80 therms for a family of four.

Estimate the yearly gas bills (including 4 standing charges) using the rates in questions **1** and **2** above for the following situations.

**1.** 2 people in a terraced house

**2.** 2 people in a detached house

**3.** a family in a terraced house

**4.** a family in a detached house

## 1.4 The Cost of Fuel – Electricity

If you are heating your home using electricity you may be using the 'Economy 7' tariff which means that you get cheaper electricity during the night.

An Electricity Board leaflet estimates that water heating for an average family would use about 60 units at the night rate and 5 units at the day rate each week. Storage heaters, which are usually heated up at night, would each use about 60 units of night rate electricity per week.

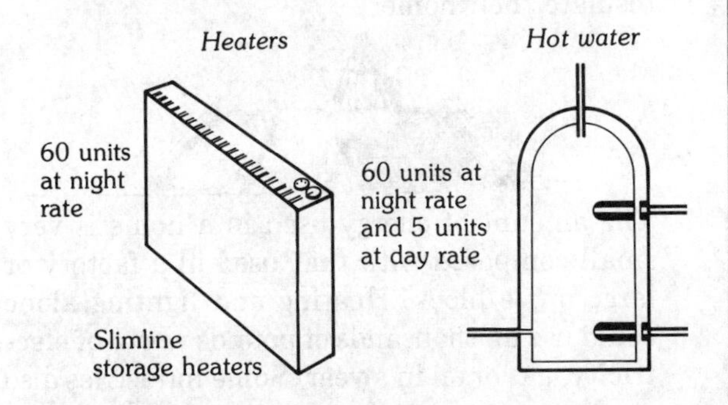


In 1984 the rates were as follows.

5.09 p per unit in the daytime
1.94 p per unit at night
plus a quarterly standing charge of £8.50.

Follow through this example to find the weekly cost (without the standing charge) for an average family who have a water heater and 4 storage heaters.

The number of daytime units would be 5, which at 5.09 p per unit costs 25.45 p.

The number of night-time units would be $60 + (4 \times 60) = 300$ units, which at 1.94 p per unit costs 582 p.

The total weekly cost would be 582 p + 25 p = £6.07 to the nearest penny.

Cooking for an average family uses about 20 units a week (normally at the day rate).

Cooker
A week's meals for a family of 4 - about 20 units


**1.** Work out the total *yearly* bills (including 4 standing charges) for an average family in each of the following cases, using the prices for 1984. Assume that the heaters are used for only 40 weeks of the year but that hot water is needed all year round.

a) hot water and 5 storage heaters

b) cooking, hot water and 4 storage heaters

c) cooking, hot water and 5 storage heaters

d) hot water and 3 storage heaters

**2.** Most people use electricity for lighting and for fridges, freezers etc. A leaflet gives the following figures.

fridge – 1 unit a day
automatic washing machine – 9 units a week
twin tub – 12 units a week
freezer – 15 units a week

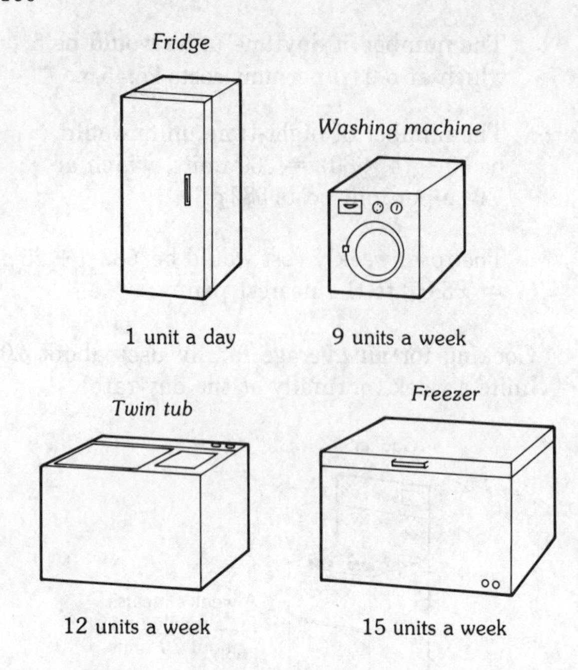


What would each of them cost
a) per week b) per year (52 weeks)
at the standard day rate of 5.09 p per unit?

**3.** All light bulbs or tubes are marked with a certain power rating such as 60, 100 or 150 watts.

Electric fires are usually rated as 1 or 2 kilowatts where

1 kilowatt = 1000 watts

A unit of electricity is a kilowatt hour so that a 100 watt bulb burning for 10 hours would use 1 unit of electricity.

The cost of lighting a house therefore depends on the ratings of the bulbs or tubes used, and on how long you leave them on.

On average, lighting a house uses about 2 units a day. How much would this cost
a) per week b) per year?

**4.** Now work out the costs per week of the following combinations using present costs per unit (your local Electricity Board can supply these).
a) fridge, twin tub, cooking and lighting
b) fridge, automatic washing machine, freezer, cooking and lighting
c) hot water, 3 storage heaters, fridge, twin tub and lighting
d) cooking, fridge, freezer, lighting, automatic washing machine and a dishwasher (5 units a day)
e) cooking, hot water, 5 storage heaters, fridge, freezer, lighting and an automatic washing machine
f) your home

## 2. Saving Energy

You can of course reduce your energy costs slightly by switching off lights which are not needed, using your automatic washing machine at night, etc., but the most important thing is to make sure that your home is properly insulated. Covering your hot water tank and insulating your loft could save up to 20% of the heating bill. Double glazing, cavity wall insulation and draught excluders could save even more. Get hold of some leaflets and work out what difference could be made to a family with a heating bill of £400 a year if they took all these steps to insulate their home.

The amount of energy used in a home is very small compared with that used in a factory or large office block. Heating and lighting alone could use up thousands of pounds worth of electricity, gas or oil in a year. Some industries also use large amounts of energy to make their products. A large chemical factory may use as much electricity as a small town.

You can do your bit to save energy at work or at school or college by shutting doors, turning off unwanted lights, turning off heaters rather than opening windows, and so on. Can you see ways in which energy is being wasted where you are now?

# Colin's and Cheryl's Energy Costs

Colin and Cheryl have just bought a house which has gas central heating, a gas fire in the living room and a gas cooker. So that they can budget properly they decide to try to estimate their gas and electricity bills for the next quarter, which will be a winter one.

They estimate that the gas fire will only be used in the evenings and at most will be on for 3 hours each day at a low setting. As they are both working the central heating will be operating for about 7 hours a day during the week and 15 hours a day at the weekend.

**1.** For how many hours will they be using the gas fire in a 13-week quarter?

**2.** If the fire uses about 1 therm every 10 hours on a low setting how many therms of gas will they be using in the quarter? (Give your answer to the nearest therm.)

**3.** Using the present cost per therm find out how much the gas fire will cost in that quarter.

**4.** For how many hours will the central heating be on during the quarter?

**5.** By using the thermostat on their central heating wisely they can get about $2\frac{1}{2}$ hours of operation per therm. How many therms will they use per quarter on the central heating?

**6.** What will be the cost of the central heating for the quarter?

**7.** A gas board leaflet estimates that they will use about 1 therm of gas per week on cooking. How much will this cost them for a quarter?

**8.** By adding together the costs in questions **3**, **6** and **7** and adding on the standing charge find their estimated gas bill for the quarter.

**9.** Bearing in mind the fact that they may have slightly under-estimated their use of gas, how much should they be putting away each month in order to save up enough money for the gas bill when it comes?

**10.** The meter reading at the beginning of the quarter was 2197 and at the end of the quarter was 2572.
How much gas have they actually used in hundreds of cubic feet? Is this more or less than they estimated they would use?
(1 therm = 100 cu ft × 1.035)

**11.** The electricity board leaflet gives an estimate of 3 units a day for lighting. How many units would this use up in a quarter?

**12.** What would be the cost of lighting for a quarter at the present rate per unit?

**13.** They decide to invest in a fridge and a twin tub washing machine. Assuming that they use their washing machine half as much as an average family how many units would they use per quarter for the fridge and the twin tub?

**14.** What would this cost per quarter?

**15.** By adding together the amounts from the answers to questions **12** and **14** and adding the standing charge estimate their quarterly bill.

**16.** How much should they be putting away each month in order to save up for their electricity bill?

# Teacher's Notes

**1.** This project deals only with gas and electricity costs. Comparisons could be made with the cost of using oil or solid fuel for heating. Information can be obtained from local branches of the major oil companies and from the Solid Fuel Advisory Service.

The question of the use of energy in industry and commerce could also be extended through visits to local firms, talks by speakers from industry, etc.

**2.** The issue of energy conservation could be extended to bring in calculations on the savings obtained by loft insulation, double glazing and so on, and the costs of doing it.

**3.** Other subject areas could be brought into this project in a variety of ways, for example

*science* production and distribution of fuels, heat values of different fuels, heat transfer, the science of conservation

*social studies* the environmental and social implications of fuel production and distribution, the future needs of society, the debate about nuclear power

*information technology* automation of production and distribution systems, production and payment of bills, heating control systems

*career development* jobs available in the energy industries, skills needed for jobs

*communications* surveys on the use of various types of fuel, interpretation and production of leaflets and posters, interviews with people working in the industry.

**4.** Various organisations such as the Gas and Electricity Boards, conservation groups and local councils will be happy to give talks on different aspects of this project and may be able to suggest places for group or individual student visits.

Chapter 13

# Getting the Order Right

Whether you are solving a mathematical problem or doing a simple job at home or at work it is often important to do things in the correct order. For example, if you light the gas under a saucepan of vegetables before you put in the water you are likely to need a new saucepan. In this chapter you will be looking at the construction of simple flow charts, which will then lead on to representing more complicated operations as flow charts. This is the beginning of the management technique known as *critical path analysis*. The ideas used in constructing flow charts can also be useful when planning ahead, either at work or at home.

## 1. Order is Important

### 1.1 Simple Flow Charts

Pretend that you are standing outside a telephone box waiting to make a call to a friend. List in order the things you will have to do between now and the time when you are next outside the telephone box after making the call. Assume that your friend is in and answers the phone.

Check your list with other people in the group.

The following simple flow chart shows the processes involved in boiling an egg.

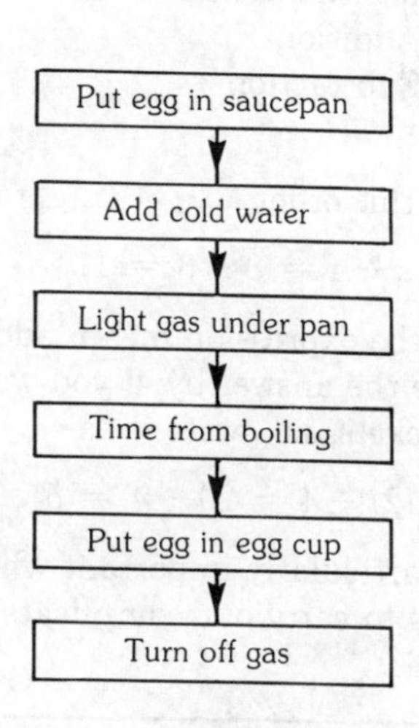


Notice that there are arrows in the flow chart to show the order of doing things.

Make up simple flow charts to show the order in the following operations.

1. writing and sending a letter
2. putting a record on your stereo system
3. getting from home to school or college
4. cooking some baked beans
5. making a cup of tea (don't forget to warm the pot)

## 1.2 Order in Number Work

What is meant by $8 + 12 \div 4$?

It could mean 'add 12 to 8 and divide the answer by 4', or 'divide 12 by 4 and add the answer to 8'.

These two alternatives give different answers, of course. It is important to remember the standard order in which operations are carried out in number work. Some people like to remember

**B** Brackets
**O** Order
**D** Division
**M** Multiplication
**A** Addition
**S** Subtraction

Following this order,

$$8 + 12 \div 4 = 8 + 3 = 11$$

To make the expression mean 'add 12 to 8 and divide the answer by 4' you would have to use brackets, i.e. write it as

$$(8 + 12) \div 4 = 20 \div 4 = 5$$

Order is particularly important when using calculators to carry out complicated calculations.

Draw simple flow charts to show how to

1. calculate the area of a circle of radius 10 cm
2. find 15% of £65
3. work out an electricity bill if 58 units have been used at 5.8 p per unit and the standing charge is £8.50
4. find the petrol cost at £2.10 per gallon for a journey of 120 miles in a car which will do 25 m.p.g.

If you have a problem getting the order correct either with or without a calculator, find the appropriate chapter in part B.

## 1.3 Basic Algebra

Algebra frightens some people but really it is just a type of shorthand which has special rules about order. Flow charts help when deciding which order to do things in, for instance the temperature conversion formula

$$C = \frac{5(F - 32)}{9}$$

can be written in flow chart form as follows.

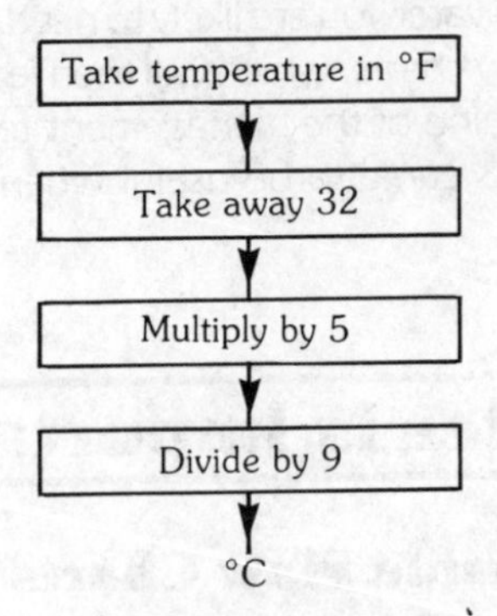


Notice that, as in BODMAS, the brackets are worked out first.

Write out flow charts for the following formulae.

1. $A = \frac{(a + b)}{2}h$
2. $V = \pi r^2 h$
3. $T = 20M + 20$
4. $C = NI + D$
5. $T = 30\frac{(I - D)}{100}$
6. $T = P\sqrt{\frac{l}{g}}$

## 1.4 Reversing the Order

Some things you do in mathematics or in day-to-day life can be reversed, for example getting to college or school. This could be written in simple flow chart form.

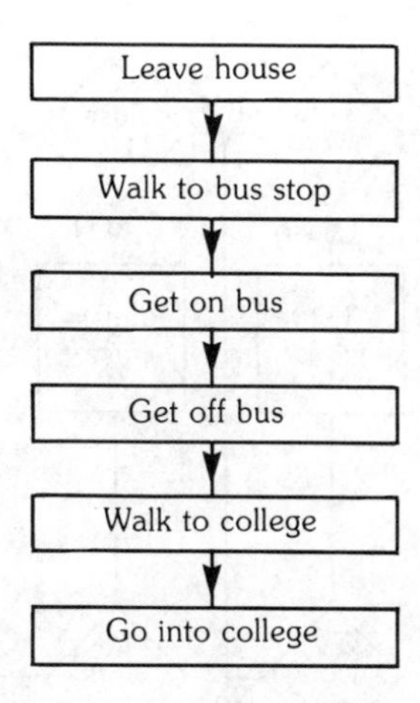


Going home might look like this.

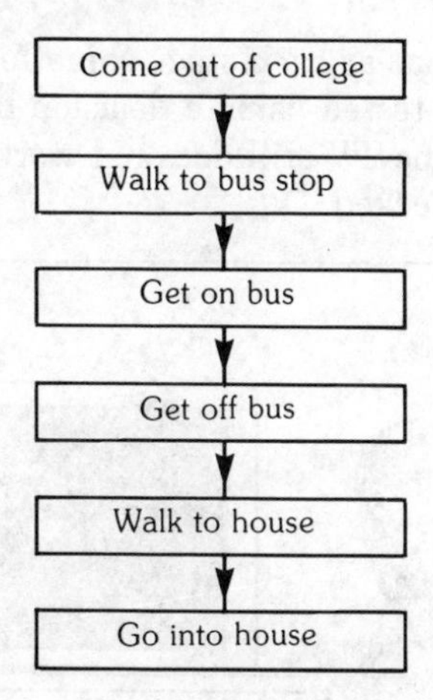


So to reverse a flow chart you *reverse the order and do the opposite thing at each stage.*

This idea is very useful in algebra. The area $A$ of a circle is given by

$$A = \pi r^2$$

which can be written in flow chart form as

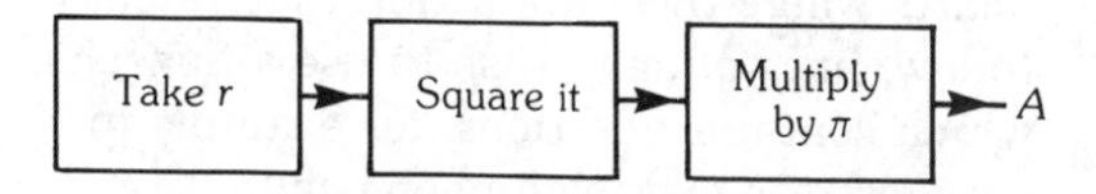


The reverse would be written

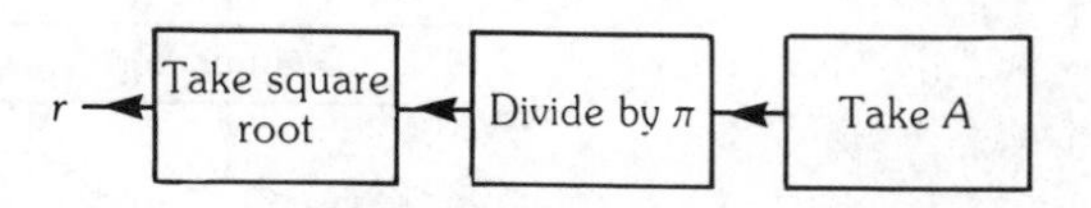


which would enable the value of $r$ to be found for a given value of $A$.

In algebra this would be written as

$$r = \sqrt{\frac{A}{\pi}}$$

---

By drawing and reversing flow charts find

**1.** $M$ in terms of $T$ from $T = 20M + 20$

**2.** $l$ in terms of $T$ from $T = 2\sqrt{l}$

**3.** $r$ in terms of $V$, $\pi$ and $h$ from $V = \pi r^2 h$

**4.** $F$ in terms of $C$ from $C = \dfrac{5(F - 32)}{9}$

**5.** $T$ in terms of $n$, $r$, and $I$ from $I = \dfrac{nrT}{100}$.

---

# 2. Complicating the Issue

## 2.1 Alternative Routes

So far you have only dealt with simple flow charts where there are no alternative paths to follow, but you may wish to use a flow chart which contains questions, for example to demonstrate making a phone call.

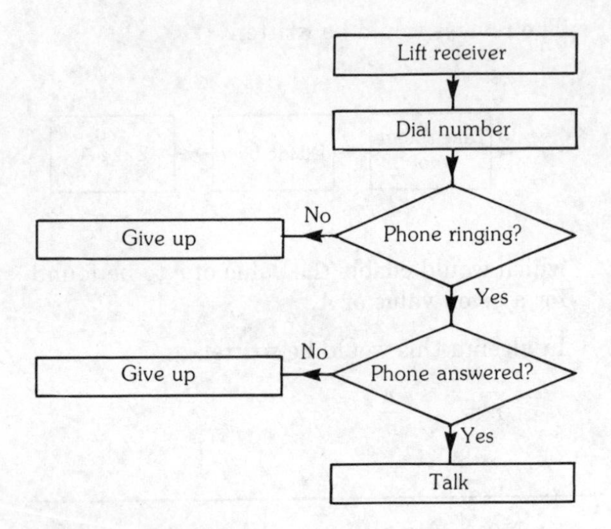


Notice that there are two boxes in the shape of a rhombus. These are called *decision boxes* and they have two possible outcomes, 'Yes' and 'No'. You may find flow charts similar to this in books on computer programming as the flow chart can be a useful way to set out the steps of a computer program before actually writing it.

Write flow charts using decision boxes for the following.

**1.** baking a cake (checking to see if it is done)

**2.** starting a car (check to see if it is in gear)

**3.** going out of the house (have you got your keys?)

## 2.2 Checking for Faults

The flow chart can give us a useful way of writing down a method when looking for a fault in something which is not working, for example your TV not giving a picture or any sound. You might run through the following system of checks.

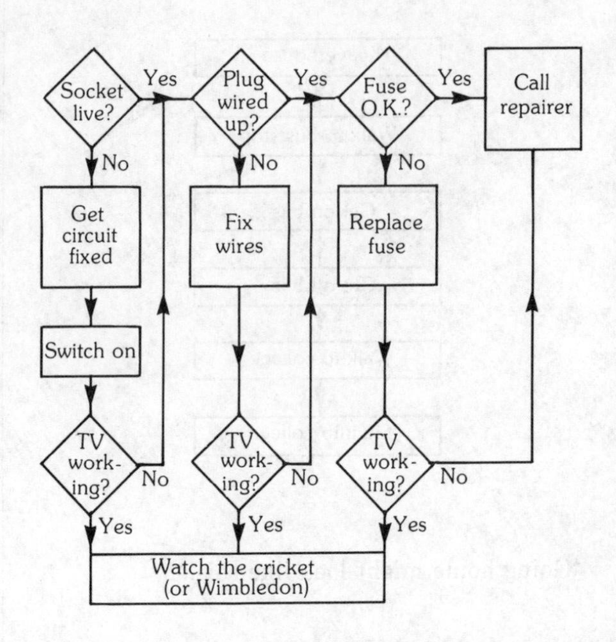


Notice that at each stage something is checked, tested using a decision box and then, if the TV still does not work, something else is checked.

A car manual gives a few useful tips for drivers whose cars will not start in the morning.

(a) If the car has been out in the rain, spray a drying agent on the wires.

(b) Check the battery water level and top up if low.

(c) Check the sparking plugs and clean if dirty.

(d) Call out the garage.

Write a flow chart to show the method of carrying out these checks in the order given.

## 2.3 Number Patterns

Flow charts with loops can also be used to produce number patterns or sequences, as shown here.

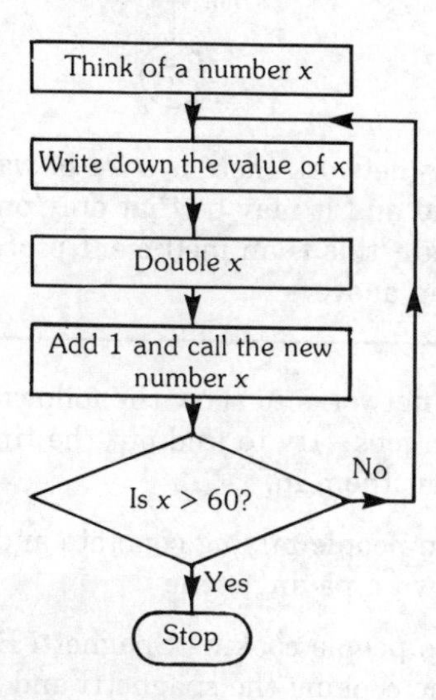


If you start with the number 2 this gives the sequence 2, 5, 11, 23.
Notice that with this flow chart the last number calculated, 95, is not written down.
Can you see a way of changing the flow chart so that the last number calculated is written down?

**1.** Write down the sequences given by the following flow charts.

a)

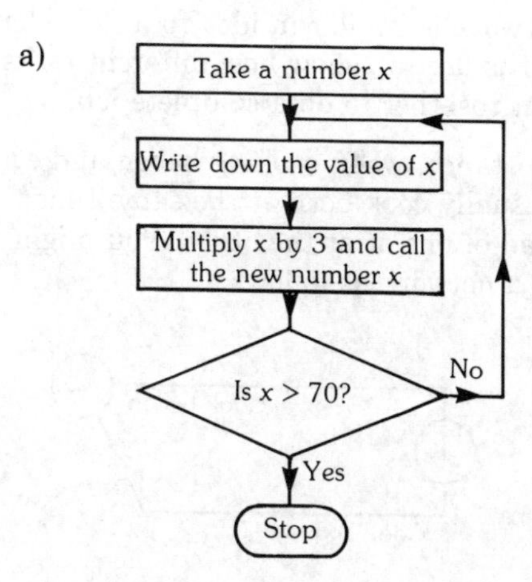


b)

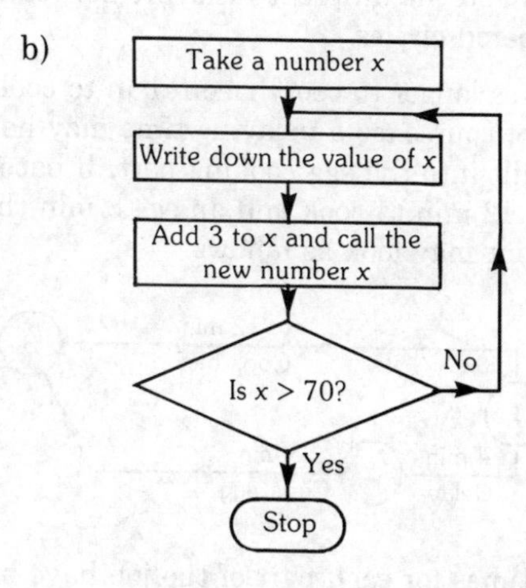


**2.** Make up flow charts which would give the following sequences.

a) 2, 4, 6, 8, 10, 12, 14

b) 1, 6, 11, 16, 21, 26

c) 1, 4, 9, 16, 25, 36

d) 2, 6, 14, 30, 62

# 3. Networks

## 3.1 Drawing a Network

A network is similar in idea to a flow chart and it is used to show how different tasks can be put together to do a complete job.

For instance, when cooking bacon and egg you usually cook both at the same time, instead of one after the other. You might draw a network as follows.

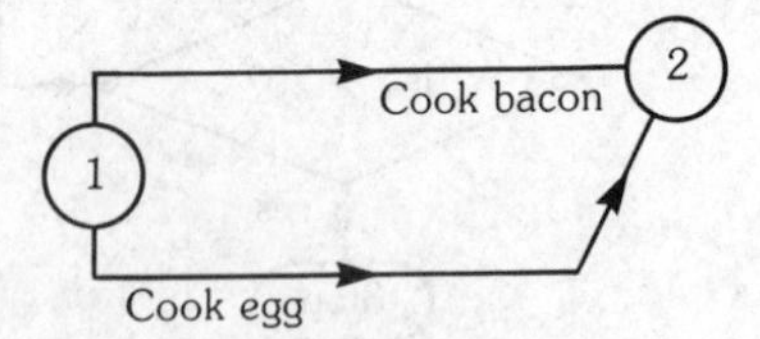


Notice that circles have been used instead of boxes and the different tasks are between the numbered circles.

It takes longer to cook bacon than to cook an egg, of course, so a delaying time may have to be built into the egg cooking part. If bacon takes 12 min to cook and an egg 8 min the network may look as follows.

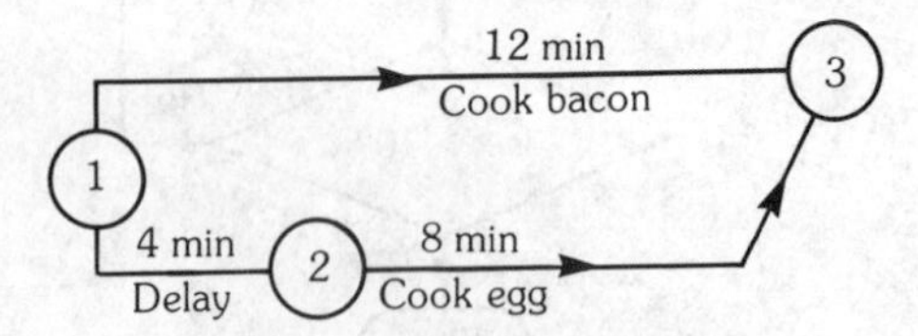


The times for each part of the job have been put into the diagram.

In other cases, for example when 2 people are making different parts of a product, the delay may not be necessary if the first person to finish is allowed to wait for the second or if the product is being made in large numbers, one after the other. For instance, when two people are making tea, one might make the tea and the other set out the cups and saucers, as follows.

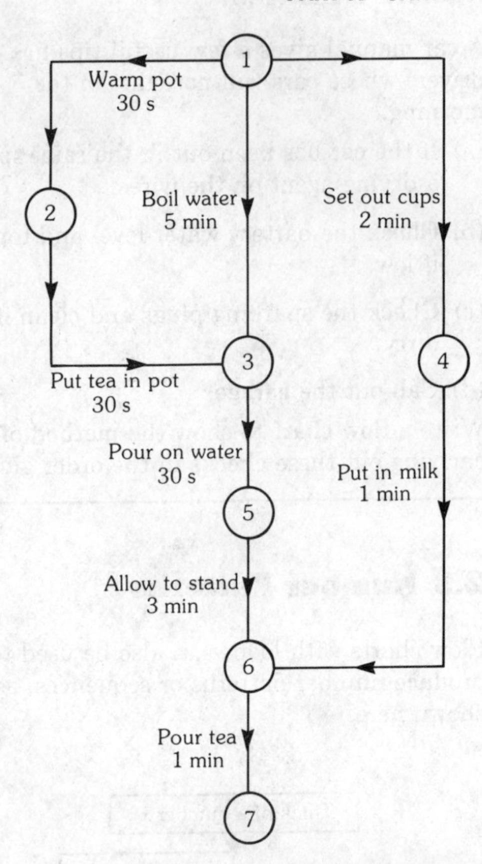


In this network there is a lot of waiting around and it may be that only one person is needed – this is an inefficient process as written above.

---

Draw networks to show the following two-person jobs – try to find out the times needed and put them in.

**1.** two people mixing concrete and cementing down a paving stone

**2.** two people cooking Spaghetti Bolognaise – one cooking the spaghetti and the other the sauce

**3.** two people in a house getting ready in the morning – dressing and using the single bathroom

**4.** going through the checkout at the supermarket and paying by cheque

## 3.2 Critical Path Analysis

In all of the networks you have just drawn you will have found some waiting time for one or both of the people concerned. At work time costs money, so an employer will want to keep waiting time to a minimum. In order to do this he or she may draw a network of the different jobs to look for the *critical path*. This is the part of the job that takes longest.

In the tea-making project the critical path is the path where the water is boiling. The person setting out the cups has $5\frac{1}{2}$ min waiting time. The whole process can be made more efficient by getting a kettle which boils more quickly, or giving people another job to do while they are waiting. Can you think of other alternatives?

In the following networks the different stages are numbered and the times for each part given. Write down the total time for each route through the network and find the critical path.

**1.** four routes (times in min)

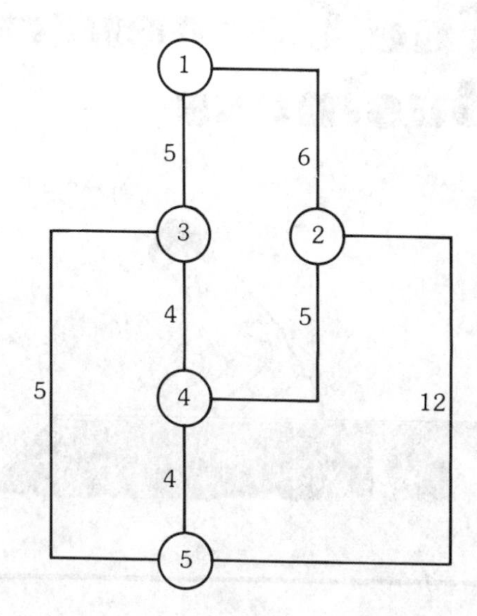


**2.** five routes (times in hours)

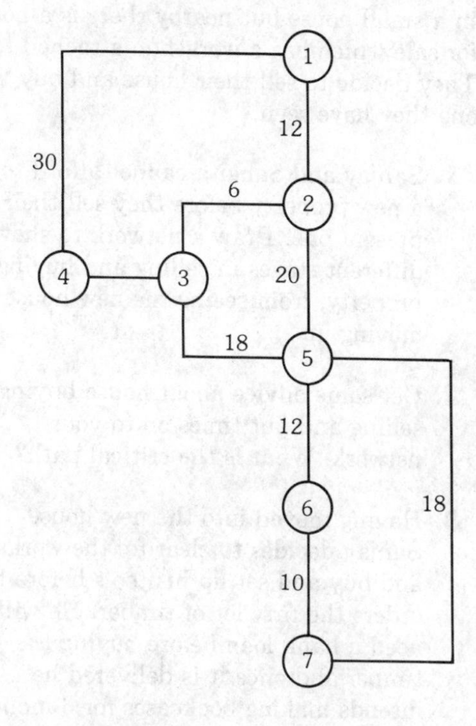

# The Carpentry Business

Sanjay and his wife Sangita have decided that they want to set up their own business and as Sanjay has always been good at carpentry they decide to start producing bookcases made to order. At present they live in a small house but nearby there is a house for sale which has a workshop attached to it. They decide to sell their house and buy the one they have seen.

**1.** Sanjay and Sangita cannot afford to buy a new property before they sell their present one. Draw a network to show the different stages in selling and buying property, from seeing the new house to moving in.

**2.** Get some advice about house buying and selling and put times on to your network. What is the critical path?

**3.** Having moved into the new house Sanjay decides to clear up the workshop and buy and set up his tools before he orders the first lot of timber. He will need a bank loan before buying the timber and once it is delivered he intends making bookcases for 6 months in his spare time before leaving his job with a building firm. Once he starts work Sangita will be organising the advertising side.
Draw a network for them both to show this process, from moving in to Sanjay leaving his job.

**4.** Put times on to your network. Sangita will have a lot of waiting time – how can she be useful to the business in that time?

**5.** Sanjay and Sangita have decided on the following responsibilities once their business becomes full-time for Sanjay.

**Sanjay**
- measuring up at customer's house
- drawing plans and cutting timber
- making bookcases
- doing first stages of finishing
- delivering to customer

**Sangita**
- taking and recording orders
- ordering materials
- making up invoices and keeping accounts
- doing final stages of finishing, e.g. staining, varnishing, etc.

Look at this list and see if you can add anything which has been missed out.

**6.** A customer rings to order a bookcase for which materials will have to be ordered specially. Make up a network to show the complete process for Sanjay and Sangita for this bookcase.

**7.** How long should the whole job take from order to delivery?

**8.** Business is going well and Sanjay decides to take on a trainee, Chris. Chris will be at college one day a week and with Sanjay for the other four days. Of course Sanjay does not let Chris do much on his own at first. Assuming that Chris is still at the learning stage, rewrite your network from question **6** above to include Chris. How much time will he save Sanjay in the early days?

**9.** Chris settles in well and Sanjay decides to keep him on and give him the job of measuring up, making plans, finishing the bookcases and delivering. Using the times you have used before and assuming that they have a good stock of timber, draw a network for Sanjay, Sangita and Chris for making and delivering 3 bookcases. (Assume that the orders came in on the same day and that they had no other work to finish off.)

**10.** Can you see any ways in which this little firm could be made more efficient?

# Teacher's Notes

**1.** The ideas used in flow charts can lead to a lot of discussion about methods of thinking and working, whether at college or school or in a work situation. Networks and critical path analysis have applications in a wide range of subjects and can be usefully incorporated into most vocational areas.

**2.** Links with other subjects could include

*science* the use of formulae for calculations, flow charts for laboratory experiments

*social studies* the structure of business, job descriptions and responsibilities in business

*career development* the section on the carpentry business in this chapter could be used as a link with discussion on working for yourself

*information technology* computerisation of jobs, identifying time saving methods by using computer technology

*communications* the whole chapter is as much an exercise in communications as in numeracy and could be used by a variety of staff.

**3.** Whatever vocational area a student is working in or intending to work in there will be a variety of tasks to be performed by one or more people. The production of networks for these tasks would encourage students to think about the efficient use of their time.

**4.** Visits to small businesses and discussion there could be a useful link with this chapter. A study of processes at school or college may also bring interesting results.

**5.** Staff involved in the teaching of management studies at a school or college can be of great help to students who wish to pursue the ideas introduced here.

Chapter 14

# How Many Cornflakes? Volume, Weight and Surface Area

Many things which you buy, from cement to orange juice, are bought by weight or volume. For example, you can buy 1 l of orange juice or 6 kg of cement. Other items such as cornflakes may be sold in sizes – small, family size, giant, etc.

The basic units of weight and liquid volume have been covered in chapter 1. Here you will be looking at the connection between volume and weight, how to calculate the volumes of standard shapes and the reasons why these shapes are used for packaging. This work will give you a better understanding of the ideas of shape and size and will also be useful for students who are likely to work in packaging or selling goods.

## 1. Measuring Volume and Weight

### 1.1 Standard Shapes

Before you can work out volumes you need to know the names of some standard shapes.

The most common 3-dimensional shape you will meet is the rectangular box. This may be a shoe box, a cupboard, a speaker, a room, etc. The proper name for such a shape is a *cuboid*.

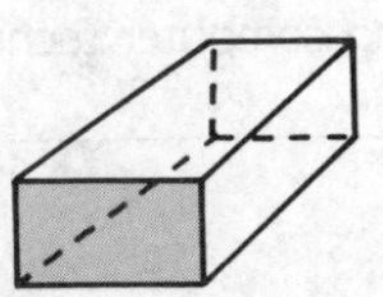

If all the edges of a cuboid are the same length then it is called a cube, for example a die or some stock cubes.

In the kitchen the most common shape is likely to be a *cylinder*, usually containing tinned fruit, baked beans, custard powder, etc.

The cylinder and cuboid are really special cases of a *prism*. This is a shape which has a constant *cross-section*, i.e. wherever you cut at right angles to the length you get the same shape. Simple examples of this are a basic ridge tent (triangular prism) or a box-shaped house with a sloping roof – what name is given to the cross-section in this case?

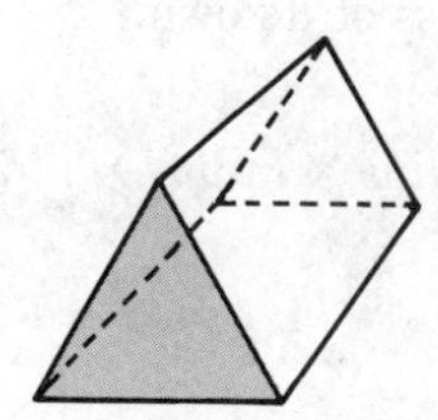

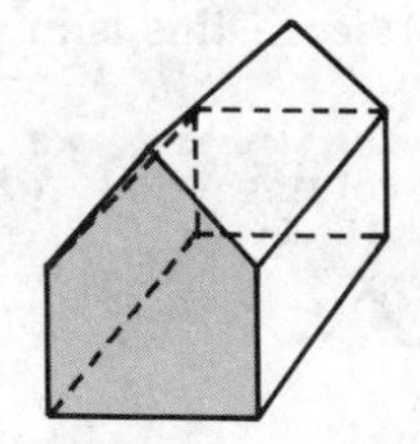

Another shape which you can probably find around you quite easily is the *sphere* – cut it in half and you have a hemisphere.

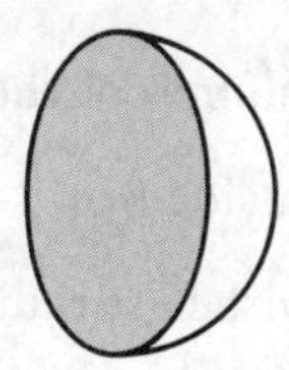

The last main type of standard solid which you are likely to meet is the *pyramid* – a base with corners attached to a point above the base.

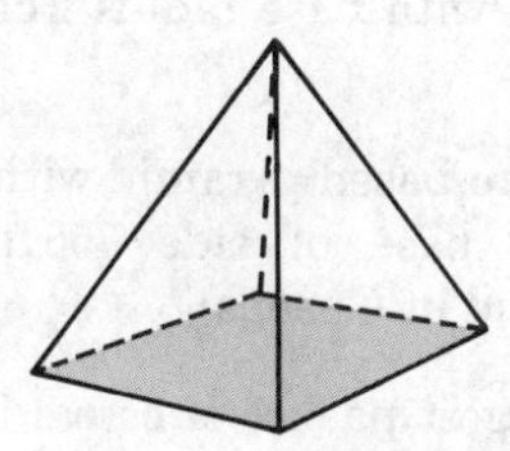

Most people think of pyramids as having a square base as shown but the base can in fact be any shape, for example a triangular base (*tetrahedron*)

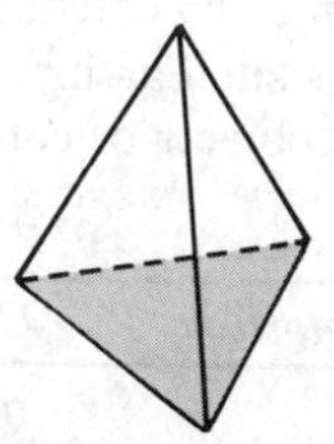

or the special case where the base is circular – the *cone*.

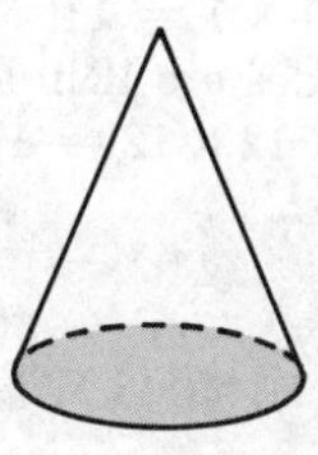

If you are going to work with 3-dimensional shapes it is important to be able to draw them so that they look 3-dimensional to you – notice that edges which are at the back of the figure are drawn as broken lines.

---

**1.** Copy the drawings above of a cuboid, a cylinder, a triangular prism, a tetrahedron and a cone – make sure that you draw them as three-dimensional objects.

**2.** Pick two or three simple objects in the room where you are now, such as a book, a desk, or a pencil, and draw them.

**3.** Draw the following as 3-dimensional objects.
a) a portable radio
b) a glass
c) a bottle
d) a house
e) a cone with a hemisphere on top (like an ice cream)
f) a cuboid with a pyramid on top (like a church tower and spire)

---

## 1.2 Units of Volume

The volumes of liquids are generally measured in litres, pints or fluid ounces, as seen in chapter 1. To calculate areas in chapter 5, units such as $in^2$, $ft^2$ or $m^2$ were used. For solid objects or containers, cubed units are used.

| *Unit of length* | *Unit of volume* |
|---|---|
| m | $m^3$ |
| in | $in^3$ |

Now if a cube has sides each 1 ft long then the volume is $1 \times 1 \times 1 = 1\,ft^3$. You could also say that the sides are 12 in long giving a volume of $12 \times 12 \times 12 = 1728\,in^3$.

so $1\,ft^3 = 1728\,in^3$
and similarly $1\,m^3 = 1\,000\,000\,cm^3$
$(100 \times 100 \times 100)$

These are the most commonly used units of volume.

To convert from imperial to metric or metric to imperial use the fact that 1 m is about 3.3 ft, so

$1\,m^3$ is about $35.4\,ft^3$

**1.** Write the following in $ft^3$.
a) $5\,m^3$ b) $6.8\,m^3$ c) $11.2\,m^3$

**2.** Write the following in $m^3$.
a) $45\,ft^3$ b) $102\,ft^3$ c) $49.6\,ft^3$

## 1.3 Volumes of Standard Shapes

The different shapes in section 1.1 can all be put into 3 basic types. This is probably the easiest way to remember how to calculate their volumes.

*Prisms* including the cube, cuboid and cylinder.

volume = area of cross-section × length

or $V = Al$

The cross-section is the shaded area in section 1.1 and its area is calculated using the methods in chapter 5.

*Pyramids* including the cone and tetrahedron.

volume = one third of base area × height

or $V = \frac{1}{3}Ah$

Again, the base has been shaded in the diagrams in section 1.1.

*Sphere* – this is in a class of its own.

volume = $\frac{4}{3} \times \pi \times$ radius × radius × radius

or $V = \frac{4}{3}\pi r^3$

**1.** Write the formula for the volume of a sphere in the form of a flow chart.

Calculate the volumes of the following.

**2.** a cube with side 8 in

**3.** a cylinder with base radius 4 cm and height 12 cm

**4.** a tent in the form of a triangular prism of width 1.2 m, height 1.6 m and length 2.5 m

**5.** a cone with base radius 2 cm and height 12 cm

**6.** a square-based pyramid with height 481 ft and a base of side 755 ft (the Great Pyramid of Cheops)

**7.** the largest packet of cereal in your house

**8.** a medium sized baked bean tin

**9.** a standard sized football

**10.** the room you are in (taken as a cuboid)

## 1.4 Weight and Density

The contents of a large packet of breakfast cereal may weigh 500 g, not much more than the contents of a tin of baked beans, even though the tin is a lot smaller. This is due to the fact that baked beans have a much higher *density* then cereals. To calculate density use

$$\text{density} = \frac{\text{weight}}{\text{volume}}$$

(strictly mass rather than weight)

So the units of density are g/cm$^3$, oz/in$^3$ etc.

Many packets and tins also have an air space but as this varies a lot from tin to tin it is difficult to take it into account.

Find the volume and density of the following.

**1.** a 400 g tin of strawberries of height 10.3 cm and base radius 3.6 cm

**2.** a 439 g tin of baked beans of height 10.8 cm and base diameter 7.2 cm

**3.** a 213 g tin of tomatoes of height 5.8 cm and base radius 3.6 cm

**4.** a 170 g tin of milk of height 5.6 cm and base diameter 6.2 cm

**5.** a 500 g packet of cereal which has width 20 cm, height 29.2 cm and depth 6.9 cm

Compare your answers to question **5** above with those obtained from other cereal packets. Make up a table showing the weight of an average dishful of different cereals – you will have to somehow find out the volume of an average dish.

# 2. Shape and Size

## 2.1 Surface Area and Nets

Having worked out the volumes of some basic shapes it may sometimes be useful to know the surface area. This is needed, for instance, to find the amount of material required to make the shape.
In order to work out the surface area it is often best to imagine the object flattened out to form what is called a *net*, for example with a cylindrical tin:

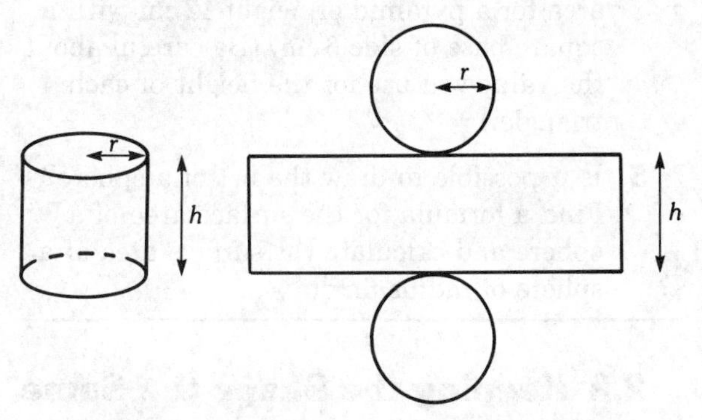


To find the total surface area you have to add together the areas of the two circular ends and the area of the curved surface. On the net all the measurements needed are known apart from the width of the rectangle. Since this rectangle comes from opening out the curved surface of the tin then its width must equal the circumference of the tin, so that

$$\text{width of rectangle} = 2\pi r$$

Putting this together,

$$\text{surface area of cylinder} = \underbrace{2\pi rh}_{\text{curved surface}} + \underbrace{2\pi r^2}_{\text{ends}}$$

For the strawberry tin in question 1, section 1.4

$$\begin{aligned}\text{surface area} &= 2 \times 3.14 \times 3.6 \times 10.3 \\ &\quad + 2 \times 3.14 \times 3.6 \times 3.6 \\ &= 232.9 + 81.4 = 314\,\text{cm}^2 \text{ to the nearest cm}^2\end{aligned}$$

**1.** Work out the total surface area of the tomato tin in question 3, section 1.4.

**2.** Draw a net for the cereal box in question 5, section 1.4 and calculate the total area of cardboard used, if the flaps at top and bottom are 4 cm deep, and there is a 2 cm overlap for gluing.

**3.** Draw a net for a triangular prism whose cross section is an equilateral triangle of side 5 cm and whose length is 8 cm. Calculate the total surface area.

**4.** Draw a net and calculate the total surface area for a pyramid of height 12 cm with a square base of side 6 cm. (Be careful about the value you use for the height of each triangle.)

**5.** Is it possible to draw the net of a sphere? Find a formula for the surface area of a sphere and calculate the surface area of a sphere of radius 6.2 cm.

## 2.2 Keeping the Shape the Same

In chapter 11 the idea of *similar figures* was mentioned. The word similar is not used just for 2-dimensional figures, it also applies in 3 dimensions. If 2 shapes are similar then one can be turned into the other by using an enlargement with the required scale factor – in this section this will be referred to as a *length scale factor*.

If a square of side 1 unit is enlarged with scale factor 2 then the sides each become 2 units long. The area of the square changes from $1 \times 1 = 1$ unit$^2$ to $2 \times 2 = 4$ unit$^2$ giving an *area scale factor* of 4.

If the length scale factor is 3, what will the area scale factor be?

What is the general result for a length scale factor $k$?

Now for 3 dimensions, consider a cube of side 1 unit which is enlarged with length scale factor 2. The original volume is 1 unit$^3$ and the new volume is $2 \times 2 \times 2 = 8$ unit$^3$.

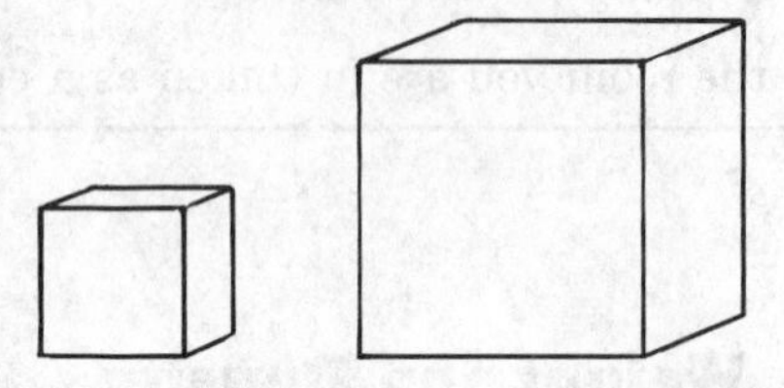

By looking at other cases you should be able to come up with the results below.

> if length scale factor is $k$
> then area scale factor is $k^2$
> volume scale factor is $k^3$

These results do not apply just to squares and cubes but to any 2- or 3-dimensional similar figures. For example

a statue is to be made 4 times as large as a model
the length scale factor is therefore 4
the area scale factor is $4 \times 4 = 16$
the volume scale factor is $4 \times 4 \times 4 = 64$

Therefore the surface area of the statue will be 16 times that of the model and its volume 64 times that of the model.

**1.** Find i) the area scale factor ii) the volume scale factor for each of the following length scale factors.
a) 5 b) 10 c) 100 d) $\frac{1}{2}$

**2.** A 'giant size' tin of beans is similar to a small one and twice as tall. If a small one contains $5\frac{1}{2}$ oz of beans how much should the giant size contain?

**3.** If a tin 12 cm high contains 15 oz of fruit what would have to be the height of a similar tin for it to contain 32 oz of fruit?

**4.** For 2 tins to be similar the ratio of height to base diameter should be the same for each tin. Carry out an investigation to see if you can find two tins which are similar.

**5.** Measure the radius and height of a tin and work out how large a similar tin would have to be to hold twice as much. Is this a sensible size?

### 2.3 The Best Shape

If you look at tins of fish, tomatoes, milk, fruit, etc. you will see that food tins come in a variety of shapes and sizes. Sometimes there may be good reasons for a particular shape because of the type of contents – can you think of any?

It may be that tins are made a particular size to save metal. This can be investigated by working out the surface area.

**1.** The volume of the 439 g tin of baked beans in section 1.4 was about 440 $cm^3$. The same volume could be obtained by using cans of a variety of shapes such as in the table below.

| $r$ (cm) | 2.4 | 3.6 | 4.2 | 4.8 | 5.4 |
|---|---|---|---|---|---|
| $h$ (cm) | 24.3 | 10.8 | 7.9 | 6.1 | 4.8 |

Work out the surface area for each of the shapes in this table.

**2.** Draw a graph of surface area against radius using the information from question 1.

**3.** From your graph estimate the radius which would give you the least surface area.
Is it the one used?

**4.** Use the formula $V = \pi r^2 h$ with $V = 439$ and your estimate of the best radius in question **3** to find the height of your ideal can. Why is this shape not used?

If you carry out an investigation similar to the above for the 213 g tin of tomatoes in question **3**, section 1.4 you may notice something about the 'best' tin shape.

If you look around a supermarket you will notice that most tins have the same base radius whatever their height. Which are the exceptions and why?

## 3. Value for Money

'Big is cheapest' is an idea you may use when buying in a supermarket. In other words it is better to buy the giant family size than the standard size. To check this out you need to work out the cost in p/l or p/g for different sizes, for example a 220 g tin of baked beans costs 16 p giving a cost of 0.073 p/g, while a 439 g tin costs 22 p giving 0.05 p/g and therefore these beans are cheaper to buy.

Of course if you live alone you may not want 439 g and may prefer to buy a small tin.

Compare the costs in p/g or p/l for the following and work out which is cheapest.

**1.** 340 g of marmalade for 29 p or 454 g for 39 p

**2.** 500 ml of paint for £2.35 or 1 l for £3.75

**3.** 930 g of washing powder for 82 p or 3.1 kg for £2.56

**4.** 410 g tin of peaches for 35 p or 822 g tin for 65 p

**5.** 500 g packet of cornflakes for 68 p or 750 g for 99 p

**6.** 200 g of coffee for £2.69 or 300 g for £3.94

# Boxes, Packets and Tins

Paul works in a supermarket and spends most of the time stacking things on shelves. Being a thoughtful chap, when he is not particularly busy at work he starts to wonder about the shapes and sizes of items on the shelves and how they are transported. Put yourself in Paul's place and follow up some of his ideas by visiting a local supermarket or cash and carry.

**1.** Many tins on the supermarket shelf are the same size, for instance baked beans, soup, peaches etc. Do they all come in boxes containing the same number of tins? If not, why not? Does weight make a difference?

**2.** Some tins are wider or narrower than usual. Are their boxes also a different size? Would it be possible to produce a box which could be used for small or family sized tins of beans (but not a mixture)? Would there be a difference in overall weight of beans per box in each case?

**3.** Packets of cereal are different from tins of beans in that they can be fitted together without gaps. Are 500 g and 750 g packets of cornflakes packed in the same size box? What are the measurements of the smallest box which could be completely filled with either 500 g packets or 750 g packets?

**4.** The shape of the box can of course also vary. There are various shapes which could hold 48 of the 500 g packets of cornflakes. Which of these shapes has the minimum surface area?

**5.** Many liquid containers are cuboids and therefore easily packed. Milk sometimes comes in containers which are prisms with a cross-section which is a pentagon. Are they designed so that they can be stacked with alternate layers upside down? Is there any wasted space in a box of these containers? If so, what percentage of the space inside the box do you think is wasted?

**6.** Tins could be stacked in a box in one of two ways as shown in these plans.

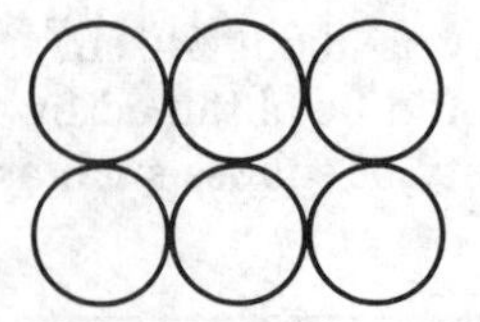
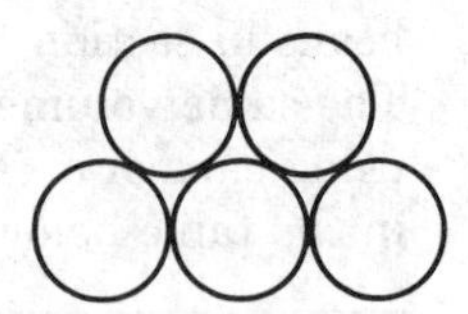

If 12 tins are to be put in a box, which of these methods would require the smallest box?
You can extend this problem by thinking about 3 rows, an odd number of tins etc.

There are endless questions which could be asked about packaging – think of some for yourself and try to answer them.

# Teacher's Notes

**1.** In this project students will be working on the mensuration of the basic solid shapes. There is obviously scope for extending it to cover more complex shapes and to include more practice on the use of formulae.

**2.** Extensions to other subject areas could include

*social studies* the environmental effects of packaging and the use of materials such as paper, plastics etc., architecture and building shapes

*career development* jobs in the packaging and marketing industries

*information technology* stock control systems, storage and recording methods

*science* materials used in packaging, packaging dangerous or toxic substances

*communications* survey of shapes and sizes of goods, labelling.

**3.** Any vocational topics linked to the retail or distribution trade could extend the ideas here to the display, storage and transport of goods. Various shapes are used in the construction industry and problems such as working out roof space could be brought in. In engineering it is also useful to be able to work out the quantities of materials needed. In catering and caring courses the buying of food in small or large quantities could be used to extend this project.

Chapter 15

# Home Budgeting

In earlier chapters you will have dealt with the financial problems of earning a living, buying a car, paying for accommodation, taking a holiday and paying the bills. In this chapter this all comes together when you look at the problem which everyone has of making the money go round.

## 1. Paying for what you Buy

### 1.1 Paying by Instalments

When buying something fairly expensive many people prefer to pay by instalments even though this may be more expensive in the long run. It often makes more sense to do this than to try to save up for what you want.

For each of the following items advertised in a catalogue work out a) the total cost, paying by instalments b) the extra cost of the instalment method.

**1.** a dress costing £16.99 or 85 p/week for 20 weeks

**2.** a radio costing £89.99 or £2.37/week for 38 weeks

**3.** a shirt costing £8.50 or 43 p/week for 20 weeks

**4.** a hi-fi system costing £499.99 or £6.06/week for 100 weeks

**5.** a three-piece suite costing £599.99 or £15.79/week for 38 weeks or £7.27/week for 100 weeks

You will have noticed that for the short term (38 weeks or less) plans the extra money is only a few pence. For the 100-week schemes the extra amount is very large.

**6.** For questions 4 and 5 work out the extra amount paid for the 100-week schemes as a percentage of the cash price.

## 1.2 Paying on 'Easy Terms'

Many shops selling large items such as washing machines, hi-fi systems, carpets, etc. offer the chance to pay by instalments. Now that so many people have credit cards most places do not even ask for a deposit on goods – you just sign the forms and then take the goods away.

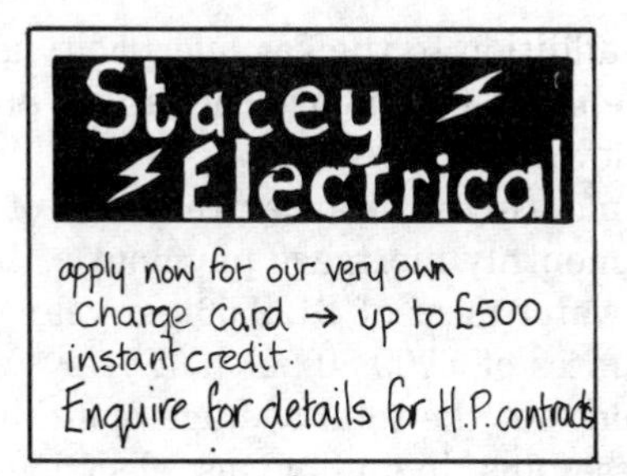

Stacey
Electrical
apply now for our very own
Charge card → up to £500
instant credit.
Enquire for details for H.P. contracts


A chain of shops operates its own credit card system on the basis that you agree on how much you can pay per month and they give you credit of up to 24 times that amount.

**1.** Make up a table to show the credit limits for agreed monthly payments of a) £5 b) £8 c) £10 d) £15 e) £25.

**2.** The company makes a service charge (interest charge) of 2.4% per month, so if at the end of the month you owe £110 they add 2.4% of £110 on to your bill.

A family buys an article for £200 agreeing to pay £10 per month. At the end of the first month they therefore owe £190 plus interest at 2.4%.

a) How much do they owe in total after 1 month?
b) Repeat the exercise above to find the amounts owed after 2 months, 3 months, 4 months and 5 months.
c) Draw a graph to show the amounts still owed after 1, 2, 3, 4 and 5 months.
d) In 5 months the family have paid out £50, a quarter of the original loan. How much of their original debt of £200 do they still owe?

If you carried on this process you would find that even after 21 payments of £10 the family still owe about £50 and that in fact they have to make 27 payments to pay off their debt completely.

**3.** Get hold of some information from various shops on their credit systems. Make up a table to compare the credit limits and monthly service charges for different schemes.

## 1.3 Taking out a Loan

Banks and finance companies are always ready to make you a loan as long as you don't have a history of not paying up. They make their money out of the interest you pay.

A finance company produced the following table showing the monthly repayments over certain periods for different loans.

| *Amount of loan (£)* | *Paying over 12 months* | *Paying over 24 months* | *Paying over 36 months* |
|---|---|---|---|
| 500 | £46.66 | £25.83 | £18.88 |
| 800 | £74.66 | £41.33 | £30.22 |
| 1000 | £93.33 | £51.66 | £37.77 |
| 2000 | £186.66 | £103.33 | £75.55 |

**1.** Make up a table showing the repayments over each of these periods for loans of a) £3000 b) £1500 c) £5000.

**2.** Work out the total amount paid to repay a loan of £1000 over a) 12 months b) 24 months c) 36 months.

**3.** Visit a local bank or finance company and see if you can obtain up-to-date figures for loans to compare with those in the table above (the monthly interest rate in the table above is about 1.8%).

**4.** Credit card companies operate on a slightly more complicated system because of the delay between buying goods and being charged for them. What monthly interest rate do they charge?

# 2. Paying the Bills

## 2.1 Paying as they Come

If you are looking after your own finances you are likely to have several regular bills coming in at various times, for example rates and water rates every 6 months, gas, electricity and telephone every quarter, mortgage repayment and credit accounts every month, insurance, rent, instalment payments monthly or weekly.

One way of dealing with these bills is to pay them as they come and to do this most people have to make sure that they put a little money away every week or month so that they can pay the bills when they arrive.

**1.** Work out the average weekly costs represented by the following bills giving answers to the nearest £1. (Take 1 month as $4\frac{1}{2}$ weeks.)

a) monthly life insurance payments of £36.29
b) monthly mortgage repayment of £63.13
c) quarterly gas bill of £120.34
d) yearly charity donation of £120
e) quarterly telephone bill of £56.49
f) half-yearly water rate bill of £64.02

**2.** In a year a family pays the following bills.

| *Gas* | *Electricity* | *Telephone* |
|---|---|---|
| £120.34 | £46.77 | £56.49 |
| £122.16 | £43.05 | £44.22 |
| £46.09 | £16.89 | £39.95 |
| £48.01 | £26.58 | £46.85 |

Work out the average weekly cost to the nearest £1 over the whole year of
a) gas b) electricity c) telephone.

**3.** In addition to the gas, electricity and telephone bills above the family pays the following.

monthly insurance premiums of £36.29
monthly mortgage payment of £63.13
water rates of £129.56 per year
rates of £209.80 per half year

Work out the total average monthly cost of these bills, including gas, electricity and telephone, to the nearest £10.

**4.** The bills above are for last year. By adding an inflation figure of 5% work out how much the family should be putting away per month this year.

## 2.2 Budget Accounts

Most banks now offer an opportunity to cope with bills by a system of monthly payments. You tell them what you expect to have to pay next year and they transfer $\frac{1}{12}$ of the year's total from your current account each month to a special budget account. You then pay your bills out of that account.

The family in questions **2** and **3** from the previous exercise decides to open a budget account for next year's bills.

**1.** In addition to the bills already dealt with in the previous exercise they wish to include the following.

car insurance £110
road fund licence £100
TV licence £52
holiday £400

Find the total yearly cost of these items plus the items in questions **2** and **3** of the previous exercise.

**2.** Add a 5% inflation figure to the answer to question **1** above.

**3.** The banks make a charge for this service. One bank charges £38 for the first £500 paid out per year plus £1 for each additional £50 or part of £50. Work out the charges for total bills of
a) £600 b) £850 c) £2100 d) £1642.

**4.** Work out the service charge which will be made for paying the total amount worked out in question **2**.

**5.** By adding the service charge in question **4** to the total from question **2** and dividing by 12, work out the monthly amount which will need to be put into the budget account.

## 2.3 Paying as you Use it

Many organisations who are likely to present you with large bills, such as your local council, offer a system of payments by instalments as an alternative.

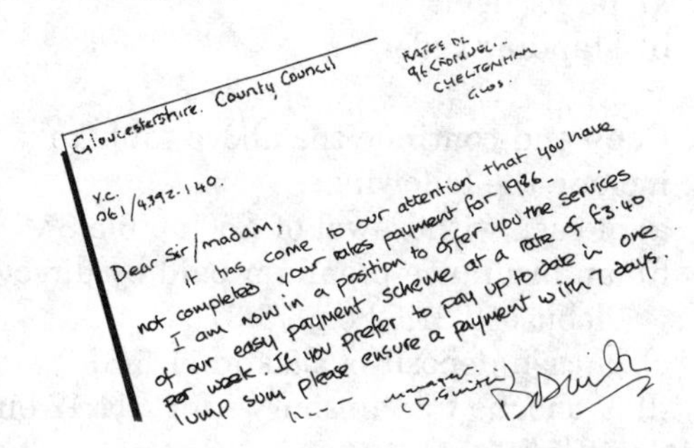
Gloucestershire. County Council

Y.C.
061/4392-140.

Dear Sir/madam,
it has come to our attention that you have not completed your rates payment for 1986.
I am now in a position to offer you the services of our easy payment scheme at a rate of £3.40 per week. If you prefer to pay up to date in one lump sum please ensure a payment with 7 days.

manager (D. Smith)

**1.** The water rate bill for a house one year is £131.33. If the account is paid by 1st May only £124.76 need be paid – what is the percentage discount for early payment? The account can also be settled by two payments of £64.02 in April and £64.01 in October – what is the percentage discount on the total in this case?

**2.** A local authority offers a method for the payment of rates based on paying by 10 equal instalments from May to February. If a house has a rateable value of £254 and the rate has been set at 159 p in the £, find the total rate bill for the year and the amount payable for each instalment.

**3.** Gas bills can be paid as a lump sum or through a 'budget scheme'. Based on her 1985 costs a householder is offered the chance of paying the 1986 bills at a rate of £26 per calendar month, with any balances being settled with the last monthly payment. If the 1986 bills are in fact £123.62, £113.48, £49.18 and £56.20 how much will have to be paid in the last month to settle the account?

**4.** Other organisations such as British Telecom and your local electricity board also operate budget account schemes. Find out how they work and how much you would have to pay per calendar month for the bills in your home.

# 3. Giving your Money Away

Most people have their favourite good cause which they give to on a regular basis. If the good cause is a registered charity such as Oxfam, Mencap or a local church it is possible to take out a *covenant* for a minimum of 4 years which gives the charity even more money. When you covenant money to a charity and you are a regular income tax payer the charity also gets back from the government the tax you have paid. If the basic tax rate is 30% then the amount you pay is only 70% of what the charity gets.

$$\text{amount going to charity} = \text{amount you pay} \times \frac{100}{70}$$

For example, if you covenant £1 then the charity gets $£1 \times \frac{100}{70} = £1.43$.

**1.** If the basic tax rate is 30% how much will a charity get on a covenant of
a) £10 b) £25 c) £150 d) £500?

**2.** What would a charity get for a £10 covenant if the tax rate is
a) reduced to 28% b) increased to 34%?

**3.** A tax payer agrees to covenant £2 a week to her local church. How much will the church actually receive from this covenant per year if the tax rate is 30%?

Remember that covenants only help a charity if you are paying tax and that you have to sign up for at least 4 years.

# 4. Keeping Track of your Money

Unless you are very well off the chances are that you will need to keep track of how your finances are going. Even if you are getting regular statements from your bank it might be useful to keep your own record of money going in and out of your account.

This could be done by using a standard system, for instance in a table with the following headings.

*Date Credits Debits Details Balance*

The words *credit*, *debit* and *balance* are standard terms which appear on most bank statements.

*Credits* are amounts going into your account.

*Debits* are amounts going out.

The *balance* is the amount actually in the account at any time.

The top of a new page in your personal account book might look something like this.

| *Date* | *Credits* | *Debits* | *Details* | *Balance* |
|---|---|---|---|---|
| | | | Brought forward | 162.51 |
| 2/5/87 | | 38.62 | Mortgage | 123.89 |
| 3/5/87 | 200.00 | | Cheque in | 323.89 |

**1.** If you are going to keep a check on your finances you will need to find out the meanings of the following terms.
a) direct debit
b) standing order

**2.** Copy and continue the above table to include the following.
a) a cash withdrawal of £50 on 6/5/87
b) an insurance premium paid by direct debit of £21.52 on 8/5/87
c) a cash deposit of £80 on 11/5/87
d) a cheque to Sainsbury's for £11.12 on 12/5/87
e) payment of a credit card account of £116.21 on 16/5/87
f) a cash withdrawal of £60 on 22/5/87

**3.** If you have a bank account of your own keep your own record for the next month and check that it agrees with the bank statement at the end.

# Family Finances

Time has passed and Liz and Stuart have bought a terraced house and have just moved in.

Liz has an average weekly gross income of about £95 with deductions of about £34. Stuart's monthly salary is £612 with deductions of £148 (month = calendar month).

**1.** a) How much does Liz bring home per week on average?
b) How much does Stuart bring home per month?

**2.** It is winter and Liz and Stuart are looking forward to their summer holiday, so they are trying to work out their finances over the next 6 months. How much will they bring home between them in 6 months (26 weeks) assuming that they do not get any wage increases?

**3.** The net repayment on their mortgage is £137.21 per month and the rate bill for the present year is £214.60. If they decide to pay the rates monthly what will they have to pay out in mortgage repayments and rates each month?

**4.** From information given by the previous owner of the house they estimate that during those 6 months they will have to pay gas and electricity bills of about £250 and £80 respectively, and water rate bills of about £70. Work out the average monthly expenditure on gas, electricity and water to the nearest £1.

**5.** Looking ahead Liz and Stuart have taken out insurance policies which cost £10.20 per month. They are also buying several items on easy terms which means paying out £11.60 per week. They decide to pay their mortgage, rates, water rates, gas, electricity and insurance bills through a budget account at their bank. How much is likely to be taken from their current account in those 6 months to pay these bills?

**6.** After paying these bills and their easy terms instalments how much are they likely to have left to spend per month?

**7.** You should have already worked out the cost, without spending money, of a holiday for Colin and Cheryl in chapter 10. Assuming that Liz and Stuart are going to spend a similar amount, how much should they put away in order to save the amount needed in 6 months?

**8.** Liz and Stuart also wish to put £20 a month into a building society account for emergencies. After they have paid their regular bills and saved for their holiday and for emergencies how much will they have left to spend per month?

**9.** Stuart is keen to try to get a bank loan to buy a newer car to replace their old banger. He decides to go for a loan of £2500 payable over 36 months. Use the table in section 1.3 to work out the monthly repayments on such a loan.

**10.** Once they have paid all their bills and taken out the loan for the car, can they afford to live?

# Teacher's Notes

**1.** As well as giving more practice in money skills, this topic brings together work in other chapters such as rates, taxes, percentages etc.

**2.** Students on a business studies course could link this work in with their studies of banking, credit card systems and changes in retailing.

**3.** Other subject area links could include
*information technology* the use of computers and electronic systems in the retail trade and in banking
*communications* the problems of budgeting and living on a low or fixed income
*career development* careers in banking and business services.

**4.** A visit to the headquarters of a building society, credit card company or bank could be linked in with this work.

Chapter 16

# Gathering Information

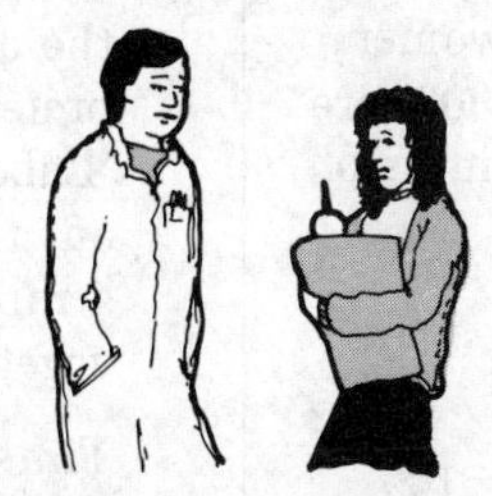

When a company wishes to see how well its product is doing or a newspaper wants to make a guess at the results of the next election, they may decide to do a survey. Surveys are used in many different situations and may be carried out by telephone, by letter, by knocking on doors or by asking questions in the street.

In this chapter you will be looking at ways of carrying out surveys and using the information obtained. This work can be applied to a lot of different subjects in your course and you may even find yourself being paid to carry out a survey at some time in the future.

## 1. The Survey

### 1.1 How Large?

What is meant by the statement 'two out of three people prefer 'Oozo' washing up liquid'? It could mean that the person making the survey asked his wife, his brother and his mother and two of them said that they preferred Oozo, just to please him. It could, on the other hand, mean that he asked 3000 people and 2000 preferred Oozo. In the first case the result is pretty meaningless but in the second we say that the answer was quite *significant*. In general, the larger the number in the survey the better.

---

Discuss with others in your group how many people you would want to question to find out the following information.

**1.** whether a particular pop group was more popular among male or female students

**2.** which political party leader is the most liked by young people

**3.** which make of car is the most popular

**4.** who is likely to win the next general election

**5.** how many people in your group are smokers

**6.** how many teenage smokers there are in the whole country

---

### 1.2 Fair Samples

Questions **5** and **6** in section 1.1 show the major problems of surveys – how many to ask and whom. For question **5** it is easy to ask every member of the group and then give an exact figure for the number of smokers. For question **6** it would be impossible to ask all teenagers and you might be tempted to use your group as a 'sample', for instance if 40% of your group smoke then you might guess that 40% of all teenagers would be smokers.

The next problem is: is your group a fair sample of all teenagers? To be a fair sample it would have to contain about as many males as females, a good spread of ages, and members of different ethnic groups. For instance, in an imaginary parliamentary constituency of 60 000 voters 60% are women, 16% are from immigrant families and 30% are under 30 years old. Therefore a fair sample of 250 people should contain

$$\frac{60}{100} \times 250 = 150 \text{ women}$$

(and so there must be $250 - 150 = 100$ men.)

$$\frac{16}{100} \times 250 = 40 \text{ immigrants}$$

$$\frac{30}{100} \times 250 = 75 \text{ under 30s}$$

It is also necessary to know, of course, what % of the under 30s are women, and so on.

---

The following questions are best answered in group discussion.

**1.** You are to carry out a survey to find out how many households in your town own computers. If you are able to take a sample of 100 homes how would you make your survey a fair one?

**2.** Would it be fair to conduct the survey in question 1 by telephone?

**3.** Would it be fair to carry out the same survey by asking 100 students at your school or college?

**4.** What would be the major problems in carrying out a survey to find out how many households in your town contained a handicapped person?

**5.** When, where and how would you carry out a street survey to find out how many people in your town like Mars bars?

---

## 1.3 What, When and Where?

In discussing the questions in the previous section you will probably have worked out for yourself some basic rules about surveys and samples, for instance it is important to make the questions clear and easy to answer – preferably a one-word answer such as 'yes', 'Labour', 'Woolworths' etc. You must carry out the survey in a place where you will meet a range of people – don't ask questions about vegetarianism outside a butcher's shop.

Timing is important. On a weekday morning the majority of shoppers are housewives, at lunchtime they are working men or women, on Saturday they are families.

Pick your people carefully so that you get a variety of types, ages, backgrounds etc. – unless you want to question only ladies in their 80s, of course.

---

Work on these questions individually, in small groups or in a larger discussion as you think best.

**1.** You wish to find out people's attitude to the present government. Write down a question to ask them.

**2.** Six of you are to carry out a street survey in your town on where women buy their clothes. Work out where you should each stand to do the survey.

**3.** You are to carry out a survey on where men buy their clothes. List some advantages and disadvantages of a street survey at the following times.
a) 8.30 a.m. on Monday
b) 12.30 p.m. on Wednesday
c) 3.00 p.m. on Saturday
When would you choose?

**4.** You are planning a survey to find out people's attitudes to a current topic of discussion, for example abortion, hanging etc. In a sample of 100 people of voting age how many would you choose from each of the following groups?
a) women over 60
b) teenagers
c) people of West Indian origin
d) women with young children

**5.** Plan and carry out a street or college survey on a topic connected with your course. Write a report about your methods and results.

# 2. Using the Information

## 2.1 The Tally System

When carrying out a survey it is important to be able to get the results quickly and accurately. A good way of doing this is to use the tally system.

In a simple survey of 40 people the answers were yes (Y) or no (N) as follows.

Y Y N N Y N N Y N N
Y N Y Y Y N N Y Y N
N Y N N Y Y Y N Y N
Y Y Y N N Y N Y N N

You could just count up how many Ys and Ns, or alternatively you could go through the list once and write down a mark for each Y or N, grouping the marks by crossing every 5th one as below.

Y ~~||||~~ ~~||||~~ ~~||||~~ ~~||||~~ 20

N ~~||||~~ ~~||||~~ ~~||||~~ ~~||||~~ 20

**1.** Use the tally system to find the number of students for each grade in an examination given the following grades.

A E U C B D E C C B
E U B C C D E D A C
U E D D C C B A B C

**2.** Use the tally system to find the number of words of different lengths in the introduction to this chapter.

**3.** Use the tally system to find the number of times each letter of the alphabet is used in this sentence.

## 2.2 The Mean

Suppose that two similar surveys have been carried out in different towns and you wish to compare the results. If you just have tables of results to compare, this is quite difficult. It is often useful therefore to have only a couple of *statistics* for each group which you can compare.

One useful statistic would be the average value which is called the *mean*.

If you have a set of numbers it is easy to find the average, for example the heights of 5 people are

1.8 m 1.6 m 1.72 m 1.66 m 1.59 m

The average height is therefore

$$\frac{1.8 + 1.6 + 1.72 + 1.66 + 1.59}{5}$$

$$= \frac{8.37}{5}$$

$$= 1.67\,\text{m}$$

In question **2** of the previous exercise you will have made up a table showing the number of words of length 1, 2, 3 etc. This is called a *frequency table.*

Suppose that the results for another passage (100 words long) were

| *Word length* | 1 | 2 | 3 | 4 | 5 | 6 | 7 |
|---|---|---|---|---|---|---|---|
| *Frequency (number of words)* | 8 | 10 | 12 | 22 | 22 | 18 | 8 |

The mean word length would then be found by

$$\text{Mean} = \frac{1 \times 8 + 2 \times 10 + 3 \times 12 + 4 \times 22 + 5 \times 22 + 6 \times 18 + 7 \times 8}{100}$$

$$= 4.26$$

**1.** Find the mean value for each of the following sets of numbers.
a) 1 1 5 7 9 10
b) 2 2 3 3 2 3 6 8
c) 101 101 105 107 109 110
d) 0.2 0.2 0.3 0.3 0.2 0.3 0.6 0.8

**2.** A survey of the number of children in 20 households in a street gave the following results.

2 1 2 0 3 5 1 2 6 1 3
0 1 0 1 2 2 4 3 1

Find the mean number of children per house.

**3.** Find the mean age of your group in years and months using ages given to the nearest month.

**4.** Choose two books – a classic and a modern novel. In each book pick a passage of about 100 consecutive words and find the mean word length for each book. Is there a significant difference?

**5.** Find the mean height of the male members of your group and their mean shoe size. Do the same for the female members. Is the ratio of mean height : mean shoe size the same for males and females?

## 2.3 Modes and Medians

Suppose that a survey similar to the one in question **4,** section 2.2, using 30 consecutive words, has given the following results.

**Book A**

| *Word length* | 1 | 2 | 3 | 4 | 5 | 6 | 7 |
|---|---|---|---|---|---|---|---|
| *Frequency* | 2 | 4 | 8 | 10 | 3 | 1 | 2 |

**Book B**

| *Word length* | 1 | 2 | 3 | 4 | 5 | 6 | 7 |
|---|---|---|---|---|---|---|---|
| *Frequency* | 4 | 6 | 5 | 7 | 3 | 4 | 1 |

The mean is a good measure of the 'centre' of the distribution of results but it has to be calculated. It would have been quite different if you had used a passage in the book with a lot of long words. Two simpler measures of the centre are the mode and the median.

**Mode**

The mode (or modal value) is the value which has the highest frequency – the most popular value. In the example above the modes are 4 for book A and 4 for book B. The mode is easy to find but does not tell you much about the two books.

**Median**

The median is the middle value when a set of values is arranged in ascending or descending order, for example 5 students have heights 5 ft 8 in, 5 ft 10 in, 5 ft 7 in, 6 ft and 5 ft 11 in.

When arranged in order these are 5 ft 7 in, 5 ft 8 in, 5 ft 10 in, 5 ft 11 in, 6 ft. The median is the middle one, i.e. 5 ft 10 in. For an even number of people, the median would be half-way between the middle two. When working with a frequency table the median has the same definition but it is harder to find.

For the 30 words from book A on the previous page, the middle value is between the 15th and 16th. There are 14 words with 3 letters or less and 24 with 4 or less. The 15th and 16th words when arranged in length order must therefore have 4 letters each, so the median is 4.

---

**1.** Find the mode and median for the following sets of numbers.
a) 1 1 5 7 9 10 6
b) 2 2 3 3 2 8 6
c) 5 5 8 2 6 4
d) 1 8 7 5 4 2
e) 2 1 2 0 3 5 1 2 6 1
3 0 1 0 1 2 2 4 3 1

**2.** Find the mode and median for the ages of people in your group which you collected for the previous exercise.

**3.** Carry out a survey of 30 cars in a car park to find their ages according to the registration letters. Find the mean, median and mode for this information.

---

## 2.4 The Range

It is not always enough to find a measure of the centre of a set of figures, you sometimes also need to know how spread out the values are. There are several ways of doing this but the easiest is the range.

range = highest value − lowest value

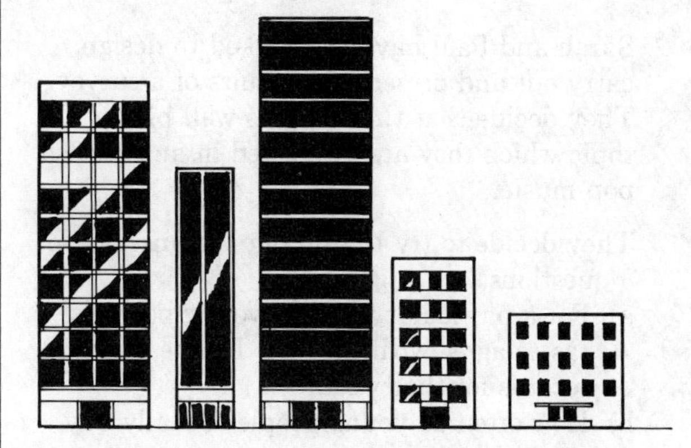

Two different sets of values may have the same mean but quite different ranges, for instance

50 51 52 – mean 51, range 2
10 51 92 – mean 51, range 82

---

**1.** Find the range for each of the following sets of figures.
a) 1 1 5 7 9 10 6
b) 2 2 3 3 2 8 6
c) 1 8 7 5 4 2

**2.** Find the range for each of the following sets of figures.
a) the ages of people in your group
b) the heights of the female members of your group
c) the heights of the male members of your group

# Popular Music

Sarah and Paul have been asked to design, carry out and present the results of a survey. They decide that they may as well pick a topic which they are interested in and choose pop music.

They decide to try to find out the answers to 3 questions.

a) For a particular week do young people's tastes agree with the Top Ten best selling records for that week?
b) How often do young people actually buy singles?
c) What sort of groups make an impact on older people?

**1.** Design questions for a) and b) – try them out on a small group of people to see if they make sense.

**2.** Choose your sample size for questions a) and b) and work out how to make that sample a fair one.

**3.** Carry out your survey recording the answers carefully.

**4.** Think about how you can present your information, along with the Top Ten for this week.

**5.** Find the mode for question a) – i.e. the most popular record. Find also the mean number of singles bought per month from your answers to question b). What is the range for this information?

**6.** For question c) you will obviously need a different sample. How large should it be? How will you choose the sample?

**7.** Design a suitable question for the survey to answer question c).

**8.** Collect this information and present it in a suitable way. Have you taken any account of age differences in the sample?

**9.** The published Top Ten shows the records which are selling best that week. How are your questions related to that? Have you really compared like with like?

**10.** Is there any evidence from your surveys that the published Top Ten influences people in their choice of favourites? Are the record companies and DJs just telling us what we should like?

# Teacher's Notes

**1.** In many courses which include mathematics, students are expected to obtain and use their own information. This could easily lead on to carrying out surveys and interpreting the information obtained.

**2.** Links with other subject areas could include

*science* surveys of the effects of pollution, the impact of science on society, etc.

*social studies* surveys of community concerns and needs

*career development* job surveys, analyses of local employers and products; the relevance of various college or school courses

*information technology* data storage and analysis of survey results

*communications* the writing and testing of questionnaires, interpretation of information, communicating with strangers.

**3.** Surveys could come into any vocational area and enable students to draw sensible conclusions about their work and its relevance.

**4.** Local market research firms may be willing to talk about their work and universities and polytechnics will probably be happy to provide speakers on this topic.

Chapter 17

# The World Around Us – Representing Information

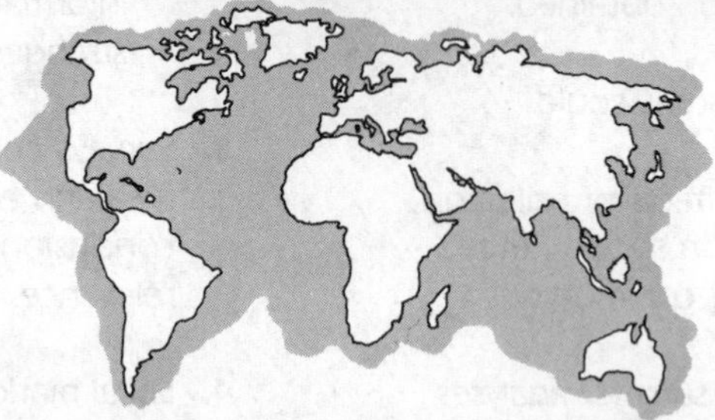

Quite often in newspapers, on the radio or on TV you are presented with statistics, such as '$3\frac{1}{2}$ million unemployed people', '4 out of 5 cannot tell margarine from butter', 'the Liberals got 30% of the votes'. Often these numbers mean little because you do not know the full story. Some people are frightened by numbers and cannot deal with them. In this chapter you will be looking at ways of turning numbers into diagrams by looking at comparisons between life in the U.K. and in other countries.

## 1. Tables and Diagrams

### 1.1 Frequency Tables and Bar Charts

In chapter 16 you will have worked on methods of collecting information through surveys and samples and on methods of using that information. One of the first things you may wish to do with information collected from a survey is to turn it into a *frequency table*. A frequency table shows the number of times any one item occurs.

In a college a group of 40 students were asked their ages on their last birthday. The answers are given below.

| | | | | | | | | | |
|---|---|---|---|---|---|---|---|---|---|
| 16 | 18 | 16 | 17 | 16 | 18 | 22 | 19 | 17 | 16 |
| 17 | 16 | 16 | 16 | 18 | 19 | 21 | 19 | 17 | 17 |
| 16 | 16 | 17 | 18 | 17 | 16 | 17 | 16 | 19 | 18 |
| 16 | 17 | 16 | 16 | 16 | 17 | 17 | 19 | 21 | 18 |

The tally system used in chapter 16 can be used to turn this into a frequency table as shown below.

| *Age* | | *Frequency* |
|---|---|---|
| 16 | ~~IIII~~ ~~IIII~~ ~~IIII~~ | 15 |
| 17 | ~~IIII~~ ~~IIII~~ I | 11 |
| 18 | ~~IIII~~ I | 6 |
| 19 | ~~IIII~~ | 5 |
| 20 | | 0 |
| 21 | II | 2 |
| 22 | I | 1 |
| | | 40 |

Do not forget to check your total at the end.

The simplest type of diagram to use to display this is called a *bar chart*. A bar chart is a set of horizontal or vertical bars, one for each value, and the height of each bar shows the frequency. The ages of the students from the above example are shown on the bar chart below.

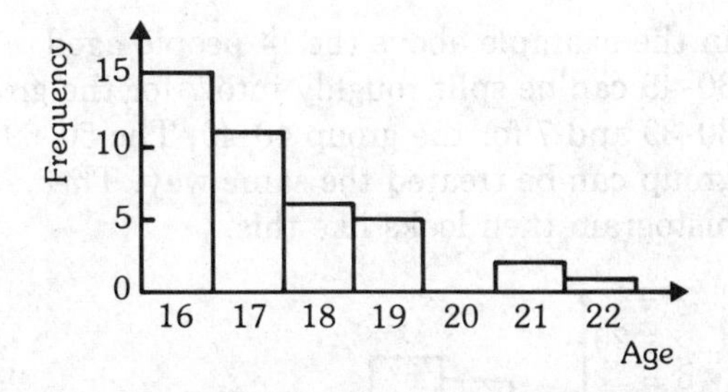


Sometimes the bars may refer to a group of values rather than a single value. A 'grouped frequency table' may be produced from the original information and used to draw a bar chart.

---

Draw bar charts to illustrate the following information.

**1.** The table shows the estimated number of child deaths per year from measles in different countries.

| *Brazil* | *Bangladesh* | *Mexico* |
|---|---|---|
| 34 000 | 173 000 | 57 000 |
| *Ethiopia* | *Vietnam* | *Algeria* |
| 60 000 | 46 000 | 25 000 |

**2.** The table shows the populations per square mile in various countries.

| *Argentina* | *Kenya* | *UK* |
|---|---|---|
| 24 | 65 | 593 |
| *Jamaica* | *Nigeria* | *Ethiopia* |
| 491 | 186 | 63 |

**3.** The table shows the average yearly rainfall in mm in various cities around the world.

| *Moscow* | *New York* | *Delhi* |
|---|---|---|
| 53 | 109 | 72 |
| *London* | *Timbuktu* | *Lagos* |
| 62 | 22 | 182 |

**4.** The table shows the life expectancy at birth in various countries.

| *Argentina* | *Kenya* | *UK* |
|---|---|---|
| 70 | 55 | 73 |
| *Jamaica* | *Nigeria* | *Ethiopia* |
| 71 | 49 | 40 |

**5.** A group of 20 students at a college were asked which countries they were born in. The answers were as follows.

UK Greece Jordan Hong Kong
UK UK Hong Kong Malaysia UK
Jordan South Africa UK Hong Kong
UK Hong Kong UK UK Iran
Iran UK

Make up a frequency table from this information and draw a bar chart.

**6.** A sample of 40 people were asked how many years they thought it would take for the world's population to double if it carried on growing at the present rate. The answers were as follows.

| | | | | | | | | |
|---|---|---|---|---|---|---|---|---|
| 10 | 100 | 20 | 25 | 5 | 20 | 30 | 90 | 95 |
| 70 | 2 | 12 | 25 | 45 | 80 | 70 | 50 | 65 |
| 25 | 40 | 16 | 35 | 60 | 75 | 80 | 90 | 10 |
| 15 | 30 | 60 | 20 | 35 | 15 | 25 | 40 | 35 |
| 70 | 75 | 100 | 5 | | | | | |

Draw up a grouped frequency table from this information using groups 1–20, 21–40, 41–60 and so on. Draw a bar chart.

---

## 1.2 Pictograms and Histograms

The bar chart is not the only way of illustrating information – another useful, and more artistic method is to use a *pictogram*.

In a pictogram a simple picture is used to show a certain number of units and the frequencies are made up by using whole or part pictures. For instance, in the example in section 1.1 you might use a picture of a birthday cake to represent 2 students. The beginning of the pictogram would look like this.

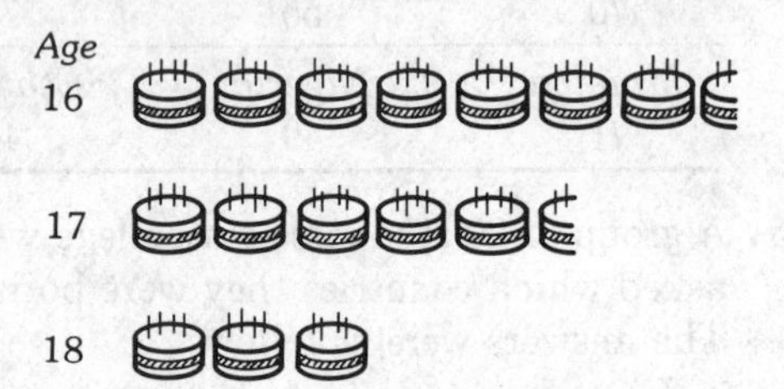


At the bottom you would write ≡ 2 students.

The pictogram is a visual representation, and gives scope for you to show your artistic talents.

---

Think up some good symbols and draw pictograms to show the information in the questions in section 1.1.

---

Now suppose that a survey is carried out of the ages at their last birthday of people in a particular street and the results are put into a grouped frequency table, as shown below.

| *Age* | 0–9 | 10–19 | 20–29 | 30–49 | 50–69 |
|---|---|---|---|---|---|
| *Frequency* | 8 | 15 | 18 | 14 | 12 |

This information is shown in the following bar chart.

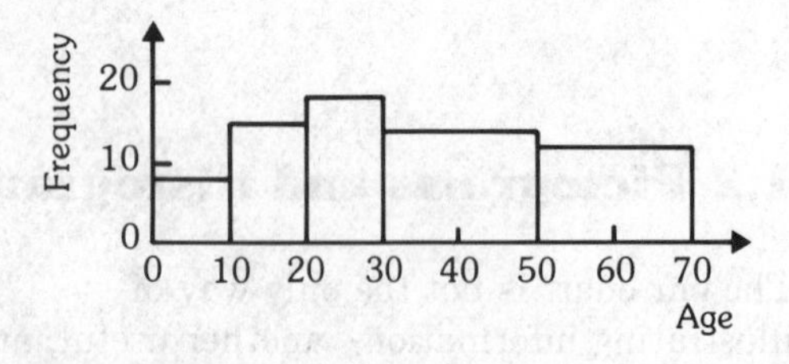


Notice that the first three bars are 10 years wide and the last two 20 years wide. This gives the impression that a large proportion of the street are 30 or over. In fact only 26 out of 67 are 30 or over. The problem is that the right-hand two bars are wider. In a *histogram* the *areas* of the bars represent the frequency. This means that if a bar is twice as wide it must be half the height.

In the example above the 14 people aged 30–49 can be split roughly into 7 for the group 30–39 and 7 for the group 40–49. The 50–69 group can be treated the same way. The histogram then looks like this.

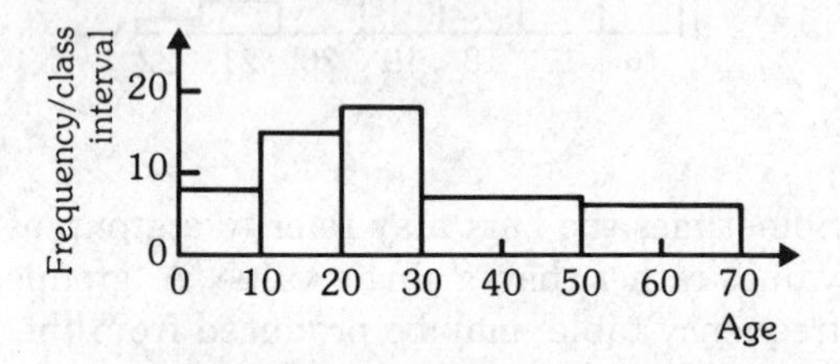


This gives a much better picture of 'people in the street'.

---

**1.** The infant mortality rates per thousand live births in countries in Africa are recorded in the following table. (Frequency, here, is number of countries.)

| *Rate* | 0–40 | 41–80 | 81–120 |
|---|---|---|---|
| *Frequency* | 3 | 3 | 22 |
| *Rate* | 121–160 | 161–240 | |
| *Frequency* | 19 | 5 | |

Taking 40 as the standard bar width draw a histogram to show this information. (For comparison the infant mortality rate in the UK is 12.1.)

**2.** The populations of countries in Africa (except Nigeria) are given in millions in the following table.

| *Population* | 0–5 | 5–10 | 10–15 |
|---|---|---|---|
| *Frequency* | 25 | 15 | 3 |
| *Population* | 15–20 | 20–30 | 30–50 |
| *Frequency* | 1 | 4 | 4 |

Draw a histogram using 5 million as the standard bar width. (For comparison the population of the UK at this time was 56 million.)

### 1.3 Pie Charts

Bar charts, pictograms and histograms are easy to draw but they do not give a good diagram when they contain only 3 or 4 bars. For these cases a *pie chart* may be better. In a pie chart a circle is drawn and divided up into sectors so that the angle of each sector represents the frequency.

In the exercise in section 1.1 you made up a table of the countries of birth of 20 students. This table could be simplified as follows.

| *Country* | UK | Hong Kong | Others |
|---|---|---|---|
| *Frequency* | 9 | 4 | 7 |

To draw the pie chart you need to work out the angle of each sector as follows. The total angle at the centre of a circle is 360°.

*for the UK* angle $= \frac{9}{20} \times 360 = 162°$

*for Hong Kong* angle $= \frac{4}{20} \times 360 = 72°$

*for others* angle $= \frac{7}{20} \times 360 = 126°$

(as a check, $162 + 72 + 126 = 360$)

The pie chart would then look like this.

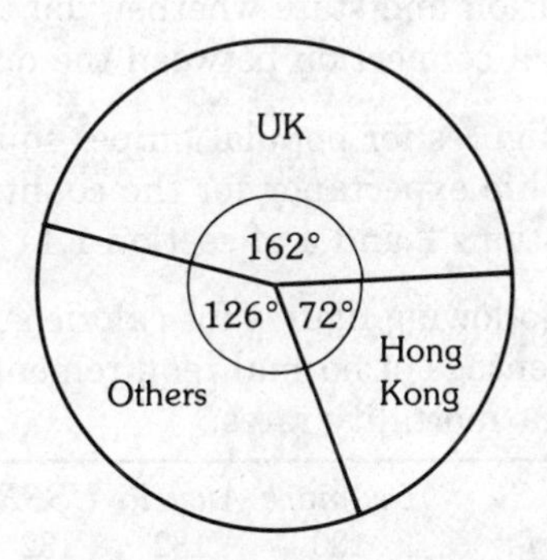


Notice that the pie chart is very useful for showing the rough proportions of different groups, but it is difficult to read exact figures from it.

**1.** The populations of the 25 small (population less than 5 million) countries of Africa can be divided up as follows.

| *Population* | Less than 1 million | 1–2 million | Over 2 million |
|---|---|---|---|
| *Frequency* | 12 | 5 | 8 |

Use a pie chart to show this information.

**2.** Tea is exported from various countries. The percentages are as follows.

| *India* | *Sri Lanka* | *China* | *Other countries* |
|---|---|---|---|
| 28% | 25% | 9% | 38% |

Draw a pie chart to show this information.

**3.** Jobs can be divided up very roughly into three types. The percentages of people in each of these for rich and poor countries in the 1970s were as follows.

| | *Agriculture* | *Industry* | *Services* |
|---|---|---|---|
| *Rich countries* | 10% | 38% | 52% |
| *Poor countries* | 63% | 15% | 22% |

Show this information on two separate pie charts.

**4.** In 1983/4 the major voluntary aid organisations raised the following amounts to the nearest million US $.

| *Action Aid* | *Cafod* | *Christian Aid* | *Oxfam* | *Save the Children* |
|---|---|---|---|---|
| 9 | 7 | 13 | 28 | 18 |

Use a pie chart to show this information.

# 2. Using Graphs

### 2.1 Line Graphs

In earlier chapters you will have drawn graphs which gave straight lines or smooth curves, for example conversion graphs.

In this section graphs will be used to show changes in time. They are not necesarily straight or smooth but are made up of line sections.

In each of the following cases choose suitable scales, mark your points on graph paper and join them with a series of straight lines.

**1.** The table shows the percentages of the population aged 55 or over in Northern Europe.

| 1950 | 1960 | 1970 | 1980 | 1990 | 2000 |
|---|---|---|---|---|---|
| 20% | 23% | 25% | 25% | 23% | ? |

**2.** For Japan the figures for over 55-year-olds are

| 1950 | 1960 | 1970 | 1980 | 1990 | 2000 |
|---|---|---|---|---|---|
| 10% | 12% | 15% | 18% | 22% | ? |

**3.** Youth unemployment as a % of total youth labour force in OECD (Organisation for Economic Cooperation and Development) countries increased as follows.

| 1979 | 1980 | 1981 | 1982 | 1983 | 1984 |
|---|---|---|---|---|---|
| 11.3 | 13.1 | 14.8 | 17.3 | 19.3 | 19.5 |

**4.** The following figures in millions show the increase in world tourism from 1958 to 1983.

| 1958 | 1963 | 1967 | 1973 | 1983 |
|---|---|---|---|---|
| 55 | 93 | 140 | 215 | 286 |

## 2.2 Scatter Diagrams

Sometimes you may wish to find out if there is any connection between two particular sets of figures, for example is there a connection between age of death of smokers and the average number of cigarettes smoked per day?

There are some quite complicated methods of looking at this problem but one simple method is to use a scatter diagram.

In a scatter diagram scales for the two sets of figures are put on a pair of axes and each pair of figures is marked by a cross.

| *Age at death* | 65 | 72 | 55 | 68 | 70 | 60 |
|---|---|---|---|---|---|---|
| *Number of cigarettes* | 20 | 20 | 40 | 10 | 15 | 15 |

The scatter diagram would look like this.

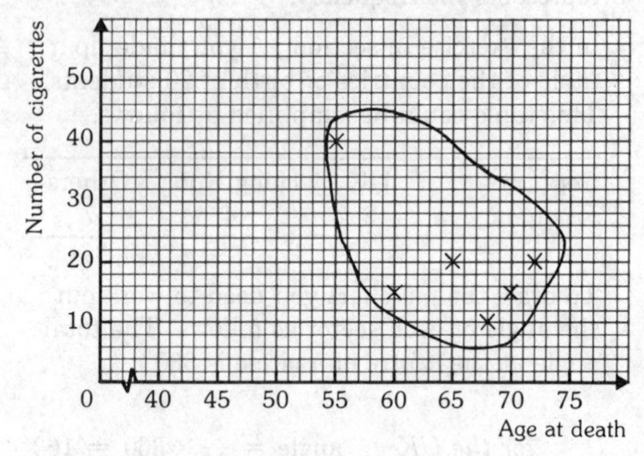


In the diagram the points have been surrounded by a ring. If this ring is long and thin then we might conclude that there is some connection. If the ring is fairly circular then no connection has been shown. Remember, however, that the scale chosen can distort a scatter diagram.

Draw scatter diagrams to show the following information and state whether you think there is a connection between the quantities.

**1.** the figures for population per square mile and life expectancy for the countries in questions 2 and 4 of section 1.1

**2.** the following figures for calorie supply as a percentage of normal requirements and infant mortality rates.

| | *Denmark* | *Austria* | *USSR* | *Jamaica* |
|---|---|---|---|---|
| *Calories* | 130 | 132 | 132 | 114 |
| *Mortality* | 8 | 13 | 33 | 28 |

| | *Jordan* | *India* | *Kenya* |
|---|---|---|---|
| *Calories* | 97 | 90 | 88 |
| *Mortality* | 68 | 122 | 86 |

# A Publicity Campaign

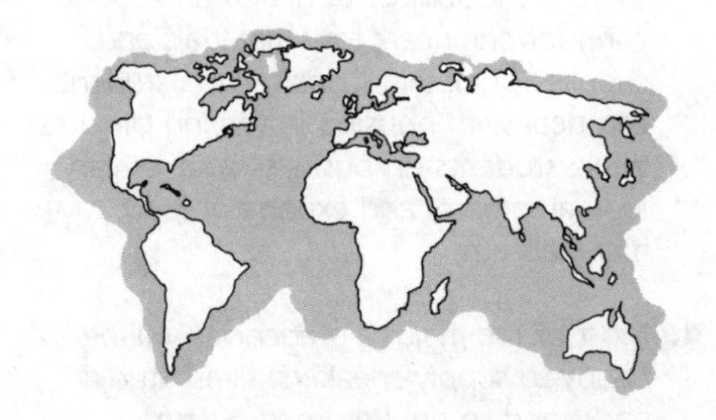

The Oxfam group which Sanjay and Debbie belong to wishes to mount a publicity campaign to show the differences between people around the world. They wish to produce a display for their local library and decide that pictures are better than tables of figures.

They first of all get together the following figures for countries around the world.

1. Find a map of the world, trace it on to a sheet of paper and mark the positions of these countries on your map.
2. Check that you understand the meanings of each of the headings in the table. Write down their meanings.
3. Look for possible connections between the different columns of figures. If you think there are connections show these by using scatter diagrams.
4. Look for major differences between countries. Draw bar charts or pictograms to show these differences.
5. One reason often given for poverty is that the population is too high. Is this true? Draw a diagram (or more than one) to show whether this is true or not.
6. Think about other causes of poverty. Can you get figures to show whether these are really causes or not? If you can, draw some diagrams to illustrate them.
7. Put your work together into a display using a variety of diagrams.
8. Make up a title for your display.

| | *Life expectancy* | *Infant mortality* | *Population (millions)* | *Population /mile²* | *Calorie supply* | *GNP* |
|---|---|---|---|---|---|---|
| Argentina | 70 | 41 | 29 | 24 | 127 | 2560 |
| Dominican Republic | 61 | 67 | 6 | 268 | 94 | 1338 |
| El Salvador | 64 | 53 | 5 | 519 | 94 | 636 |
| Kenya | 55 | 86 | 19 | 65 | 88 | 432 |
| Thailand | 61 | 54 | 51 | 225 | 103 | 769 |
| Ethiopia | 40 | 146 | 31 | 63 | 74 | 142 |
| Jamaica | 71 | 28 | 2 | 491 | 114 | 1182 |
| Mexico | 66 | 55 | 76 | 86 | 120 | 2250 |
| Morocco | 57 | 106 | 23 | 108 | 109 | 869 |
| Nigeria | 49 | 134 | 84 | 186 | 99 | 873 |
| Vietnam | 54 | 99 | 57 | 378 | 93 | 175 |
| Liberia | 54 | 153 | 2 | 42 | 107 | 536 |
| UK | 73 | 12 | 56 | 593 | 131 | 8950 |
| Sweden | 75 | 7 | 8 | 48 | 117 | 14 500 |
| Italy | 73 | 14 | 56 | 499 | 144 | 6830 |
| Saudi Arabia | 54 | 112 | 10 | 11 | 119 | 12 720 |

# Teacher's Notes

**1.** From the mathematical point of view this chapter is mainly about the use of diagrams in statistics. There are, however, many interesting implications arising from the topic and a whole range of ways of extending it.

**2.** Other subject areas can be included in a variety of ways, for example

*science* nutrition, basic food items such as salt and their production, minerals and raw materials, agriculture

*social studies* relationships with developing countries, climate, imports and exports

*career development* careers in technology, opportunities to work in developing countries

*information technology* impact of information technology on the developing countries, appropriate technology

*communications* life in the third world, aid agencies, attitudes to problems of other countries, media coverage.

**3.** Most vocational areas could link with the work in this chapter. For caring courses there is the subject of health and social care, for engineers raw materials and supplies of metals. Construction students can deal with housing in varying climates while students on business courses can look at imports and exports of food, raw materials etc.

**4.** Most of the major aid agencies will be happy to supply speakers, films, study packs and so on. Returned overseas volunteers can be contacted through organisations like VSO and IVS. Many areas have Development Education Centres run either by local authorities or voluntary organisations. Some local authorities also employ specialists in this field.

# Chapter 18

# What are the Chances?

If you always knew what was going to happen to you life would be very boring – you need to take chances sometimes to make life more interesting. However, before you take a chance it is often useful to know what the outcome is likely to be. It may be a great feeling leaping over double decker buses on a motorbike but, as most people know that for them it would mean almost certain death, they don't try it.

Similarly a firm cannot afford to take too many chances over new products or new factories without knowing whether they are likely to succeed. In short, you need to have methods for working out how likely something is to happen – this is called probability.

## 1. Basic Probabilities

### 1.1 Probabilities and Odds

You often hear people saying things like

'the odds for that horse were 10 to 1'

'there's a 90% chance that I will get the job'

'his chances of recovery were fifty–fifty'

'9 out of 10 people prefer Zonk to any other make'

These are all statements about chance and therefore about probabilities. In mathematical work the *probability* of something happening is written down not in terms of 'odds' or 'chances' but as a fraction between 0 and 1. For example, odds of 10 to 1 would be written as a probability of winning of $\frac{1}{11}$ and you would write

$$\text{p(win)} = \frac{1}{11}$$

a 90% chance would be written as $\frac{90}{100} = \frac{9}{10}$

so, $\text{p(getting job)} = \frac{9}{10}$

a fifty-fifty chance would be $\frac{50}{100} = \frac{1}{2}$

so, $\text{p(recovery)} = \frac{1}{2}$

In theory a probability of 0 should mean that something cannot happen at all and a probability of 1 should mean that it is certain to happen. You will come across this later.

---

Write the following as probability statements.

**1.** 'odds of 11–2'

**2.** 'odds of 5–4 *on*'

**3.** 'a 20 to 1 chance'

**4.** '95% certain'

**5.** 'a 1 in 4 chance'

**6.** '1 in 1000 babies'

**7.** 'a chance in a million'

**8.** 'no chance'

---

## 1.2 Finding Probabilities

In the last section you were turning statements about odds and chances into probabilities. This was easy because each statement contained some numbers, but what about statements like

'there's not much chance'
'I'm almost certain'
'I'm not sure'

You sometimes need to be able to turn these into probabilities, especially if you want to compare statements. For example, which statement gives the higher probability, 'I'm almost certain' or 'I'm very sure'?

There are three main ways of finding probabilities.

**a) The Equally Likely Method**

The study of probability started among gamblers because there it is normally expected that all possibilities are equally likely to happen. For example

tossing a coin – heads and tails are equally likely
throwing a die – 1, 2, 3, 4, 5 and 6 are equally likely
cutting a pack of cards – each card is equally likely to turn up

Since equally likely possibilities must have equal probabilities then

p(getting a head on a coin) $= \frac{1}{2}$

p(getting a 6 on a die) $= \frac{1}{6}$

p(getting a King of Clubs from a pack of playing cards) $= \frac{1}{52}$

---

Find the probabilities of the following.

**1.** getting the double 6 from a set of dominoes

**2.** winning first prize in a raffle if 500 tickets are sold

**3.** your phone number ending in a 3

**4.** your sister having the same birthday as you

**5.** getting double 6 when throwing a pair of dice

**6.** suffering from lung cancer in the future if you are a heavy smoker

---

**b) The Past Experience Method**

The last question in the previous set of questions cannot be answered using the equally likely method because heavy smokers seem to be more likely to suffer from lung cancer than non-smokers. The only way to find this probability is to estimate it from the information you have. For instance, if in a survey of 1000 heavy smokers 400 had lung cancer then we might say

p(heavy smoker getting lung cancer)

$$= \frac{400}{1000} = \frac{2}{5}$$

Turn the following statements into probability statements.

**1.** 'of 800 people interviewed 600 thought that Labour would win'

**2.** 'out of 280 children born this year 60 had some form of handicap'

**3.** 'out of 600 teenage motorcyclists 420 were injured in accidents in their first year of riding'

**4.** 'last year 60 candidates out of 160 obtained at least grade C in GCSE'

**5.** 'out of 800 cases of rabies in people none were cured'

**6.** 'Burnley have won all of their 8 matches so far'

### c) Guesswork

Questions **4, 5** and **6** of the previous section bring out certain problems when finding probabilities.

In question **4** the answer $\frac{3}{8}$ applies to the 'average' candidate – it does not tell you what *your* chance of getting grade C is, as you may be better or worse than the average candidate.

The answer 0 in question **5** suggests that surviving rabies is impossible – a statement which no doctor would make.

The answer 1 in question **6** suggests that Burnley will certainly win their next match, but nothing in sport is certain.

In all these cases you have to guess at probabilities based on the information available.

**1.** Write down the probabilities which you think go with the following words.
a) sure b) fairly sure c) almost certain
d) unlikely

**2.** Write down estimates of the probabilities of the following.
a) Manchester United winning the league this year
b) there being a change of government at the next election
c) England winning the next World Cup
d) this week's Top Ten number 1 being number 1 next week

Compare your answers to these questions with others in your group – there may be quite big differences.

# 2. Combining Probabilities

## 2.1 Several Possibilities

In section 1.1 you worked on finding the probabilities of one outcome out of several possibilities. This method can also be extended to more difficult cases, for instance the probability of getting a 5 or 6 when throwing a die.

You are interested in 2 outcomes out of a possible 6, so

$$p(5 \text{ or } 6) = \tfrac{2}{6} = \tfrac{1}{3}$$

When a card is picked from a shuffled pack,

$$p(\text{card is a club}) = \tfrac{13}{52} = \tfrac{1}{4}$$

Find the probabilities of the following.

**1.** picking an ace from a shuffled pack of cards

**2.** picking a domino with a 6 on it from a full set

**3.** scoring 3 or more when throwing a die

**4.** being picked from a group of 30 to go on a trip if there are only 12 places available

**5.** winning one of the 20 prizes in a raffle if 1000 tickets are sold

**6.** getting a number ending with 0 or 5 from a book of draw tickets numbered 1 to 100

**7.** picking a blue ball from a bag containing 4 red, 3 blue, and 3 yellow balls

**8.** getting a total of 9 when throwing two dice

## 2.2 Possibility Spaces

In question 8 of the previous section you were asked to find the probability of getting a total of 9 when throwing two dice.

One way of tackling this problem is to draw a *possibility space diagram.* On die 1 there are 6 possible scores, marked as shown, and there are also 6 possible scores for die 2. This gives a total of 36 possible scores. A total of 9 can be obtained in the 4 combinations 3,6, 4,5, 5,4 and 6,3. These have been ringed on the diagram below.

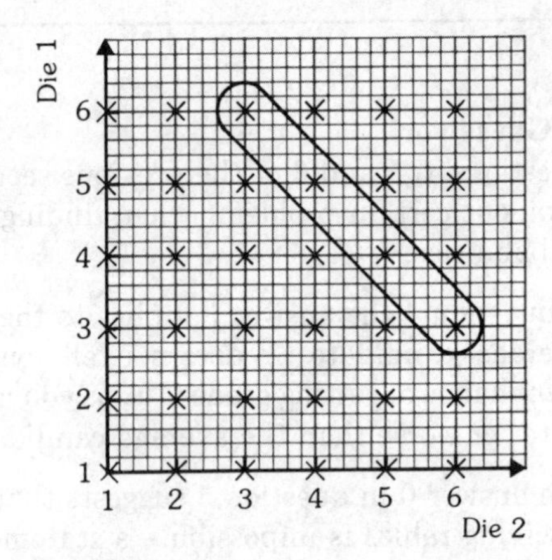


$$p(\text{total of } 9) = \tfrac{4}{36} = \tfrac{1}{9}$$

This method can be used for any similar situation as long as the different outcomes are equally likely to happen, as in throwing a die and tossing a coin together, which will give 12 possible outcomes.

If two fair dice are thrown, find the probabilities that

**1.** a total of 4 is obtained

**2.** the total score is 11

**3.** both dice show the same score

**4.** at least one of the dice shows a 6.

If a die is thrown and a coin tossed, find the probabilities of getting

**5.** a 6 and a head

**6.** a 6 or a head or both.

Three coins are tossed. By trying to imagine a 3-dimensional possibility space diagram, or by using some other method, find the probabilities of getting

**7.** a head and 2 tails

**8.** 3 heads.

---

## 2.3 Expected Values

'Knowing my luck, if I buy a ticket for a raffle I don't expect to win anything.' This is the sort of statement that people often make. In probability work the word 'expectation' or the phrase 'expected value' is used in a more mathematical way.

A school has organised a raffle and has spent £100 on prizes. If 2000 tickets are sold then on average each ticket will win $\frac{\pounds 100}{2000} = 5\,\text{p}$ worth of prizes.

The expected value of the prize per ticket is then 5p. If tickets are sold for 10 p each then there is an expected loss per ticket of $10\,\text{p} - 5\,\text{p} = 5\,\text{p}$, so on average every ticket loses 5 p and the school wins 5 p per ticket.

> in general, the expected value is an average value

Expected values can also be worked out using probabilities. For example, in a game involving throwing 2 dice there is a prize of 50 p for a total score of 12 or 10 p for a total of 10 or 11.

$$\text{p(total of 12)} = \tfrac{1}{36}$$
$$\text{p(total of 10 or 11)} = \tfrac{5}{36}$$

(from the possibility space diagram)

So expected winnings per game are

$$50 \times \tfrac{1}{36} + 10 \times \tfrac{5}{36} = \tfrac{100}{36} \simeq 3\,\text{p}$$

If an entry fee of 5 p is charged for the game then on average *you* would expect to lose $5\,\text{p} - 3\,\text{p} = 2\,\text{p}$ per game.

---

In the dice game above what would you expect to lose per game if

**1.** the entry fee is 5 p and there is a prize of £1 for a total score of 12

**2.** the entry fee is 5 p and there is a prize of 20 p for getting a double

**3.** the entry fee is 10 p and there is a prize of 20 p for getting a double or a total of 11 or more?

In a 'pick-a-straw' game the straws are numbered from 1 to 300. What would you expect to lose per go if

**4.** the entry fee is 5 p and a 10 p prize is given for numbers ending in 0 or 5

**5.** the entry fee is 10 p and a prize of 50 p is given for numbers ending in 0?

A bag contains 100 balls. Fifteen have the number 2 on them, four the number 5 and one the number 50. The rest are blank. If a ball is drawn out of the bag the drawer gets a prize of the same number of pence as the number on the ball.

**6.** What is the expected value of the drawer's winnings for one ball?

**7.** If you were the 'banker', how much would you charge as an entry fee for this game?

**8.** With this entry fee how much should the bank make per 100 players?

# Prices and Prizes

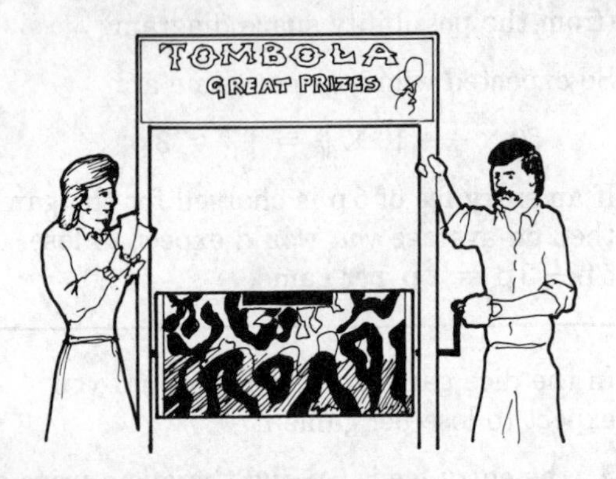


The Oxfam group to which Sanjay and Debbie belong is holding a Spring Fête. Several organisations are providing stalls but Debbie and Sanjay have been asked to hunt around for a few sideshows and organise them. Obviously the sideshows are to make a profit but they want to make them reasonably fair by making sure that a good proportion of the takings is given back in prize money – they decide that the prizes should be about 60% of takings.

**1.** Using this 60% figure what should the expected winnings be for a game costing a) 5 p b) 10 p?

**2.** Look back at the 'pick-a-straw' game in section 2.3. Do either of the suggested systems for that fit the 60% condition? If not, suggest a better system.

**3.** The tombola contains 60 prizes, some bought and some donated, with a total estimated value of £50. If 600 tickets numbered 1 to 600 are being used and a prize is given to every ticket which ends in a 0, what should the price per ticket be?

**4.** In a 'Lucky Dip with a Difference' there are to be 100 identical parcels, 90 containing a stone and 10 containing items worth about £2 each. What would be a fair charge for the Lucky Dip?

**5.** In a football game people are given 3 balls to kick through a small hole and are to be charged 10 p a go. Watching this earlier Debbie noticed about 5% of people could get three balls in, 10% only two and 10% only one. Work out a fair prize system for this game using the 60% rule.

**6.** The children's art competition costs 10 p per entry. There are to be 3 prizes worth about £1.50 each. How many entries are needed for the 60% rule to work?

# Teacher's Notes

**1.** This chapter has dealt with only the basic ideas of probability in order to give students an introduction to the principles involved. This could of course be extended to cover tree diagrams and more complex problems.

**2.** Links with other subject areas are not as varied with this topic but could include
*science* basic genetics
*communications* interpretation of statements about chance
*social studies* an analysis of different types of gambling.

**3.** One area where there could be vocational applications is in caring courses where probability could link in with a study of infectious diseases. In a business-oriented course probability could be brought into decision-making processes.

**4.** As a last resort you could always get students gambling and working out odds – in a friendly way, of course.

Chapter 19

# You and Yours

This chapter is about people, but it also contains quite a lot of numeracy work. It brings together work you have already been doing on units and various types of graphs, gives you opportunities to carry out your own investigations and introduces several important issues which affect young people. The work in this chapter will also be useful for anyone going into the manufacturing or retail sides of the clothing industry.

## 1. Buying Clothes

### 1.1 Metric and Imperial Units

In earlier chapters you have had to work with both metric and imperial units, but usually not both together. When buying clothes it is often necessary to be able to convert from one set of units to another. The two standard units you are likely to need are

the inch (in) for the imperial system
the centimetre(cm) for the metric system

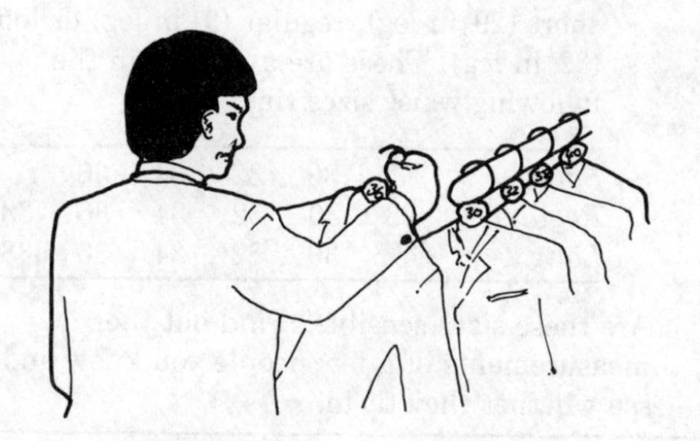

The relationship between these which is generally used is

$$1\,\text{in} = 2.54\,\text{cm}$$

For example, $40\,\text{in} = 40 \times 2.54$
$= 101.6$
$= 102\,\text{cm}$ to the nearest cm

**1.** Make up a table showing the number of cm (to the nearest cm) for different numbers of in from 20 to 60 in steps of 5 in.

**2.** Use your table from question 1 to draw a graph of cm against in from 20 in to 60 in.

**3.** Use your graph to find
a) the number of cm (to the nearest cm) equivalent to
i) 32 in ii) 38 in iii) 46 in.
b) the number of in (to the nearest in) equivalent to
i) 100 cm ii) 120 cm iii) 136 cm.

**4.** When buying clothes you need to know certain measurements. Make a list of the following measurements for yourself, both in inches and in centimetres.

| | |
|---|---|
| for a male | chest, waist, inside leg, height |
| for a female | waist, hips, bust, height |

**5.** Carry out a survey of local clothes shops to see which mark sizes in metric units, which in imperial units and which in both. Is there a difference between men's and women's shops in this respect?

## 1.2 Children's and Women's Sizes

Clothes for young children are classified by height on the assumption that children are not short and fat or tall and thin but are all 'average'. Women's clothes, on the other hand, are sold by using a mysterious size system which usually starts at 10 and goes up in 2s to 24 or even 30 for some styles of clothes. Again, if you are a funny shape you are in trouble as no size will be quite right.

**1.** A catalogue gives the following table for children's sizes.

| | | | | |
|---|---|---|---|---|
| *Height (cm)* | 104 | 110 | 116 | 122 |
| *Chest (in)* | 22 | 23 | 24 | 25 |
| *Waist (in)* | 22 | 22 | 23 | 23 |
| *Height (cm)* | 128 | 134 | 140 | 146 |
| *Chest (in)* | 26 | 27 | 28 | 29 |
| *Waist (in)* | 23 | 23 | 24 | 24 |

On the same graph paper draw graphs of chest size against height and waist size against height using the above figures.

**2.** Use a couple of children you know as 'guinea pigs' and see if they fit in with the above scheme. You will find out that measuring children accurately is not easy!

**3.** The same catalogue gives the following table of women's sizes.

| *Size* | 10 | 12 | 14 | 16 | 18 | 20 | 22 | 24 |
|---|---|---|---|---|---|---|---|---|
| *Bust (in)* | 32 | 34 | 36 | 38 | 40 | 42 | 44 | 46 |
| *Hips (in)* | 34 | 36 | 38 | 40 | 42 | 44 | 46 | 48 |

On the same graph paper draw graphs of bust and hip measurements against size. Do you fit this scheme?

**4.** This system of women's sizes may not be universal. Visit a couple of shops and see if their systems agree with the one above. Note down any differences.

## 1.3 Men's Sizes

If you are male you may not have wanted to go out and investigate women's clothes shops so this is your chance to shine. In general, to buy clothes for yourself you need to know four numbers – the sizes for your neck, chest, waist and leg length.

**1.** Men's shirts are usually sold by neck size starting at $14\frac{1}{2}$ in and going up to about 18 in. Some shirts also give the chest size (or range of sizes) for which they are designed. Carry out a survey of this and make up a table of collar size and chest size. Draw a graph of average chest measurement against collar size – does it give a straight line?

**2.** Trousers are sold by waist size and leg length. In a catalogue trousers are sold as short ($29\frac{1}{2}$ in leg), regular (31 in leg) or long ($32\frac{1}{2}$ in leg). These are available in the following waist sizes (in).

| | | | | | | |
|---|---|---|---|---|---|---|
| *Short* | | 30 | 32 | 34 | 36 | |
| *Regular* | 28 | 30 | 32 | 34 | 36 | 38 |
| *Long* | | 30 | 32 | 34 | 36 | 38 |

Are these sizes sensible? Find out the measurements of a few people you know and see whether they fit in.

### 1.4 Shoe Sizes

Most of the time you buy shoes in sizes like $5\frac{1}{2}$ or 8 or 4. Sometimes, however, you pick up a pair of shoes and find that they are size 40 or size 36. This is because there are two main systems of sizes in use, the first is the British system and the second the continental system.

**1.** The following table shows approximate equivalents between British and continental shoe sizes.

| *British* | 4 | 5 | $6\frac{1}{2}$ | 8 | $9\frac{1}{2}$ |
|---|---|---|---|---|---|
| *Continental* | 37 | 38 | 40 | 42 | 44 |

Mark these values on a piece of graph paper using a range of 2 to 10 for British sizes. Draw the best straight line you can through these points. What are the equivalent continental sizes (to the nearest whole number) for the following British sizes?

a) $2\frac{1}{2}$ b) 6 c) $8\frac{1}{2}$

**2.** A foot of length 9 in requires a size 4 shoe whereas an 11 in foot needs a size 10. Assuming that the relationship between length and size gives a straight line show this information by drawing another line on your graph from question 1.

Get someone to measure your foot length when you are standing up straight. Check from your graph that you are wearing the correct size of shoes.

**3.** Carry out two separate surveys of shoe sizes of males and females among your fellow students. For each survey draw a bar chart showing the sizes worn. If you know someone who works in a shoe shop you could try to find out how they decide which sizes to stock.

## 2. Looking After Yourself

### 2.1 Keeping in Shape

It is quite common for people to worry about their shape or size. They think that they are too tall, too short, too fat or too thin. You can't do much about your height, of course, but you may be able to do something about your weight. The first thing to do is to find out if you really are the right shape.

**1.** A slimming club produces a leaflet showing the 'ideal' weight for women of small, medium and large frame.

| *Height* | | *Weight* Small frame | | Medium frame | | Large frame | |
|---|---|---|---|---|---|---|---|
| *ft* | *in* | *st* | *lb* | *st* | *lb* | *st* | *lb* |
| 4 | 10 | 7 | 2 | 7 | 9 | 8 | 5 |
| 5 | 0 | 7 | 6 | 8 | 1 | 8 | 11 |
| 5 | 2 | 8 | 0 | 8 | 7 | 9 | 3 |
| 5 | 4 | 8 | 6 | 9 | 1 | 9 | 11 |
| 5 | 6 | 9 | 0 | 9 | 9 | 10 | 5 |
| 5 | 8 | 9 | 9 | 10 | 3 | 10 | 13 |
| 5 | 10 | 10 | 3 | 10 | 11 | 11 | 8 |

Using the same axes and scales draw 3 graphs to show weight against height for the 3 different frames. *Remember that there are 12 in in 1 ft and 14 lb in 1 st, and be careful with your scales.*

What weight would be ideal for
a) a small-framed woman of height 5 ft 1 in
b) a medium-framed woman of height 5 ft 5 in?

**2.** A table for men gives the following information.

| Height | | Weight | | | | | |
|---|---|---|---|---|---|---|---|
| | | *Small frame* | | *Medium frame* | | *Large frame* | |
| *ft* | *in* | *st* | *lb* | *st* | *lb* | *st* | *lb* |
| 5 | 4 | 8 | 13 | 9 | 7 | 10 | 2 |
| 5 | 6 | 9 | 6 | 10 | 0 | 10 | 12 |
| 5 | 8 | 10 | 0 | 10 | 9 | 11 | 7 |
| 5 | 10 | 10 | 9 | 11 | 3 | 12 | 1 |
| 6 | 0 | 11 | 3 | 11 | 12 | 12 | 10 |
| 6 | 2 | 11 | 11 | 12 | 8 | 13 | 6 |

Draw a similar graph to the one in question 1 and estimate the ideal weights for the following.

a) a medium-framed man of height 5 ft 9 in
b) a medium-framed man of height 6 ft 4 in

**3.** See where you fit in on one of the graphs above. If you think you are overweight don't rush into a crash diet without taking proper advice. Think about the fat-producing foods which you could *easily* cut out. Exercise will also do you good.

**4.** Energy, whether going in, in the way of food, or going out through exercise, is measured in kilocalories, also called Calories, or, increasingly, in kilojoules (kJ).

The average man needs an energy intake of about 11 000 kJ per day while a woman needs about 9000 kJ. This sounds a lot but a single egg contains about 350 kJ. Get a calorie chart from somewhere (for instance a slimming club) and work out your energy intake for yesterday. (1 Calorie is about 4.2 kJ.)

**5.** The chart below shows the energy used up in half an hour for various types of exercise.

| *Exercise* | *Energy kJ* |
|---|---|
| Slow walking | 400 |
| Brisk walking | 650 |
| Jogging | 1000 |
| Badminton or dancing | 650 |
| Squash | 1000 |
| Swimming | 850 |
| Watching TV | 200 |

How many kJ did you burn up yesterday doing these activities? Work out a weekly programme of exercise to use up 3000 kJ.

## 2.2 Healthy Living

Being overweight can be a nuisance or it can seriously damage your health. Taking drugs or smoking or drinking heavily is, however, much more likely to carry you off to an early grave. In this section you will have a chance to practise some statistical work and learn a few nasty facts.

**1.** The numbers of people (to the nearest 10) per million of the population who died from lung cancer from 1961 to 1981 are given below.

| *Year* | *Males* | *Females* |
|---|---|---|
| 1961 | 870 | 140 |
| 1966 | 970 | 180 |
| 1971 | 1060 | 220 |
| 1976 | 1110 | 280 |
| 1979 | 1120 | 310 |
| 1980 | 1170 | 330 |
| 1981 | 1090 | 330 |

Show these two sets of figures on the same graph and comment on the trends. Try to find out figures for the years since 1981 and add them to your graph. Has the trend changed since 1981?

**2.** As you will already know quite a few young people take up smoking while still at school. In national surveys the % of regular smokers among schoolchildren was found to be as follows.

| | *1982* | | *1984* | |
|---|---|---|---|---|
| | *Boys* | *Girls* | *Boys* | *Girls* |
| *4th years* | 19 | 15 | 17 | 24 |
| *5th years* | 26 | 28 | 31 | 28 |

Illustrate this information in a suitable way and comment on the results.

**3.** 'My grandad smokes 30 cigarettes a day and he's still alive at the age of 86', you might say, but 40% of heavy smokers die before retirement age compared with 15% of non-smokers. Make up an anti-smoking poster using this information.

**4.** Alcohol is another 'socially acceptable' drug which can have serious effects on health. One way of measuring how much a person drinks is by the number of 'standard drinks' he or she has. A 'standard drink' is the equivalent of half a pint of ordinary beer, a single measure of a spirit or a glass of wine.

Dave and Alan are friends who go out for a drink together about 4 times a week. Usually Dave has about 3 pt of beer and a double whisky while Alan has 3 pt of a special beer (2 standard drinks per $\frac{1}{2}$pt) each time they go out. Work out how many standard drinks they each have in an average week.

**5.** A man who takes more than 50 standard drinks a week or a woman drinking more than 35 is possibly heading for alcohol problems. Put this information on a poster.

# The Generation Gap

Sanjay's father owns a small clothing factory making men's trousers and women's skirts and trousers. He makes certain styles for the 'younger market' and others for older people and wonders whether he should make different sizes for the two styles. Sanjay has offered to do some market research for him.

1. Put yourself in Sanjay's position and start with people of your own age. Carry out a survey in small groups of 50 males and 50 females between the ages of 16 and 25. In each case find out the waist size and, in the case of the males, leg length.
2. This information must now be put in a useful form. Make up frequency tables showing the different waist sizes for each sex.
3. How are you going to use the information you have about leg length for men?
4. Now carry out the same exercise for groups of 50 men and 50 women in the over-25 age group and make up frequency tables.
5. You now have 2 sets of frequency tables and they have to be compared to see whether there is much difference. Work out the average or mean value for each of the 4 tables. Is there any difference between sizes for people over and under 25?
6. Another way to compare the figures would be to make up a bar chart for each table. Do the bar charts differ with age?
7. Are two age ranges enough? You could split up the over 25s into 25–50 and over 50 and compare the two groups.
8. Sanjay's father wants some advice. What would you advise him to do? How would you suggest he organises the sizes for the different styles?
9. 50 is quite a small sample for this sort of work. Compare your group's findings with that of another group. Do they come up with the same answers?
10. You need not stop here. You could talk to a buyer for a local shop and find out how he or she decides which sizes to buy. Does this agree with your survey results?

# Teacher's Notes

**1.** This chapter reinforces and extends earlier work done on units in a very relevant way. It provides a good introduction to various topics which could be the basis of surveys and further statistical work.

**2.** Links with other subject areas include
*science* the effects of drugs, nicotine, alcohol on the body, and the effects of not being an ideal shape or size
*information technology* equipment used to monitor personal fitness and health, stock control methods in shops
*communications* surveys of sizes, shop stocks, health problems, effects of drugs, etc.
*social studies* the 'drug problem', health education, smoking and drinking habits
*career development* work in the retail and manufacturing clothing trades.

**3.** Caring courses could use this as part of a project on health education in general. Students on a retail course could link this in with their work via clothing shops.

**4.** Local Health Education departments are usually willing to send speakers to groups and they can supply large quantities of information. A buyer for a local shop may be able to explain the shop's policy for stock control of various sizes.

Chapter 20

# Putting it Together

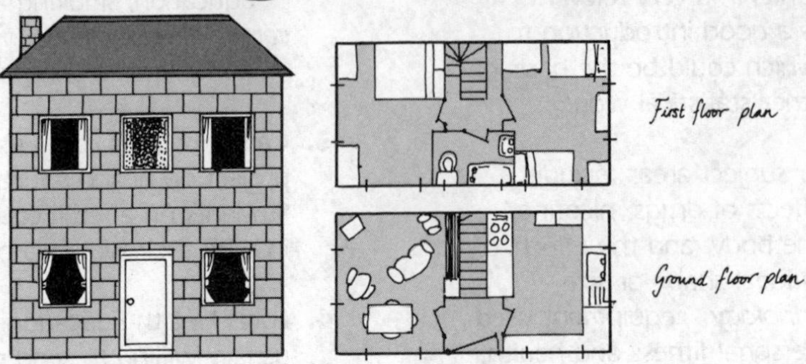


In the work you have done so far you have had to use mathematics in a variety of ways, some real and some artificial. In this chapter you will be putting some of these ideas together to do something practical. The end product from this chapter is a plan for producing something useful. You need not stop there, of course. Once you have the plan, with the right skills, you can go on to make the product. If you don't like the project suggested, use your imagination and plan something different along similar lines.

## 1. Planning the Project

### 1.1 Market Research

Large hand-made toys are very expensive and may be outside the price range of playgroups, nursery schools, children's centres, etc. If you have a lot of children around, things like dolls' houses, forts, farms, garages and toy cookers are very useful. The aim is to obtain toys at a low cost.

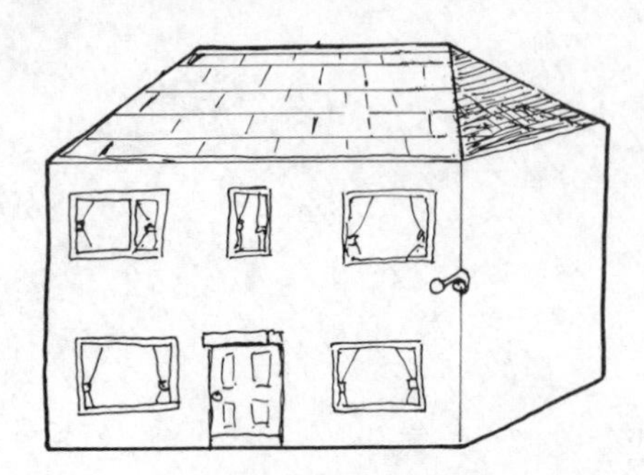

Start, by working in groups, to carry out your market research into the following areas.

**1.** Visit playgroups, etc. to find out what sort of large items they would find useful. Think in terms of toys which several children can play with as a group. If you cannot get to many places, use the skills you have learnt about samples to choose which ones to visit.

**2.** From your research decide on a suitable toy and visit shops to see what can be bought and how much it would cost.

**3.** From your combined efforts you should have enough information to be able to decide on one or more products to design.

### 1.2 Deciding on the Size

If you are designing something like a play kitchen you will need to know something about the sizes of the children who will be using it.

If the project is a farm or a dolls' house you need to decide which size animals or dolls it is for and carry out some measurements.

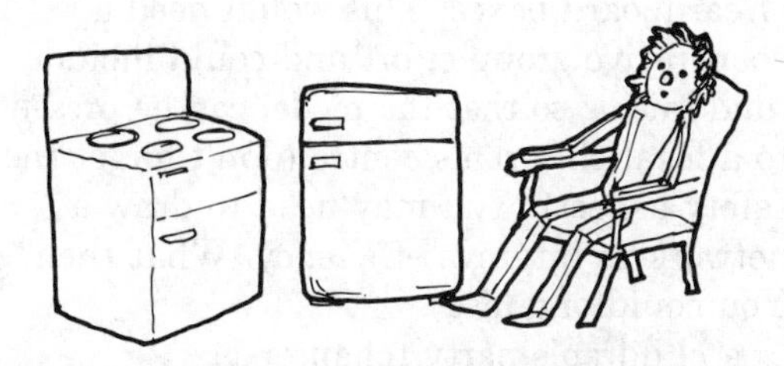

Now you need to get back to the research work.

1. Measure a sample of children and work out the average size and the range of sizes.
2. If bought equipment is to be used, for instance dolls, animals or saucepans, you need to go back to the shops and do some measuring.
3. Decide on the overall size and shape of your project.
4. Present your findings and conclusions in a report. Don't forget to include some statistical diagrams, and if you can draw well include an artist's impression.

### 1.3 The Right Scale

If you are designing a model of something bigger, like a house, then you need to get the scale correct.

1. Decide on the sort of house or other structure you are going to make and get approximate measurements for the real thing.
2. Having decided what overall size you want the model to be you can now work out the scale.

$$\text{scale} = \frac{\text{height of model}}{\text{height of real thing}}$$

3. Start getting suitable measurements for doors, windows etc. and scale them down for the model.
4. You can now draw a full-sized ground plan and elevations of your model.

## 2. Costing the Basic Work

This is a very important part of your work because if you get it wrong either your product will be too expensive to buy or you will lose money on it.

1. Firstly decide on the materials you are going to use for the basic work. Think about thicknesses, finishes etc.
2. Now comes the complicated part. You have to work out how much you will need of each type of material, how you can avoid too much waste, whether or not you can buy large sizes and cut them, and so on. This is a measuring and planning exercise which will involve a lot of numeracy work.
3. The next step is to go out again and cost materials. Visit various places to compare prices.
4. You should now be in a position to present a costing for the project. Work out the various items and present a report on your work.

## 3. The Extras

Once you have designed the basic article it is time to think about improvements. How should the design be changed – how can it be made more useful, does it need decoration, etc?

**1.** As a group, discuss possible improvements and add them to your plan.

**2.** Safety is important in toys. Are there any changes which need to be made so that children cannot be injured by the finished product?

**3.** Work out colour schemes, patterns, etc. If you are building a dolls' house, for instance, it would be good to design and print suitable scale wallpaper.

**4.** Models may need furniture. Can these be made or must they be bought? Carry out some costings. What about energy saving? Could you make your model 'energy-efficient'?

## 4. What Next?

Having got this far it would be a pity, if you have the skills, money, and equipment, not to make a prototype – even if it is only made out of cardboard boxes. This would need a cooperative group effort and could include fund-raising so that the model can be presented to a local children's centre (don't forget the safety aspect). You may need to draw a network to improve efficiency. What then? You could organise

a children's party (chapter 1)
an outing for the children (chapters 2, 9)
decorating the centre (chapters 5, 11)
taking bets on how long your toy will stay intact (chapter 8)

There are a lot of possibilities for extending this project in various ways and it may lead some of you into a little manufacturing business. How could you get help to set this up and keep it going?

# A Final Comment

Most people can get by in life using very little mathematics. To do this, however, you do have to trust the experts like the kitchen designer, the travel agent or the supermarket cashier. You also have to make a lot of guesses such as how much paint to buy for the ceiling – this can lead to expensive mistakes. By using some mathematics as you have in this and previous chapters you can be more efficient, more accurate and tackle more difficult problems.

Think about it.

# PART B

## Introduction to Part B

You may find some of the questions in part B rather easy. This is intentional. They have been designed both to increase your speed and accuracy and to show the number of different ways the questions can be worded. For example 'increase by', 'find the sum of' and 'what is the total if . . .' all imply that addition should be used. Knowing what the words mean is half the battle in understanding mathematics.

For some of these exercises you will need a calculator. Common sense should dictate when and how to use a calculator and the answer should *always* be checked to see whether it is sensible. Try not to be totally dependent on your calculator. You should aim to be able to add, subtract, multiply and divide whole numbers, decimals and simple fractions without a calculator. Chapters 1 to 3 provide practice at this.

It would be sensible to use a calculator for long multiplication and long division and when complicated numbers are involved. Some calculators can even deal with fractions.

Chapter 1

# Working with Whole Numbers

## Signs

| | |
|---|---|
| + plus | used for *addition* |
| − minus | used for *subtraction* |
| × times | used for *multiplication* |
| ÷ divided by | used for *division* |

## Place value

5104 means 5 thousands 1 hundred 0 tens 4 units

| | |
|---|---|
| 10 000 | ten thousand |
| 100 000 | one hundred thousand |
| 1 000 000 | one million |
| 10 000 000 | ten million |
| 100 000 000 | one hundred million |
| 1 000 000 000 | one billion* |

*This is the American billion which is often used on television for Government figures about the economy. The English billion is one thousand times greater but is seldom used now.

# Addition

25 water-colour pictures, 18 oil paintings and 34 pen-and-ink drawings make a *total* of 77 pictures altogether. The *sum* of 25, 18 and 34 is 77.

**Example**

Add 5004, 67 and 319.

```
5004     line up the units
  67     on the right
 319 +
----
5390
   2
```

## EXERCISE 1

Add

**1.** 4, 9 and 17

**2.** 15, 23 and 11

**3.** 62, 133 and 667

**4.** 56, 71, 65 and 134

**5.** 603, 71 and 9.

Find the sum of

**6.** 52 + 17

**7.** 314 + 205

**8.** 412 + 517 + 16

**9.** 6041 + 32 + 451

**10.** 7205 + 3500 + 75

**11.** Speed practice

| A | B |
|---|---|
| 6 + 9 = | 18 + 8 = |
| 5 + 7 = | 11 + 4 = |
| 8 + 3 = | 12 + 9 = |
| 4 + 8 = | 18 + 7 = |
| 9 + 5 = | 17 + 9 = |
| 2 + 8 = | 14 + 8 = |
| 5 + 3 = | 13 + 6 = |
| 4 + 7 = | 15 + 5 = |
| 8 + 5 = | 17 + 8 = |
| 9 + 3 = | 16 + 5 = |

| C | D |
|---|---|
| 7 + 23 = | 50 + 75 = |
| 9 + 22 = | 25 + 50 = |
| 8 + 25 = | 75 + 25 = |
| 3 + 29 = | 25 + 25 = |
| 6 + 24 = | 80 + 20 = |
| 7 + 28 = | 250 + 60 = |
| 9 + 29 = | 350 + 90 = |
| 3 + 22 = | 425 + 75 = |
| 5 + 25 = | 155 + 25 = |
| 8 + 26 = | 380 + 35 = |

**12.** a) 7189
3062 +

b) 657 893
908 759 +

c) 4562
9686
7509 +

d) 4505
63
171
18 +

**13.** What are the total costs for these shopping lists? (The figures have been rounded to the nearest whole number.)

| a) £ | b) £ | c) £ | d) £ |
|---|---|---|---|
| 3 | 4 | 25 | 35 |
| 5 | 2 | 13 | 18 |
| 8 | 1 | 17 | 11 |
| 1 | 7 | 6 | 4 |
| 2 | 14 | 3 | 6 |
| 1 | 5 | 35 | 17 |
| 1 | 11 | 2 | 3 |
| 2 | 2 | 11 | 9 |
| 1 | 2 | 24 | 11 |
| 5 | 1 | 5 | 14 |

**14.** Write down half a million in figures and add this to fifty thousand.

**15.** Add five hundred and forty-seven to six hundred and four, giving your answer in figures.

**16.** Add one million, five hundred thousand to two million, forty thousand. Give your answer in figures.

**17.** A debt of 345 billion pounds is further increased by 165 billion pounds. How much is now owing?

**18.** The figures below are part of a computer print-out. They have been totalled across and down but it is known that the totals are wrong. Write down the correct totals. What do you notice?

| | | | | | | *Totals* |
|---|---|---|---|---|---|---|
| | 5567 | 3410 | 1980 | 1187 | 8009 | 20 118 |
| | 4521 | 1008 | 877 | 500 | 3777 | 10 648 |
| | 3452 | 6611 | 1993 | 5509 | 405 | 17 935 |
| | 405 | 750 | 250 | 560 | 300 | 2230 |
| *Totals* | 13 910 | 11 744 | 5065 | 7721 | 12 456 | 51 036 |

**19.** Give the dimensions of AB, BD and AE in the diagram below.

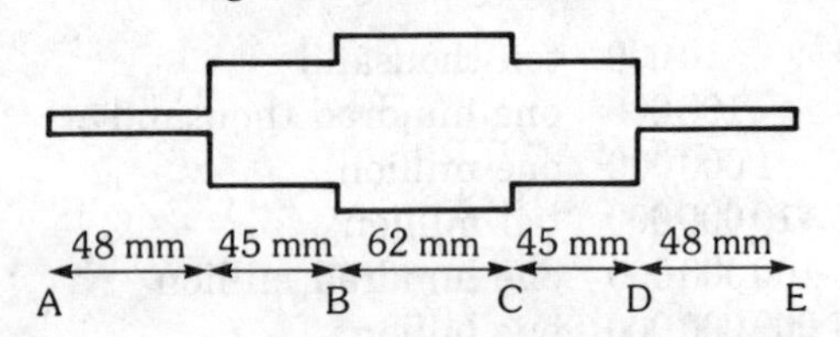


**20.** A band wishes to purchase some new instruments. Cornets cost about £200, trumpets £350 and tubas £575. How much does the band need to buy 1 cornet, 1 trumpet and 1 tuba?

**21.** a) Increase 170 by 40.
b) Find the sum of 300, 450 and 600.
c) Find 8 + 6 + 9 + 2.
d) What is 14 + 7 + 19 + 9?
e) What are the next two numbers of the sequence 148, 157, 166, 175 . . . ?

**22.** A factory's workforce is increased by 1800. If there were 6500 employees before how many are there now?

# Subtraction

A length of 56 cm is sawn from a plank that is 98 cm long. What is the length of the piece that is left? The difference between 98 and 56 is 42 so the answer is 42 cm.

---

### Example

Subtract 314 from 5062.

```
4 1 51
5062     Either cross off at the top
 314 -
----
4748
----

5062     or pay back at the bottom
 314 -
1 1
----
4748
----
```

---

## EXERCISE 2

**1.** Speed practice

| A | B |
|---|---|
| 9 − 7 = | 8 − 1 = |
| 7 − 2 = | 9 − 3 = |
| 8 − 4 = | 5 − 1 = |
| 8 − 3 = | 4 − 3 = |
| 6 − 2 = | 7 − 1 = |
| 9 − 5 = | 6 − 0 = |
| 8 − 6 = | 9 − 6 = |
| 5 − 3 = | 7 − 3 = |
| 7 − 4 = | 5 − 4 = |
| 5 − 2 = | 4 − 2 = |

| C | D |
|---|---|
| 12 − 9 = | 10 − 7 = |
| 14 − 8 = | 15 − 6 = |
| 16 − 8 = | 10 − 1 = |
| 12 − 5 = | 14 − 5 = |
| 15 − 7 = | 13 − 6 = |
| 12 − 4 = | 15 − 8 = |
| 10 − 4 = | 10 − 8 = |
| 11 − 9 = | 14 − 6 = |
| 13 − 5 = | 11 − 4 = |
| 11 − 8 = | 17 − 9 = |

Subtract

**2.** 14 from 29

**3.** 63 from 82

**4.** 571 from 974

**5.** 609 from 7082

**6.** 7003 from 9281.

Work out

**7.** 981 − 335

**8.** 4698 − 732

**9.** 5000 − 18

**10.** 7083 − 2994

**11.** 10 456 − 9007.

**12.** Take fifty thousand from five hundred thousand, giving your answer in figures.

**13.** Subtract half a million from a billion, giving your answer in figures.

**14.** What is six hundred and fifty-three minus eighty-seven in words?

**15.** Work out the following and check the answers by adding the bottom two lines, as shown by the first example.

```
a) 5329     b) 3361     c) 5078
    807 -       203 -      1469 -
   ----        ----        ----
   4522
   ----        ----        ----

   Check
    807
   4522 +
   ----
   5329
   ----

d) 76 809     e) 7000
   14 992 -         31 -
   ------         ----

   ------         ----
```

**16.** One figure in each of these answers is wrong. Copy each question, correct the figure and put a ring around the corrected figure.

a)
$$\begin{array}{r} 7009 \\ 3990\,- \\ \hline 3010 \\ \hline \end{array}$$

b)
$$\begin{array}{r} 7894 \\ 6055\,- \\ \hline 1849 \\ \hline \end{array}$$

c)
$$\begin{array}{r} 50\,000 \\ 10\,009\,- \\ \hline 49\,991 \\ \hline \end{array}$$

d)
$$\begin{array}{r} 98\,762 \\ 14\,231\,- \\ \hline 84\,931 \\ \hline \end{array}$$

**17.** 756 people attend a certain college. 498 of these are men. How many are women?

**18.** It is 40 miles from Dorchester to Salisbury via Blandford Forum, and 23 miles from Blandford Forum to Salisbury.

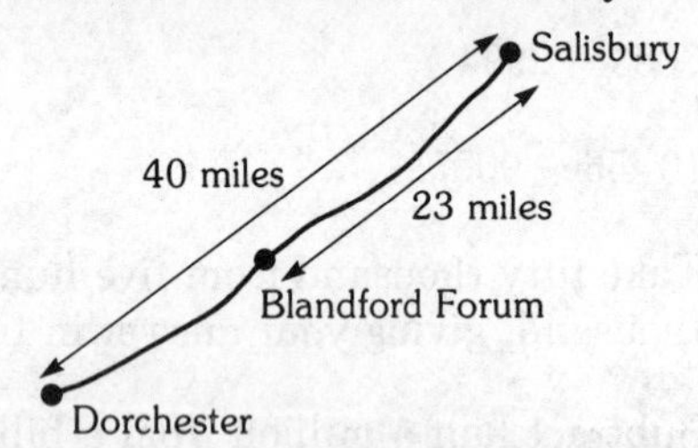


How far is it from Dorchester to Blandford Forum?

**19.** Find the difference between
a) 427 and 1500 b) 178 and 1630.

**20.** A workforce of 34 000 is reduced by 15 000. How many are left?

**21.** Attendance at a football match was 21 800 one week but only 19 760 the following week. By how many had the attendance decreased?

# Multiplication

There are $9 \times 8 = 72$ tiles in this pattern.

The *product* of 9 and 8 is 72.

The *multiples* of 7 are 7, 14, 21, 35, and so on.

Multiples of 2 are called *even* numbers. All other whole numbers are *odd*.

*Squares* $8 \times 8$ may be written $8^2$.

*Cubes* $5 \times 5 \times 5$ may be written $5^3$.

Note that $6 \times 0 = 0$, not 6.

## Examples

**1.** Short multiplication

$$\begin{array}{r} 563 \\ 4\,\times \\ \hline 2252 \\ \hline \end{array}$$

**2.** Long multiplication

$$\begin{array}{r} 563 \\ 74\,\times \\ \hline 2252 \\ 39410 \\ \hline 41662 \\ \hline \end{array}$$

## EXERCISE 3

**1.** Speed practice

| A | B |
|---|---|
| 5 × 4 = | 8 × 3 = |
| 9 × 2 = | 9 × 6 = |
| 5 × 3 = | 6 × 5 = |
| 6 × 9 = | 7 × 7 = |
| 3 × 7 = | 3 × 9 = |
| 8 × 7 = | 5 × 2 = |
| 9 × 4 = | 5 × 9 = |
| 8 × 6 = | 6 × 7 = |
| 7 × 5 = | 4 × 4 = |
| 9 × 9 = | 7 × 9 = |

| C | D |
|---|---|
| 8 × 8 = | 4 × 8 = |
| 7 × 3 = | 8 × 0 = |
| 2 × 9 = | 2 × 6 = |
| 9 × 3 = | 7 × 6 = |
| 6 × 6 = | 9 × 7 = |
| 2 × 7 = | 8 × 4 = |
| 5 × 5 = | 9 × 5 = |
| 3 × 8 = | 7 × 8 = |
| 8 × 9 = | 9 × 8 = |
| 5 × 8 = | 6 × 8 = |

Multiply

**2.** 52 by 3

**3.** 46 by 5

**4.** 128 by 2

**5.** 485 by 7

**6.** 926 by 8.

Work out

**7.** 305 × 3

**8.** 516 × 9

**9.** 438 × 7

**10.** 6008 × 8

**11.** 9867 × 9.

**12.** Find the number of pills in these packets. If each pill costs 3 p find the cost in pence.

a) b) c)

**13.**

| a) 673 | b) 512 | c) 6980 |
|---|---|---|
| 24 × | 24 × | 56 × |

| d) 7886 | e) 5329 | f) 4128 |
|---|---|---|
| 109 × | 89 × | 307 × |

**14.** Evaluate

a) $5^2$ b) $9^2$ c) $3^3$ d) $7^3$

**15.** What do 3 packets of sweets at 18 p a packet cost?

**16.** Work out the cost of 8 suites of furniture at £636 each.

**17.** A profit of £14 is made on each washing machine sold by a certain company. If the company sells 7135 washing machines how much profit has it made?

**18.** If a man earns £7850 a year how much will he earn in 25 years?

**19.** Find the cost of 350 coats if each costs £16.

**20.** A shop sells about 85 pairs of shoes each working day. How many pairs are sold in three weeks? (Take the working week to be 6 days.)

**21.** 13 dress patterns have to be cut from a roll of cloth. Each pattern needs 175 cm. Will 2300 cm of cloth be sufficient? If so, how much will be left over?

**22.** a) Work out i) $7 \times 10$ ii) $7 \times 100$.
b) Write down the rule for multiplying whole numbers by
i) ten ii) a hundred.

**23.** From the list of numbers 2, 6, 9, 12, 15, 19, 23, 25 write down
a) the even numbers
b) the odd numbers
c) the multiples of 3
d) the multiples of 5
e) the square numbers.

**24.** Write down the product of
a) 9 and 70 b) 203 and 40.

**25.** Work out
a) 35 multiplied by 20 b) 250 squared.

# Division

A meal for eight people costs £104. If the bill is shared equally how much will each person pay? 104 divided by 8 is 13. They each pay £13.

104 divided by 8 may be written as

$8\overline{)104}$, $\frac{104}{8}$, $104 \div 8$ or 104/8.

Division is concerned with sharing into equal parts. 3 and 7 are *factors* of 21 because they divide exactly with no remainder.

*Prime numbers* have no factors apart from themselves and 1. (1 is not a prime number). The prime numbers are 2, 3, 5, 7, 11, 13, . . .

The *square root* sign is written $\sqrt{\phantom{x}}$. The square root of 36 is 6 because 6 multiplied by itself is 36.

## Examples

**1.** Short division
Divide 761 by 3.

253 remainder 2
$3\overline{)761}$

**2.** Long division
Divide 761 by 13.

```
     58 remainder 7
13)761
   65
   ---
   111
   104
   ---
     7
```

$1 \times 13 = 13$
$2 \times 13 = 26$
$3 \times 13 = 39$
$4 \times 13 = 52$
$5 \times 13 = 65$
$6 \times 13 = 78$
$7 \times 13 = 91$
$8 \times 13 = 104$

Take care when there is a zero in the middle of the number, for example

801
$5\overline{)4005}$

## EXERCISE 4

**1.** Speed practice

| A | B |
|---|---|
| $12 \div 4 =$ | $16 \div 2 =$ |
| $15 \div 3 =$ | $10 \div 5 =$ |
| $27 \div 9 =$ | $21 \div 7 =$ |
| $35 \div 7 =$ | $63 \div 7 =$ |
| $42 \div 6 =$ | $40 \div 5 =$ |
| $56 \div 8 =$ | $24 \div 6 =$ |
| $81 \div 9 =$ | $56 \div 7 =$ |
| $63 \div 9 =$ | $64 \div 8 =$ |
| $28 \div 7 =$ | $49 \div 7 =$ |
| $48 \div 8 =$ | $40 \div 8 =$ |

| C | D |
|---|---|
| $45 \div 9 =$ | $20 \div 5 =$ |
| $35 \div 5 =$ | $54 \div 6 =$ |
| $30 \div 6 =$ | $36 \div 9 =$ |
| $42 \div 7 =$ | $45 \div 5 =$ |
| $72 \div 8 =$ | $72 \div 9 =$ |
| $90 \div 10 =$ | $18 \div 6 =$ |
| $54 \div 9 =$ | $48 \div 8 =$ |
| $32 \div 4 =$ | $24 \div 4 =$ |
| $14 \div 7 =$ | $12 \div 2 =$ |
| $36 \div 6 =$ | $24 \div 3 =$ |

**2.** a) Share 84 by 12.
b) How many 4s are in 32?
c) Is 54 exactly divisible by 8?
d) Which numbers divide exactly into 63?
e) Is 6 a factor of 32?

**3.** a) Divide 72 by 9.
b) How many 100s are in 3600?
c) What is the square root of 25?
d) What is the remainder when 65 is shared by 9?
e) List all the factors of 12.

**4.** a) $2\overline{)52}$
b) $5\overline{)68}$
c) $7\overline{)50}$
d) $8\overline{)100}$
e) $9\overline{)708}$

**5.** a) $206 \div 8$
b) $516 \div 3$
c) $771 \div 5$
d) $8003 \div 8$
e) $6874 \div 7$

**6.** a) 120/3
b) 250/5
c) 512/4
d) 100/5
e) 351/9

**7.** a) $\frac{561}{6}$
b) $\frac{3500}{5}$
c) $\frac{6231}{7}$
d) $\frac{8809}{9}$

**8.** A fence 882 cm long is partitioned into 9 equal sections. How long is each section?

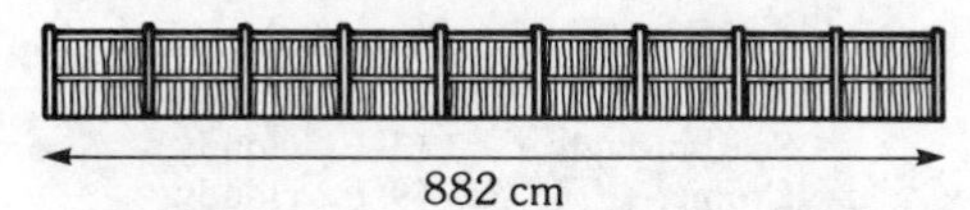


**9.** A bill of £175 is to be shared equally between 7 people. How much must each pay?

**10.** A piece of copper tubing 480 cm long is divided into 3 equal lengths. What is the length of each piece?

**11.** Give the least number of 5 lb weights needed to make a total weight of over 72 lb.

**12.** 7 peaches cost 98 p. Give the cost of each peach. Is 98 a prime number?

**13.** How many 6 cm lengths can be cut from a length of garden wire that is 75 cm long? How long is the remaining piece?

**14.** a) $50\overline{)800}$ b) $108 \div 36$ c) 985/15
d) $452 \div 30$

**15.** The bill for 47 text books is £141. How much does each book cost?

**16.** There are 875 students in a college. How many on average are there in each class if there are 25 classes?

**17.** Colin finds he can travel 800 miles on 25 gallons of petrol. How many miles per gallon is this?

# Order of Operations

If you work out 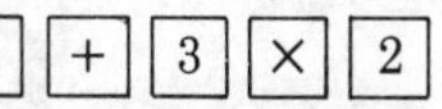
on some calculators you will get the answer 11, and on others the answer 16. Which is correct?

On all *computers* the answer would be printed as 11 because the convention is

> multiplication and division must be done before addition and subtraction

For simple calculators great care must be taken to see that the operations +, −, ×, and ÷ are carried out in the correct order.

The more expensive and advanced calculators allow you to use *brackets*. On simple calculators brackets are not available. Using the rule 'brackets must be worked first', $(5+3) \times 2$ equals 16 and $5 + (3 \times 2)$ is 11. Brackets can be used to avoid ambiguity (that is to make sure the meaning is clear).

## Examples

**1.** $5 + 2 \times 7$
$= 5 + 14$
$= 19$

**2.** $16 - 12 \div 4$
$= 16 - 3$
$= 13$

**3.** $(8 - 3) \times 2 - 7$
$= 5 \times 2 - 7$
$= 10 - 7$
$= 3$

## EXERCISE 5

Use the rules (a) do brackets first (b) do $\times$ and $\div$ before $+$ and $-$ to find the results a computer would give for

**1.** $6 + 3 \times 4$

**2.** $20 - 7 \times 2$

**3.** $(17 - 8) \times 5$

**4.** $9 \div 3 + 6$

**5.** $9 + 8 \times 5$

**6.** $24 - 4 \times 5$

**7.** $(13 - 6) \times 4$

**8.** $7 \times 2 + 8 \times 3$

**9.** $15 - 36 \div 12 + 2$

**10.** $50 - 25 \div 5$

**11.** $(6 + 2) \div (5 - 3)$

**12.** $(5 \times (4 + 2) + 2 \times (4 - 1)) \div 12$

**13.** Work out the price of 7 items at £34 plus 3 items at £45.

**14.** If a farmer fills twelve 55 kg bags from 700 kg how many kg will be left over?

**15.** A garage charges £78 for spare parts plus 12 hours labour at £8 an hour. How much is this altogether?

**16.** To cook a chicken the recipe book suggests 20 min per lb plus an extra 20 min. How long should a 3 lb chicken take?

**17.** If tea costs 18 p a cup and coffee 24 p find the cost of 3 teas and 2 coffees.

**18.** Printing a newsletter costs £4 for the master copy plus £1 for every 100 copies. What would 800 copies cost?

**19.** To change temperatures in Fahrenheit to Celsius you take away 32, divide by 9 and then multiply by 5. Change 50°F into °C.

**20.** A machine takes 40 min to set up and then 2 min to produce each article. What is the total time taken to produce 50 articles?

## EXERCISE 6 NEWSPAPER CUTTINGS

**1.**

Birmingham
50 miles
Northampton
64 miles
65 miles
Oxford
56 miles
LONDON

How far is it from Birmingham to London via Northampton? How much further would it be to travel via Oxford?

**2.**

**British resorts**

| | | |
|---|---|---|
| Scarborough | 63°F | cloudy |
| Cromer | 59°F | cloudy |
| Margate | 61°F | cloudy |
| Folkstone | 63°F | bright |
| Weymouth | 73°F | sunny |
| Brighton | 64°F | sunny |
| Jersey | 75°F | sunny |
| Tenby | 75°F | sunny |
| Anglesey | 66°F | bright |
| Blackpool | 72°F | sunny |

By how many degrees was the temperature at the warmest resort above the temperature at the coolest?

**3.**

**Fortune Theatre**
*The Closed door*

Seats still available.
Stalls £5, Circle £7
Upper Circle £4

How much would 18 seats in the stalls cost? How much more expensive would it be to take 18 seats in the circle? Would it be possible to take the same number of seats in the upper circle for £80?

**4.**

If Cheryl buys 2 tubes of toothpaste, a bottle of shampoo and 3 bars of soap how much has she saved on the old prices?

**5.**

**Project Leader**
£7500 per annum

Information Systems

A major established International Group.

Big Company Benefits.

How much is this salary per month?

**6.**

**Rare Bank Note sold for £2700**

A rare bank note fetched £2700 at an auction yesterday. Oil paintings were sold for £51 000 and a rare book fetched £30 000. The remaining contents of the house sold for £541 000.

What was the total amount received for the house and contents?

**7.**

**PROFIT ON SALE OF 52 ITEMS £156**

How much profit is this per item?

**8.**

**Song Writer Leaves £150 928**

Ian James, author and composer of over 1000 songs, left £150 928 (£166 627 gross) in his will published yesterday.

The tax is found by deducting the amount left in the will from the gross amount (assuming that the song writer had no outstanding debts). How much tax was paid?

**9.**

The members of the evening classes were asked if they owned a calculator. The results are shown in the table below.

| | *Own a calculator* | *Do not own a calculator* |
|---|---|---|
| Men | 245 | 52 |
| Women | 71 | 184 |

How many men owned a calculator? How many members of the evening classes didn't own a calculator? How many people were in the evening classes altogether?

**10.**

£17 return Cardiff to London
Leaves Cardiff 9.30 a.m.
Children 5–14 half fare.
Returns to Cardiff by 8.30 p.m.

What is the return fare from Cardiff to London for 3 adults and 2 children aged 8 and 10?

**11.**

Deane School has 38 classes with about 35 in each class. It was debated yesterday whether this school could take a further 85 children.

How many children are there in Deane School? How many would there be with a further 85 children? How many on average would there be in each class if the extra children were accepted?

**12.** Those earning between £60 and £80 a week will find their wages increased by about £4 while those earning between £80 and £100 will benefit from an increase of £5.

Find the difference in the new wages of employees who used to earn £67 and £85 respectively.

## EXERCISE 7

**1.** a) $\begin{array}{r} 41 \\ 27+ \\ \hline \end{array}$ b) $\begin{array}{r} 352 \\ 121- \\ \hline \end{array}$

c) $\begin{array}{r} 42 \\ 2\times \\ \hline \end{array}$ d) $5\overline{)565}$

**2.** a) $304 + 15$ b) $68 - 8$ c) $7 \times 9$ d) $24 \div 3$

**3.** a) Increase 11 by 5.
b) Decrease 16 by 9.
c) Calculate 14 times 2.
d) Divide 99 by 11.

**4.** a) How many 9s are in 81?
b) Find the remainder when 25 is divided by 6.

**5.** 4 is a factor of 12.
a) Write down two other factors.
b) Is 12 a prime number?

**6.** a) Correct 621 to the nearest hundred.
b) Write 462 correct to the nearest fifty.

**7.** a) Does 7 go exactly into 49?
b) How many 13 p bars of sweets could you buy for 50 p and how much change would be left?

**8.** $12 \times ? = 48$. What is the missing number?

**9.** How many hundreds are there in 1500?

**10.** Work out a) $350 \div 5$ b) 200/10 c) $3^3$ d) $9^2$

**11.** a) $\begin{array}{r} 5215 \\ 128 \\ 1609+ \\ \hline \end{array}$ b) $\begin{array}{r} 3052 \\ 198- \\ \hline \end{array}$

c) $\begin{array}{r} 184 \\ 15\times \\ \hline \end{array}$ d) $\begin{array}{r} 125 \\ 204\times \\ \hline \end{array}$

**12.** Write down 52 304 in words.

**13.** a) $16 + 8 \times 4$ b) $12 - (20 - 15)$

**14.** What is the square root of a) 25 b) 400?

# End Test

**1.** Add 105 to 23.

**2.** Take 12 from 49.

**3.** Increase 14 by 50.

**4.** Decrease 24 by 5.

**5.** Find the product of 8 and 7.

**6.** $9 + 9 + 9 + 9 =$

**7.** How many 7s are in 56?

**8.** Find the sum of 619 and 327.

**9.** 3 and 5 are factors of 15.
Write down two factors of 21.

**10.** Write 78 correct to the nearest ten.

**11.** Is 35 a multiple of 7?

**12.** Work out
a) $216 \div 8$ b) $\frac{400}{25}$
c) $8^2$ d) $6^3$

**13.** Subtract 14 from 92.

**14.** Find the remainder when 213 is shared by 6.

**15.** a) $\begin{array}{r} 47 \\ 53 + \\ \hline \\ \hline \end{array}$ b) $\begin{array}{r} 95 \\ 57 - \\ \hline \\ \hline \end{array}$

c) $\begin{array}{r} 23 \\ 6 \times \\ \hline \\ \hline \end{array}$ d) $9\overline{)288}$

**16.** $14 \times ? = 70$. Find the missing number.

**17.** Is 72 exactly divisible by 9?

**18.** How many 10s are in 4200?

**19.** Multiply 179 by 8.

**20.** a) $\begin{array}{r} 54\,198 \\ 17\,385 \\ 4219 + \\ \hline \\ \hline \end{array}$ b) $\begin{array}{r} 706\,103 \\ 39\,418 - \\ \hline \\ \hline \end{array}$

c) $\begin{array}{r} 172 \\ 35 \times \\ \hline \\ \hline \end{array}$ d) $\begin{array}{r} 184 \\ 103 \times \\ \hline \\ \hline \end{array}$

**21.** a) $6 + 3 \times 2$ b) $15 - (8 - 3)$
c) $17 - 36 \div 9$ d) $5 + 9 - 6$
e) $5 \times 4 - 9 \div 3 + 2$
f) $(5 - 2) \times 4 + 10 \div 5$

**22.** a) $21\overline{)785}$ b) $50\overline{)75\,000}$

**23.** Find the difference between half a million and twenty thousand.

**24.** Find the remaining amount when 100 g salt is poured into eight 12 g boxes.

**25.** $16 - ? = 9$. What is the missing number?

**26.** A deposit of £52 plus 9 instalments of £13 are paid on hire purchase. How much is paid altogether?

**27.** Write 60 040 in words.

**28.** How many 16 p stamps could be purchased for 96 p?

**29.** Find the difference between 732 and 1400.

**30.** From the list

16 19 24 28 34 40

write down
a) a prime number
b) the square root of 1600
c) a multiple of 7
d) the even numbers
e) a perfect square
f) a number divisible by 3
g) the next number of the sequence 1, 2, 3, 5, 8, 13, 21 . . .
h) the number of odd numbers in the list.

Chapter 2

# Using Decimal Points

## Column Headings

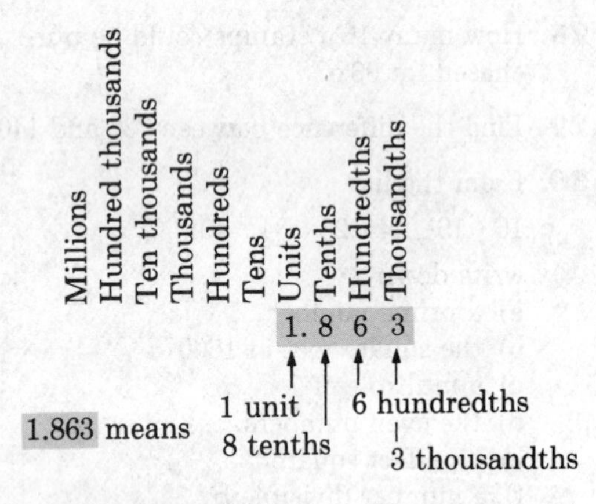


$\frac{7}{10} = 0.7$

$8\frac{3}{100} = 8.03$

and so on.

## Addition

### Example

Add 57 + 8.3 + 0.51.

57 means 5 tens and 7 units and can also be written as 57.0.

```
57.0      Line up the
 8.3      decimal points
 0.51 +
-----
65.81
-----
```

### EXERCISE 1

**1.** 41.23<br>21.05 +

**2.** 3.4<br>2.1<br>4.2 +

**3.** 16.11<br>41.56 +

**4.** 98<br>32.1<br>0.4 +

**5.** 32.7<br>20.92<br>0.23 +

**6.** 135.2 + 14.67

**7.** 157.2 + 45.3 + 2.1

**8.** 700 + 35.2 + 4.56

**9.** 98.14 + 23.5 + 1.34

**10.** 0.78 + 0.45 + 0.78

**11.** Add 45.12 and 0.89

**12.** Add 15, 0.3 and 8.9

**13.** 98.7<br>87.6<br>9.3 +

**14.** 56.98<br>3.15<br>0.27 +

**15.** Calculate the weight of a lorry plus its load if the lorry weighs 4.25 tonnes and the load is 0.37 tonnes.

**16.** Five blocks weigh 2.5 kg, 1.5 kg, 0.5 kg, 0.25 kg and 3.5 kg. How much do they weigh altogether?

**17.** A woman buys four items costing £5.13, £6.95, £0.95 and £13.50. How much does she pay?

**18.** Calculate the overall length of this block of wood.

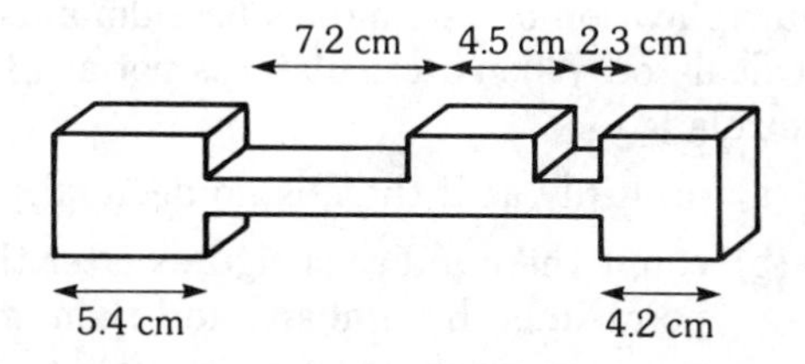


**19.** The capacities of four containers are 5.5 litres, 2.75 litres, 3.25 litres and 1.25 litres. What is the total capacity?

**20.** How much was earned in one week if the daily earnings are as shown below?

| | |
|---|---|
| Monday | £2.11 |
| Tuesday | £5.09 |
| Wednesday | £9.65 |
| Thursday | £8.97 |
| Friday | £4.51 |

**21.** Check these totals by adding across and down.

| *Weights in g* | | | *Total* |
|---|---|---|---|
| 56.71 g | 45.11 g | 67.3 g | |
| 39.87 g | 90.8 g | 56.12 g | |
| 61.78 g | 1.76 g | 3.18 g | |
| *Total* | | | |

**22.** An athlete sets himself a target of running 35 km per week. He runs 5.2 km one day and 4.8 km the next. He runs 6.4 km on the third day. How far has he run during the first three days?

If he then runs 4.7 km each day on the following four days how far has he run in one week? Has he reached his target?

# Subtraction

Carry out as for whole number arithmetic but line up the decimal points.

**Example**

Find the difference between 75.2 and 1.08.

$$\begin{array}{r} 75.20 \\ 1.08\,- \\ \hline 74.12 \\ \hline \end{array}$$

$$\begin{array}{r} 1.08 \\ 74.12\,+ \\ \hline 75.20 \\ \hline \end{array}$$ *Check by adding the bottom two lines*

Note that 75.2 was rewritten as 75.20.

## EXERCISE 2

**1.** $\begin{array}{r} 63.5 \\ 1.4\,- \\ \hline \\ \hline \end{array}$ **2.** $\begin{array}{r} 26.9 \\ 1.7\,- \\ \hline \\ \hline \end{array}$ **3.** $\begin{array}{r} 61.57 \\ 32.09\,- \\ \hline \\ \hline \end{array}$

**4.** $\begin{array}{r} 0.580 \\ 0.023\,- \\ \hline \\ \hline \end{array}$ **5.** $\begin{array}{r} 45.89 \\ 1.98\,- \\ \hline \\ \hline \end{array}$

**6.** 9.12 – 4.67

**7.** 34.5 – 6.12

**8.** 0.768 – 0.592

**9.** 6.7 – 5.65

**10.** 134.1 – 35.7

Find the difference between

**11.** 36.3 and 14

**12.** 0.56 and 0.35

**13.** 1.5 and 0.75

**14.** 56 and 34.25

Subtract

**15.** 0.27 from 3.4

**16.** 12.3 from 15.2

**17.** 64.5 from 103.9

**18.** A woman bought two items in a shop. The total cost was £12.63. She knows that one of the items was £4.99. How much was the other?

**19.** Calculate the length of AB in the diagram of the building below.

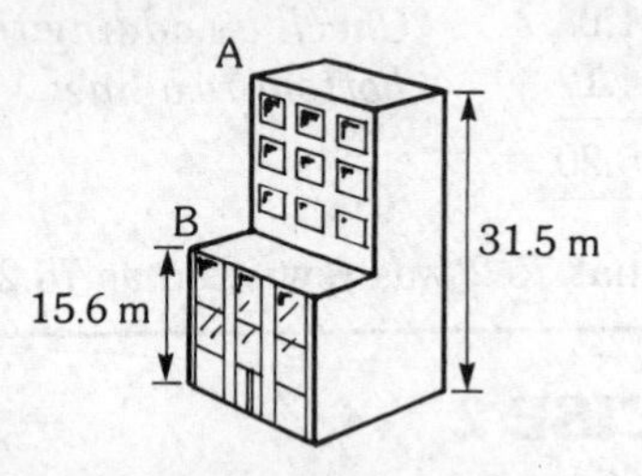


**20.** The weight of a lorry carrying steel bars is 2.57 tonnes. It is 1.95 tonnes when empty. Calculate the weight of the steel bars.

**21.** A steel bar which is 5.6 m long has a piece 3.25 m long sawn from it. What is the length of the remaining piece?

**22.** Which is longer and by how much – a copper rod which is 0.425 metres or a copper rod which is 0.35 metres?

**23.** A cup and saucer together weigh 0.25 kg. If the cup weighs 0.16 kg how much does the saucer weigh?

**24.** Calculate the lengths AB and CD in the diagram below.

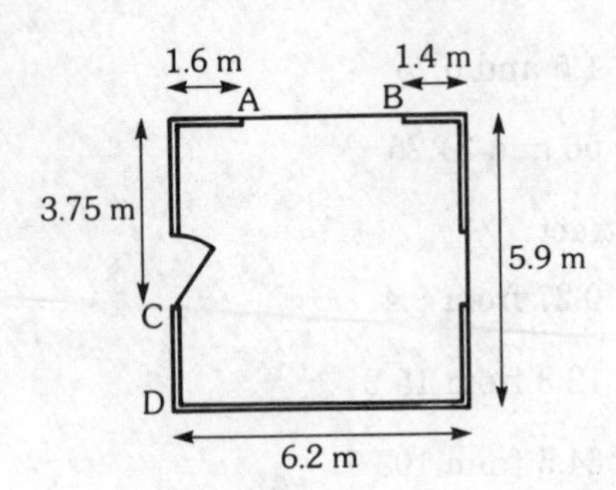


# Multiplication

Multiplication of decimals is best done using a calculator. When a calculator is not available the rule is

(a) multiply as if there is no decimal point

(b) count the number of figures after the point in both numbers, add them and put this number of figures after the point in the answer.

## Example

Multiply 5.6 by 0.03

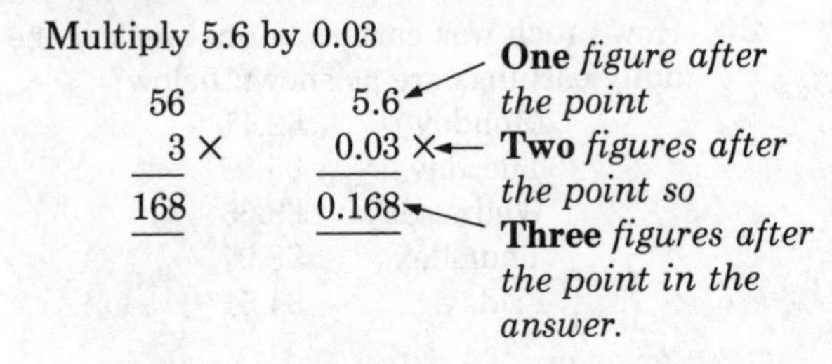


## EXERCISE 3

**1.** 3.4 × 2

**2.** 2.2 × 3

**3.** 1.21 × 4

**4.** 3.41 × 4

**5.** 1.25 × 6

**6.** 0.4 × 0.8

**7.** 0.2 × 0.4

**8.** 0.4 × 0.07

**9.** 300 × 0.08

**10.** 0.02 × 0.3

**11.** 2.13 × 0.09

**12.** 6 × 0.4

Find the product of

**13.** 1.3 and 6

**14.** 0.009 and 5

**15.** 1.8 and 0.9

**16.** 0.02 and 18

**17.** 10.7 and 0.08

**18.** Multiply a) 0.682 b) 45.3 c) 5 d) 0.0456 e) 1.26 by 10 and write down a rule for multiplying numbers by ten.

**19.** Multiply a) 0.678 b) 4.5 c) 32 d) 0.0982 e) 1.587 by 100 and write down a rule for multiplying numbers by one hundred. Comment about what to do when whole numbers are involved.

Multiply

**20.** 0.03 by 7

**21.** 7.20 by 0.3

**22.** 10.6 by 5

**23.** 3.45 by 0.2

**24.** 0.5 by 2

**25.** 0.75 by 4

**26.** 1.80 by 3

**27.** 2.75 by 0.8

**28.**
$$\begin{array}{r} 1.09 \\ 11 \times \\ \hline \\ \hline \end{array}$$

**29.**
$$\begin{array}{r} 56 \\ 4.8 \times \\ \hline \\ \hline \end{array}$$

**30.**
$$\begin{array}{r} 507 \\ 1.20 \times \\ \hline \\ \hline \end{array}$$

**31.**
$$\begin{array}{r} 32 \\ 1.08 \times \\ \hline \\ \hline \end{array}$$

**32.** Find the cost of five items each costing £5.78.

**33.** Seventy gold bars each weighing 5.6 kg are loaded into a lorry. What is the total weight of the gold bars?

**34.** A man earns £75.58 a week for 12 weeks. How much has he earned altogether?

**35.** A herd of cows yields 18.6 litres of milk per cow. What is the total amount of milk obtained if the herd consists of 35 cows?

**36.** Work out the cost of making 52 telephone calls at 5.3 pence per call.

# Division

Division of decimals is best done using a calculator. However, in case one is not available, some methods are shown in the examples below.

## Examples

**1.** Divide 18.6 by 3.

$$3\overline{)18.6}\quad = 6.2$$

*Decimal points should be in a line*

**2.** 0.186 ÷ 0.3

*Multiply both numbers by 10 or 100 until the second number is a whole number*

$= 1.86 \div 3$

$= 3\overline{)1.86}$ giving 0.62

**3.** 186.0 ÷ 0.03

*Add noughts if necessary*

$= 1\,8600 \div 3$

$= 6200$

## EXERCISE 4

**1.** 4.8 ÷ 2

**2.** 12.3 ÷ 3

**3.** 3.6 ÷ 6

**4.** 16.8 ÷ 7

**5.** 3 ÷ 5

**6.** 3.2 ÷ 0.8

**7.** 13.5 ÷ 0.05

**8.** 189 ÷ 0.9

**9.** 0.72 ÷ 0.3

**10.** 203 ÷ 0.07

**11.** Divide a) 14.2 b) 0.5 c) 80 d) 7 by 10
Write down in words the rule for dividing by ten.

**12.** Divide a) 6 b) 0.8 c) 12.3 d) 90 by 100.
Write down, in words, the rule for dividing by one hundred.

**13.** A tank contains 1560 litres of fruit juice. How many full cans each containing 0.8 litres can be filled from this tank?

**14.** How many steel bars 3.2 m long can be cut from a bar that is 48 m long?

**15.** Find the cost of a single cassette if twenty-five similar cassettes cost £53.

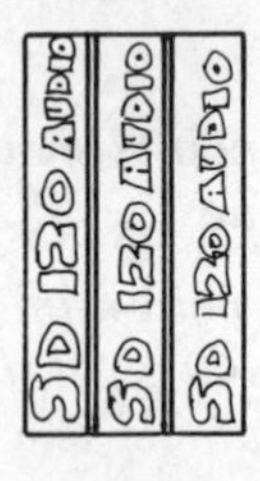


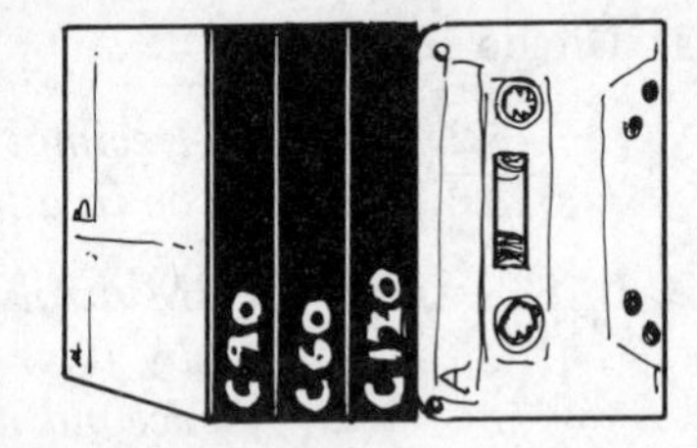


**16.** Five people share a meal costing £24. If they share the cost equally how much does each pay?

**17.** A 240 m cable is to be divided into 500 equal parts. How long will each part be?

**18.** Work out the time needed to make a single component if 270 components can be made in 135 minutes.

**19.** A pad containing 150 sheets of paper is 37.500 mm thick. If the sheets are all the same thickness then what is the thickness of each sheet of paper?

**20.** The weekly wage bill for 105 factory workers is £9712.50. How much does each earn per week if they are all paid the same amount?

# Decimal Places

Sometimes the numbers do not divide exactly. In this case the calculation may be carried out to 2 or 3 or more decimal places as required.

---

**Example**

Divide 15.3 by 7 correct to 3 decimal places.

$$\begin{array}{r} 2.1857 \\ 7\overline{)15.3000} \end{array}$$

*Answer* 2.186 correct to 3 decimal places.

Note that 15.3 has been rewritten as 15.3000. Also, the division was carried out to four decimal places. If the figure in the last place is 5 or more round up to the next figure.

---

## EXERCISE 5

Work out the following correct to 3 decimal places.

**1.** 6.5 ÷ 7

**2.** 0.32 ÷ 3

**3.** 40 ÷ 6

**4.** 51.23 ÷ 9

**5.** 789 ÷ 11

**6.** Divide 0.4 by 0.3

**7.** Divide 1.2 by 0.07

**8.** Divide 11 by 0.6

**9.** Divide 105 by 3

**10.** Divide 0.07 by 0.9

# Recurring Decimals

When the figures in the answer repeat themselves they are called recurring decimals.

**Examples**

$\frac{1}{3} = 0.33333\ldots$ and is written $0.\dot{3}$

$\frac{1}{11} = 0.090909\ldots$ and is written $0.\dot{0}\dot{9}$

$\frac{1}{7} = 0.1428571428571\ldots$

and is written $0.\dot{1}4285\dot{7}$

The dots above the figures show which digits recur (repeat themselves). When more than 2 digits recur, the dots are placed above the first and last of the recurring figures.

## EXERCISE 6

Work out the following and write them as recurring decimals.

**1.** $\frac{2}{3}$ **2.** $1 \div 9$ **3.** $11\overline{)3}$

**4.** 5/7 **5.** $5 \div 11$ **6.** 4/9

**7.** $\frac{7}{11}$ **8.** $9\overline{)8}$ **9.** $3 \div 7$

**10.** 5/13

# End Test

**1.** Add 56.3 to 11.2.

**2.** Increase 0.045 by 0.005.

**3.** Work out 5.04 minus 2.07.

**4.** Multiply 5.3 by 4.

**5.** Divide 17.6 by 8.

**6.** a) $17.23 + 15 + 9.8$ b) $18.90 - 0.27$

**7.** a) $0.5 \times 0.2$ b) $0.6 \div 0.2$

**8.** Work out $\frac{5}{6}$ correct to 3 decimal places.

**9.** Write 6/7 as a recurring decimal.

**10.** 4.3 means 4 units and 3 tenths. What does 0.801 mean?

**11.** Write $7\frac{2}{10}$ in decimal form.

**12.** Which is greater and by how much, 0.125 or 0.2753?

**13.** Place these in ascending order.
0.5, 0.125, 0.375, 0.0125

**14.** Multiply 1.5 by itself.

**15.** Express $1\frac{3}{100}$ as a decimal.

**16.** Work out 14 divided by 5.

**17.** Multiply 1.08 by 1.2.

**18.** a)
```
 23.4
  1.56
 90.0
  2.15 +
 ______

 ______
```
b)
```
 3.50
 0.86 −
 ____

 ____
```

**19.** Work out the following as decimals and place them in descending order.
$\frac{2}{5}$, $\frac{3}{8}$, $\frac{2}{9}$, $\frac{1}{4}$

**20.** Find the change from £10.00 for an article costing £3.07.

**21.** What is the total weight of three iron bars each weighing 7.2 kg?

**22.** Two items cost £4.07 and another costs £0.95. How much change will there be from £20 after buying all three?

**23.** Give the length that is left when six rods 0.45 m long are cut from a bar that is 3 m long.

**24.** Find the total weight of three copper pans each weighing 5.1 kg plus five aluminium saucepans each weighing 3.2 kg.

**25.** How many 1.8 m dress lengths can be cut from a roll of cloth that is 7.2 m long?

**26.** A consignment of nuts and bolts weighs 52 g. If there is the same number of nuts as bolts, each nut weighs 1.3 g and each bolt weighs 2.7 g, find how many nuts and bolts there are altogether.

**27.** The diagram shows part of a metal pipe. How thick is the metal?

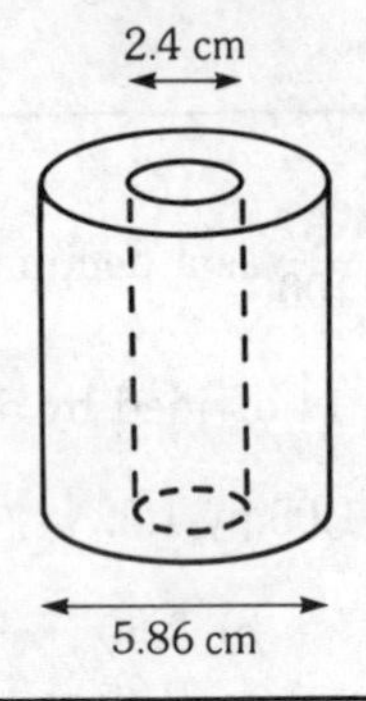


**28.** Potatoes are on sale for £0.36 per kg, carrots for £0.45 per kg and courgettes for £0.82 per kg. How much would you have to pay for 5 kg of potatoes, 3 kg of carrots and $\frac{1}{2}$ kg of courgettes?

**29.** a) What is the area of the square tile shown below? (Multiply length by width.)

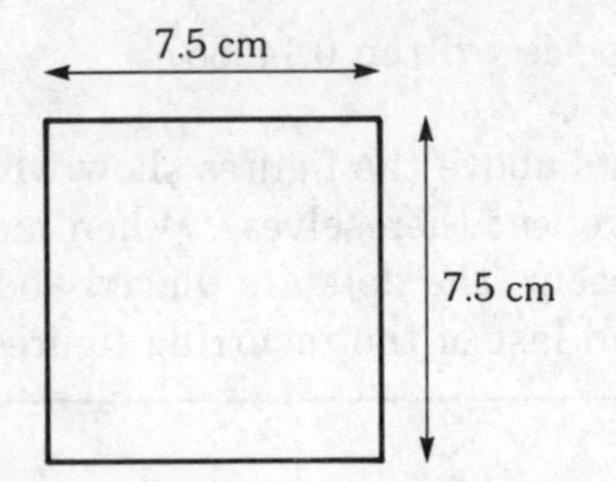


b) What would be the length (in metres) of the side of a square with area $0.09\,m^2$?

**30.** A tourist changes £37.50 into German marks. Find how many marks she gets if £1 is worth about 5 marks.

She then travels to Switzerland where 1 Swiss franc is roughly 0.8 marks. Find how many Swiss francs she would get for 176 marks.

Chapter 3

# Working with Fractions

## Changing to Equivalent Fractions

**Example**

Change $\frac{7}{8}$ to sixteenths.

$\frac{7}{8} = \frac{14}{16}$ *8 into 16 goes 2*
*2 times 7 is 14*

You can multiply top and bottom of a fraction by the same number without altering its value.

### EXERCISE 1

Fill in the missing numbers.

**1.** $\frac{1}{2} = \frac{?}{10}$ **2.** $\frac{3}{4} = \frac{?}{16}$

**3.** $\frac{5}{8} = \frac{?}{16}$ **4.** $\frac{2}{3} = \frac{?}{6}$

**5.** $\frac{4}{5} = \frac{?}{10}$ **6.** $\frac{1}{4} = \frac{?}{12}$

**7.** $\frac{3}{8} = \frac{?}{64}$ **8.** $\frac{5}{8} = \frac{?}{32}$

**9.** $\frac{3}{8} = \frac{?}{1000}$ **10.** $\frac{7}{25} = \frac{?}{100}$

**11.** $\frac{8}{50} = \frac{?}{200}$ **12.** $\frac{7}{125} = \frac{?}{500}$

## Reducing to Lowest Terms

This means 'make the fraction simpler by cancelling'.

**Example**

Reduce $\frac{12}{20}$ to its lowest terms.

$\frac{\not{12}^{3}}{\not{20}_{5}} = \frac{3}{5}$

You can divide top and bottom of a fraction by the same number without altering its value.

### EXERCISE 2

Reduce these fractions to their lowest terms.

**1.** $\frac{3}{12}$ **2.** $\frac{4}{16}$ **3.** $\frac{8}{20}$

**4.** $\frac{12}{24}$ **5.** $\frac{8}{32}$ **6.** $\frac{5}{30}$

**7.** $\frac{32}{64}$ **8.** $\frac{75}{100}$ **9.** $\frac{16}{18}$

**10.** $\frac{25}{75}$ **11.** $\frac{12}{40}$ **12.** $\frac{72}{1000}$

# Changing Top-Heavy Fractions into Mixed Fractions

### Example

Change $\frac{9}{4}$ into a mixed fraction.

$\frac{9}{4} = 2\frac{1}{4}$ *4s into 9 go 2 remainder 1*

## EXERCISE 3

Turn the following top-heavy fractions into mixed fractions.

**1.** $\frac{11}{5}$ **2.** $\frac{13}{10}$ **3.** $\frac{12}{7}$

**4.** $\frac{10}{3}$ **5.** $\frac{18}{5}$ **6.** $\frac{47}{8}$

**7.** $\frac{29}{16}$ **8.** $\frac{139}{64}$ **9.** $\frac{123}{100}$

**10.** $\frac{3003}{1000}$

# Changing Mixed Fractions into Top-Heavy Fractions

### Example

Change $3\frac{2}{5}$ into a top-heavy fraction.

$3\frac{2}{5} = \frac{17}{5}$ *3 times 5 plus 2 is 17*

Remember that the number underneath stays the same.

## EXERCISE 4

Change these mixed fractions into top-heavy fractions.

**1.** $2\frac{1}{3}$ **2.** $3\frac{1}{4}$ **3.** $1\frac{1}{2}$

**4.** $2\frac{3}{4}$ **5.** $4\frac{1}{2}$ **6.** $3\frac{7}{9}$

**7.** $9\frac{3}{8}$ **8.** $8\frac{5}{16}$ **9.** $2\frac{1}{100}$

**10.** $4\frac{273}{1000}$

# Addition

### Examples

**1.** Add $5\frac{2}{3}$ to $4\frac{1}{6}$.

Add whole numbers then fractions. You must change all the fractions to the same type. The smallest number that both 3 and 6 will go into without leaving a remainder is 6, so change to sixths.

$5\frac{2}{3} + 4\frac{1}{6} = 9 + \frac{2}{3} + \frac{1}{6}$ *Add whole numbers*

$= 9 + \frac{4}{6} + \frac{1}{6}$ *Change $\frac{2}{3}$ to $\frac{4}{6}$*

$= 9\frac{5}{6}$

**2.** Add $1\frac{1}{4}$, $\frac{1}{6}$ and $5\frac{2}{3} = 6 + \frac{1}{4} + \frac{1}{6} + \frac{2}{3}$

$= 6 + \frac{3}{12} + \frac{2}{12} + \frac{8}{12}$

$= 6 + \frac{13}{12}$

$= 7\frac{1}{12}$

## EXERCISE 5

Add the following fractions.

1. $\frac{1}{3}+\frac{1}{2}$
2. $\frac{1}{5}+\frac{2}{3}$
3. $\frac{1}{4}+\frac{2}{3}$
4. $1\frac{2}{5}+\frac{3}{10}$
5. $\frac{3}{8}+1\frac{7}{16}$
6. $4\frac{5}{8}+\frac{1}{4}$
7. $3\frac{2}{3}+\frac{5}{6}$
8. $1\frac{1}{2}+2\frac{3}{4}$
9. $5\frac{1}{2}+6\frac{3}{4}$
10. $2\frac{1}{8}+3\frac{3}{16}$
11. $\frac{3}{11}+2\frac{5}{22}+1\frac{8}{11}$
12. $\frac{5}{16}+2\frac{17}{32}+1\frac{7}{16}$

# Subtraction

### Examples

**1.** Subtract $\frac{3}{8}$ from $\frac{3}{4}$

Change to fractions of the same type, in this case eighths, since 8 and 4 both go into 8 exactly.

$$\frac{3}{4}-\frac{3}{8}=\frac{6}{8}-\frac{3}{8}$$

$$=\frac{3}{8}$$

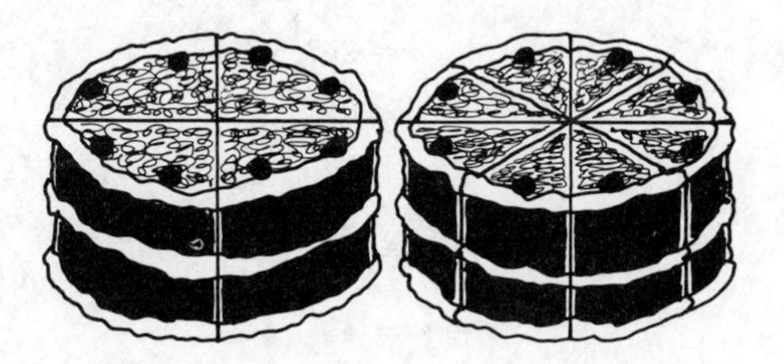

**2.** Work out $7\frac{1}{2}-5\frac{11}{16}$

$$7\frac{1}{2}-5\frac{11}{16}=2+\frac{1}{2}-\frac{11}{16}$$

*Subtract whole numbers then fractions. It is impossible to take 11 from 8 so 'borrow' from the 2 and say* $2 = 1 + \frac{16}{16}$

$$=2+\frac{8}{16}-\frac{11}{16}$$

$$=1+\frac{16}{16}+\frac{8}{16}-\frac{11}{16}$$

$$=1\frac{13}{16}$$

The diagram below shows another way of representing $1\frac{13}{16}$

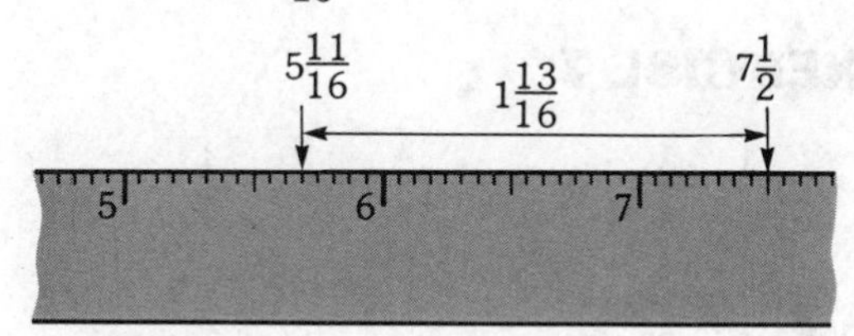


## EXERCISE 6

1. $7\frac{3}{4}-5\frac{1}{2}$
2. $\frac{5}{6}-\frac{1}{3}$
3. $1\frac{1}{4}-\frac{3}{4}$
4. $2\frac{3}{4}-1\frac{7}{8}$
5. $3\frac{1}{3}-\frac{11}{12}$
6. $2\frac{1}{2}-1\frac{7}{8}$
7. $8\frac{1}{4}-5\frac{11}{16}$
8. $3\frac{1}{8}-2\frac{1}{2}$
9. $5\frac{7}{8}-\frac{11}{16}$
10. $8\frac{1}{5}-2\frac{3}{10}$
11. $4\frac{11}{32}-3\frac{1}{64}$
12. $7\frac{13}{100}-\frac{7}{10}$

# Multiplication

If there are whole numbers, make the fractions top-heavy first. Cancel where possible. Multiply numbers on the top together and numbers on the bottom together. Change to a mixed fraction if necessary.

**Examples**

**1.** $1\frac{1}{4} \times 2\frac{2}{3} = \frac{5}{\cancel{4}_1} \times \frac{\cancel{8}^2}{3}$ *Make fractions top-heavy if there are whole numbers*

$= \frac{5 \times 2}{1 \times 3}$

$= \frac{10}{3}$

$= 3\frac{1}{3}$ *Change back to a mixed fraction*

**2.** $\frac{1}{3} \times \frac{3}{4} \times \frac{15}{16} = \frac{1 \times \cancel{3} \times 15}{\cancel{3} \times 4 \times 16}$

$= \frac{15}{64}$

## EXERCISE 7

**1.** $\frac{1}{2} \times \frac{3}{4}$

**2.** $1\frac{2}{3} \times \frac{4}{5}$

**3.** $\frac{3}{5} \times \frac{10}{27}$

**4.** $\frac{8}{11} \times 2\frac{1}{16}$

**5.** $\frac{5}{8} \times 2\frac{2}{5}$

**6.** $\frac{1}{2} \times \frac{17}{32}$

**7.** $3\frac{1}{4} \times \frac{8}{13}$

**8.** $\frac{1}{2} \times 12$

**9.** $\frac{1}{8} \times 1\frac{3}{5}$

**10.** $\left(3\frac{1}{3}\right)^2$

**11.** $5\frac{1}{8} \times 2\frac{1}{4} \times \frac{8}{9}$

**12.** $3\frac{1}{2} \times 2\frac{1}{3} \times \frac{1}{7}$

# Division

If there are whole numbers then make the fractions top-heavy. Do not cancel yet. Turn the second number upside down and multiply, cancelling where possible. If necessary change back to a mixed fraction.

**Example**

$1\frac{5}{8} \div \frac{1}{2} = \frac{13}{8} \div \frac{1}{2}$

$= \frac{13}{{}_4\cancel{8}} \times \frac{\cancel{2}^1}{1}$

$= \frac{13 \times 1}{4 \times 1}$

$= \frac{13}{4}$

$= 3\frac{1}{4}$

## EXERCISE 8

**1.** $\frac{3}{4} \div \frac{1}{5}$

**2.** $1\frac{1}{2} \div \frac{2}{3}$

**3.** $\frac{5}{8} \div 1\frac{2}{3}$

**4.** $2\frac{1}{4} \div 1\frac{1}{2}$

**5.** $3\frac{1}{2} \div 1\frac{1}{4}$

**6.** $5\frac{1}{2} \div 2\frac{3}{4}$

**7.** $8 \div \frac{1}{4}$

**8.** $3\frac{1}{2} \div 7$

# Order of Operations

Brackets must be done first, then multiplication and division and finally addition and subtraction.

**Example**

$2\left(7\frac{1}{2} + 4 \times 3\frac{1}{2}\right) = 2\left(7\frac{1}{2} + \frac{\cancel{4}^2}{1} \times \frac{7}{\cancel{2}_1}\right)$

$= 2\left(7\frac{1}{2} + \frac{2 \times 7}{1 \times 1}\right)$

$= 2\left(7\frac{1}{2} + 14\right)$

$= 2 \times 21\frac{1}{2}$

$= \frac{\cancel{2}}{1} \times \frac{43}{\cancel{2}}$

$= 43$

*Do the multiplication inside the bracket first. Then complete the addition inside the bracket. Finally remove the bracket and multiply by the number outside.*

Notice that the 2 outside the bracket means 'multiply by 2'.

## EXERCISE 9

**1.** $\left(8\frac{1}{2} - 3\right) \times \frac{1}{4}$  **2.** $\frac{1}{8} + 2 \times \frac{3}{4}$

**3.** $5\frac{3}{8} - \frac{3}{4} \div \frac{1}{2}$  **4.** $3\left(1\frac{2}{3} - \frac{1}{6}\right)$

**5.** $\left(8\frac{3}{4} - 6\frac{1}{2}\right) \times \left(2\frac{7}{16} - 1\frac{3}{4}\right)$

**6.** $8 \times \frac{3}{4} + 9 \times 1\frac{2}{3}$

**7.** $\left(8 \div \frac{1}{2}\right) \div 4$

**8.** $10\left(3\frac{1}{5} + 4\frac{17}{20} - 1\frac{3}{10}\right)$

# Turning Decimals into Fractions

### Examples

**1.** $0.8 = \frac{8}{10}$

$= \frac{4}{5}$ in its lowest terms

**2.** $1.67 = 1 + \frac{6}{10} + \frac{7}{100}$

$= 1\frac{67}{100}$

## EXERCISE 10

Express these decimals as fractions in their lowest terms.

**1.** 0.5  **2.** 1.25  **3.** 3.75

**4.** 0.125  **5.** 8.03  **6.** 0.7

**7.** 0.675  **8.** 4.2  **9.** 1.375

**10.** 0.0625

# Turning Fractions into Decimals

$\frac{4}{5}$ means $4 \div 5$. On a calculator, $4 \div 5 = 0.8$ or, writing 4 as 4.0

$$5\overline{)4.0} = 0.8 \quad \text{so that } \frac{4}{5} = 0.8$$

To turn a fraction into a decimal divide the number on the bottom (the denominator) into the number on the top (the numerator).

## EXERCISE 11

Change to decimals, correcting to 3 decimal places where necessary.

**1.** $\frac{1}{2}$  **2.** $\frac{3}{4}$  **3.** $2\frac{1}{3}$

**4.** $1\frac{1}{5}$  **5.** $\frac{1}{8}$  **6.** $4\frac{3}{8}$

**7.** $8\frac{1}{6}$  **8.** $\frac{2}{3}$  **9.** $2\frac{3}{5}$

**10.** $3\frac{1}{7}$

# Percentages and Fractions

Per cent means out of 100.

$$5\% = \frac{5}{100}$$

$$= \frac{1}{20} \text{ in its lowest terms}$$

## EXERCISE 12

Change these percentages to fractions and reduce them to their lowest terms.

**1.** 10% **2.** 25% **3.** 50%
**4.** 75% **5.** 30% **6.** 70%
**7.** 4% **8.** 150% **9.** $12\frac{1}{2}$%
**10.** $33\frac{1}{3}$% **11.** 200% **12.** $37\frac{1}{2}$%

To change a fraction to a percentage multiply by 100%.

$$\frac{1}{5} = \frac{1}{5} \times 100\%$$
$$= \frac{1}{\cancel{5}_1} \times \frac{\cancel{100}^{20}}{1}\%$$
$$= 20\%$$

## EXERCISE 13

Change these fractions to percentages.

**1.** $\frac{1}{4}$ **2.** $\frac{1}{2}$ **3.** $\frac{3}{4}$
**4.** $\frac{2}{3}$ **5.** $\frac{3}{5}$ **6.** $\frac{7}{10}$
**7.** $\frac{5}{8}$ **8.** $\frac{13}{20}$ **9.** $1\frac{1}{4}$
**10.** $\frac{3}{100}$ **11.** $2\frac{1}{2}$ **12.** $3\frac{1}{10}$
**13.** $\frac{1}{400}$ **14.** $\frac{3}{200}$

# Comparing the Sizes of Fractions

This can be done in two ways

a) changing into fractions of the same type, or
b) converting into decimals.

### Examples

**1.** Write in ascending order of size

$\frac{5}{8}, \frac{11}{16}, \frac{17}{32}, \frac{29}{64}$

Changing into sixty-fourths (since all denominators go exactly into 64)

$\frac{5}{8} = \frac{40}{64}, \frac{11}{16} = \frac{44}{64}, \frac{17}{32} = \frac{34}{64}$

and $\frac{29}{64}$ stays the same.

So, in ascending order we have

$\frac{29}{64}, \frac{17}{32}, \frac{5}{8}, \frac{11}{16}$

**2.** Alternatively, changing them all into decimals (using a calculator)

$\frac{5}{8} = 0.625, \frac{11}{16} = 0.6875, \frac{17}{32} = 0.53125,$

$\frac{29}{64} = 0.453125$

which in order of size is

0.453125, 0.53125, 0.625, 0.6875.

Turning these back into fractional form gives

$\frac{29}{64}, \frac{17}{32}, \frac{5}{8}, \frac{11}{16}$

## EXERCISE 14

By changing to fractions of the same type, write in *ascending* order

**1.** $\frac{1}{2}, \frac{2}{3}, \frac{5}{12}, \frac{3}{4}$

**2.** $\frac{7}{20}, \frac{1}{5}, \frac{3}{10}, \frac{1}{4}$

**3.** $\frac{3}{8}, \frac{9}{16}, \frac{1}{4}, \frac{5}{8}, \frac{1}{2}$

By changing to decimals, write in *descending* order

**4.** $\frac{3}{4}, \frac{5}{8}, \frac{1}{2}, \frac{7}{20}$

**5.** $\frac{1}{2}, \frac{5}{6}, \frac{3}{5}, \frac{2}{3}, \frac{1}{3}$

**6.** $\frac{3}{4}, \frac{25}{32}, \frac{47}{64}, \frac{13}{20}, \frac{5}{8}$

## EXERCISE 15 (MISCELLANEOUS)

**1.** What is the total weight of eight boxes each weighing $2\frac{3}{4}$ lb?

**2.** A window is 6 ft $8\frac{3}{4}$ in wide including the frame and the frame is $1\frac{13}{16}$ in wide as in the diagram.

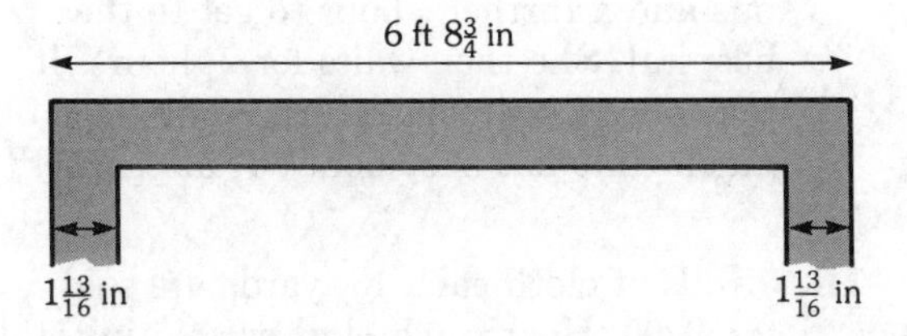


Find the width of glass in the window.

**3.** Lengths of $3\frac{5}{8}$, $2\frac{1}{2}$ and $1\frac{1}{4}$ yards have been cut from ribbon that is 12 yards long. What length is left?

**4.** Which is the shortest out of steel rods measuring $5\frac{3}{8}$, $5\frac{7}{16}$ and $5\frac{1}{2}$ inches?

**5.** If a copper pipe $13\frac{1}{2}$ feet long is divided into four equal lengths how long is each piece?

**6.** Write down the overall length of this model.

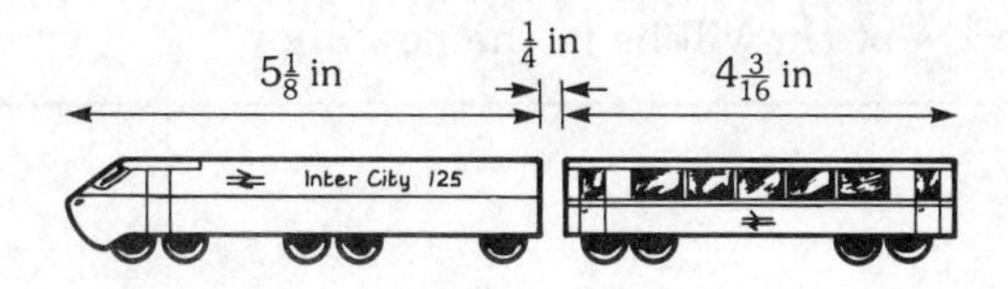


**7.** Work out the cost of making 412 zips at an average cost of $65\frac{1}{4}$p per zip.

**8.** The manager of a business allots $\frac{1}{3}$ of his profits for reinvestment and $\frac{1}{4}$ for repairs and maintenance. Out of a profit of £10 800 how much does he have left over for other uses? Sketch a pie chart to show this.

**9.** 32 out of 96 apples in a box are bad. What fraction is this in its lowest terms? Express this fraction as a percentage.

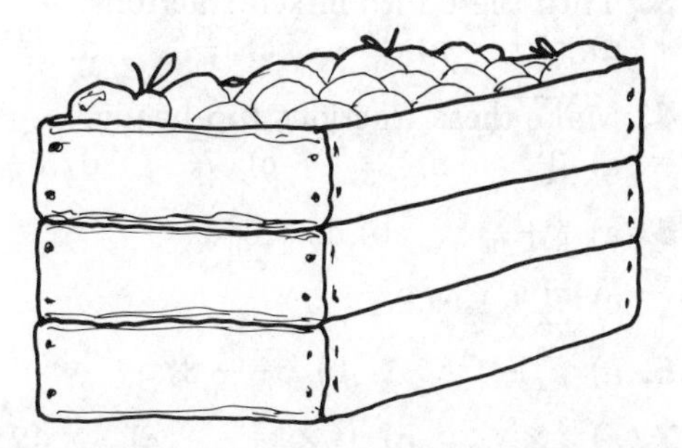

**10.** An oil container holds $35\frac{1}{2}$ litres. Seven cans each holding $1\frac{1}{2}$ litres are filled from the container. How much is left in the container?

**11.** The pie chart below shows the distribution of trees in a forest.

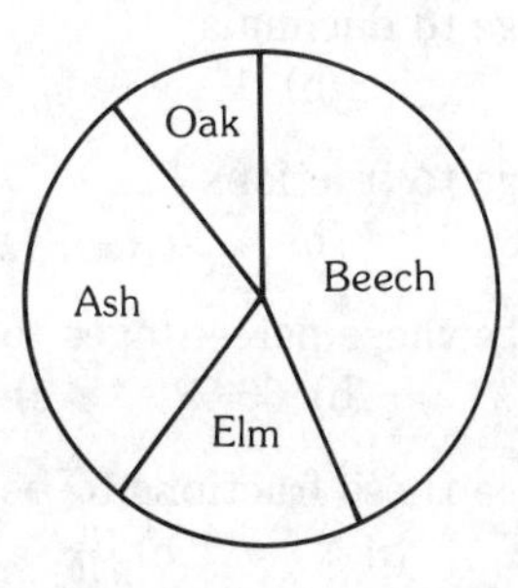


If $\frac{1}{7}$ are elm and $\frac{2}{7}$ are ash what is the remaining fraction of the forest which is oak and beech? There are three times as many beech trees as oak trees. If there are 140 trees altogether how many of them are beech?

**12.** Four-fifths of a number is 72. What is the number?

# End Test

**1.** Fill in the missing numbers.
a) $\frac{1}{2} = \frac{?}{8}$ b) $\frac{3}{5} = \frac{?}{20}$
c) $\frac{7}{8} = \frac{?}{64}$ d) $\frac{3}{8} = \frac{?}{200}$

**2.** Reduce these fractions to their lowest terms.
a) $\frac{4}{12}$ b) $\frac{6}{20}$
c) $\frac{16}{64}$ d) $\frac{75}{80}$

**3.** Turn these into mixed fractions.
a) $\frac{21}{5}$ b) $\frac{19}{6}$ c) $\frac{35}{16}$ d) $\frac{49}{10}$

**4.** Make these fractions top-heavy.
a) $3\frac{1}{2}$ b) $2\frac{2}{3}$ c) $7\frac{5}{8}$ d) $5\frac{1}{16}$

**5.** a) $\frac{3}{5} + \frac{7}{10}$ b) $2\frac{1}{2} + 3\frac{3}{8}$
c) $7\frac{1}{2} + 1\frac{3}{8} + 2\frac{7}{16}$

**6.** a) $2\frac{3}{4} - 1\frac{1}{2}$ b) $5\frac{11}{16} - 3\frac{7}{8}$

**7.** a) $\frac{3}{4} \times \frac{8}{9}$ b) $1\frac{2}{5} \times 1\frac{2}{3}$ c) $\frac{3}{5} \times 12$

**8.** a) $\frac{8}{9} \div \frac{2}{3}$ b) Divide $7\frac{1}{2}$ by $\frac{3}{5}$.
c) $8 \div \frac{1}{3}$ d) $6\frac{1}{2} \div 4$

**9.** a) $(2\frac{3}{4} + 1\frac{5}{8}) \div 1\frac{3}{4}$ b) $1\frac{2}{5} + \frac{2}{3} \times \frac{3}{4}$
c) $4(5\frac{1}{2} \div 2)$

**10.** Change to decimals
a) $\frac{11}{20}$ b) $1\frac{3}{4}$ c) $5\frac{7}{16}$

**11.** Change to fractions
a) 0.6 b) 4.25 c) 13.125

**12.** Change these percentages to fractions.
a) 20% b) $66\frac{2}{3}$% c) 140%

**13.** Change these fractions to percentages.
a) $\frac{4}{5}$ b) $\frac{1}{3}$ c) $\frac{1}{10}$ d) $\frac{1}{200}$

**14.** Write in ascending order of size
$\frac{7}{20}, \frac{1}{5}, \frac{3}{10}, \frac{4}{25}$

**15.** A litre of milk is approximately $1\frac{3}{4}$ pints. Approximately how many litres equals 7 pints?

**16.** How wide is the frame around this window?

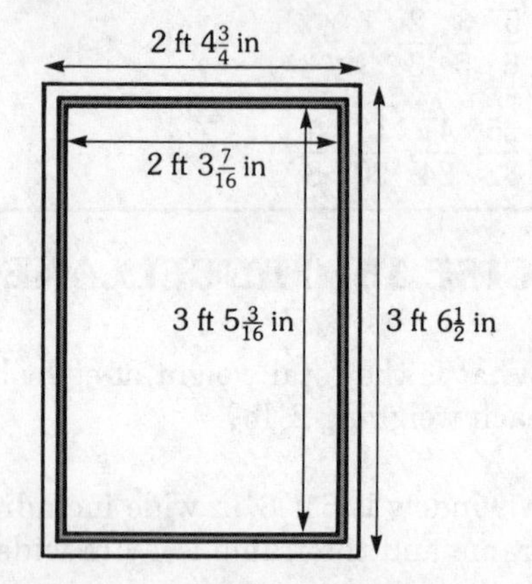


**17.** Joanna takes $1\frac{1}{4}$ hours to go to town by bus and a further $\frac{1}{2}$ hour to get to the hospital. She then waits for $2\frac{3}{4}$ hours. It takes her 2 hours to return home. How much time has she spent out altogether?

**18.** 7 rolls of cloth each $13\frac{1}{3}$ yards are sold for £200. How much cloth is sold in total and how much does each roll cost? If lengths of $1\frac{2}{3}$ yards, $1\frac{1}{3}$ yards and $3\frac{1}{6}$ yards are cut from a single roll how much remains in that roll?

**19.** A business makes profits of £25 500. The profits are shared by giving one partner $\frac{1}{3}$, a second partner $\frac{2}{5}$ and a third partner the rest. How much does the third partner receive?

**20.** Copper and tin are mixed in an alloy in the ratio $\frac{7}{10}$ copper to $\frac{3}{10}$ tin. How much copper is there in 450 g of the alloy? If 50 g of copper per 450 g of alloy are added to make a new alloy what fraction of tin will be in the new alloy?

Chapter 4

# Money Calculations

## Money as a Decimal

British currency was changed to the decimal system in 1971. There are 100 p in £1. Thus £6.10 is six pounds and ten pence, £6.01 is six pounds and one penny. Thirty-five pence could be written as 35 p or, using the decimal point, as £0.35.

### EXERCISE 1

Write out in words

**1.** £0.45 **2.** £2.70 **3.** £3.08

**4.** £4.56 **5.** £97.60.

Write using figures

**6.** twelve pence

**7.** one pound and ten pence

**8.** seventy-five pence

**9.** six pence

**10.** two hundred and five pounds fifty pence.

Write as part of a pound using a decimal point

**11.** 24 p **12.** 3 p **13.** 20 p

**14.** 17 p **15.** 8 p

## Calculating with Money

You need to be able to add, subtract, multiply and divide when totalling bills, giving change, working out the price of several similar items or sharing costs. In some situations pencil and paper or calculators might not be available.

Carry out exercise 2 saying the answers out loud, rather than writing them down.

### EXERCISE 2

**1.** Add £11.50 and £2.46.

**2.** Subtract £0.35 from £1.20.

**3.** Multiply 70 p by 3.

**4.** Divide £8 by 5.

**5.** What is twice £8.15?

**6.** Take £1.95 from £8.50.

**7.** Add 17 p to 96 p.

**8.** Multiply £4.05 by 4.

**9.** Share £6.30 by 6.

**10.** Take £2.99 from £10.

Here is an exercise for speed practice. Use pen or pencil and paper but do not use a calculator.

### EXERCISE 3 (SPEED PRACTICE)

Add

**1.** £0.67 to £3.12 **2.** £8.15 to £4.92

**3.** £3.07 to £17.12 **4.** £8.75 to £6.43

**5.** £0.51 to £9.49 **6.** £17.50 to £327.

Subtract

**7.** £2.15 from £3.89

**8.** £3.47 from £13.60

**9.** £15.92 from £26.02

**10.** £17.50 from £19.40

**11.** £0.56 from £3.49

**12.** £410.80 from £450.

Multiply

**13.** £3.45 by 6 **14.** £6.35 by 4

**15.** £19.25 by 8 **16.** £4.37 by 5

**17.** £61.82 by 7 **18.** £121.50 by 9.

Divide, giving your answer correct to the nearest penny

**19.** £16.08 by 4 **20.** £5 by 6

**21.** £166 by 8 **22.** £12 by 7

**23.** £17.44 by 9 **24.** £342.60 by 11.

# Coins and Notes

The coins in general circulation at present are 1 p, 2 p, 5 p, 10 p, 20 p, 50 p, £1 and £2. The notes are £5, £10, £20 and £50 (Scotland also still has the £1 note).

## EXERCISE 4

**1.** A shop assistant can choose which coins or notes she gives as change. A boy buys a milkshake costing 47 p. Work out two ways of giving change if he offers a) a £1 coin b) a £5 note c) a £10 note.
(Make the two ways as different as possible.)

**2.** Repeat the above for an item costing £0.25.

**3.** List three different ways of making up a) 27 p in coins b) £50 in notes.

## EXERCISE 5 (MISCELLANEOUS)

**1.** What is the cost of a sandwich at 80 p plus a cup of coffee at 37 p? I also buy a baked potato at 80 p and a cake at 45 p. What do I have to pay in total?

**2.** Joan owes me £5.18. She pays me £3.50. How much does she still owe me? She pays me a further £1.25. How much does she owe now?

**3.** Sean is working as a window cleaner. He charges 45 p per window and cleans, on average, thirty-six windows a day. How much does he earn per day? How much more would he earn if he charged 50 p per window but only cleaned 33 windows a day?

**4.** Hiring a coach costs £52. If twenty-five people share the cost how much would they each have to pay? If they received £14 towards the cost how much would they then each have to pay?

**5.** The charge for twelve sessions at an evening class is £9.60. How much is this per session? What would the new charge be if the cost of each session was increased by 10 p?

**6.** Petrol can be bought at a local garage for £2.05 per gallon. Find the price of a) 5 b) 7 c) 9 gallons and the change that would be given in each case if the bill was paid with a £20 note.

**7.** A man works 37 hours at his basic rate plus 3 hours overtime for which he is paid double his basic rate. Find how much he earns if his basic rate is a) £4.50 b) £5.20 an hour.

**8.** Sandra stays at a hotel which charges £25.75 per night (including VAT and service) and she also has to pay £8.20 for telephone charges for the week and a single charge of £6.35 for drinks. What is her total bill if she stays for seven nights?

**9.** A firm selling spare parts for cars orders seventy items at £15.60 each, two hundred and ten at £17.85 each and twenty-four at £19 each. How much is this altogether?

**10.** A hairdresser charges £4.25 for a cut and blow dry, £15.75 for a perm and £3 for a wash and set. What would be the daily takings for 12 perms, 15 wash and sets and 7 cut and blow dries?

**11.** Soup costs 45 p, a roll 25 p and a portion of butter 18 p. How much would the bill be for four people having soup, a roll and butter each?

**12.** A coach trip to London costs £18 per adult and £9 per child under 15. Find the cost for a family with 2 adults and three children aged 2, 7 and 16 given that children under 3 travel free.

# Money and Percentages

Per cent means out of 100. A 4% wage increase means that for every £100 you earn you will get an extra £4. If you earned £700 you would get £4 × 7 more. If you earned £3000 you would get £4 × 30 = £120 more.

## EXERCISE 6

How much extra would you earn if you had

**1.** a 3% increase on £500

**2.** 10% on £700

**3.** 5% on £900

**4.** 8% on £1000

**5.** 9% on £2000

**6.** 2% on £1500

**7.** 1% on £9500

**8.** 7% on £7000

**9.** 4% on £8000

**10.** $2\frac{1}{2}$% on £6000?

To find, say, 15% of £425 using a calculator you could press the keys

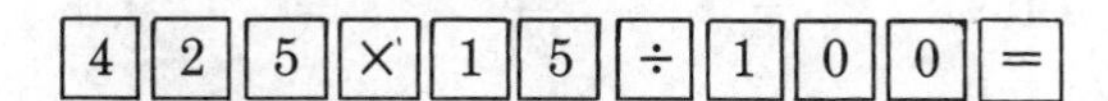


to get (63.75) on the display.

The answer is £63.75.

Sometimes you will need to be able to *round* the answer to the nearest penny, for example

(24.8275) would be £24.83

and (0.634) would be £0.63

(see part B, chapter 2, page 176)

Suppose you have no calculator. A simple way of calculating a percentage of a whole number of £s is to say 1% of £1 is 1 p. To find 6% of £40 you could write 1% of £40 is 40 p, so 6% of £40 is 6 × 40 p = 240 p = £2.40.

Use your common sense to decide whether to use a calculator or not. To work out 1%, move the decimal point back two places. To work out 10% move it back one place.

**Example**

1% of £350 = £3.50 — *1% is $\frac{1}{100}$ so divide by 100*

10% of £12.50 = £1.25 — *10% is $\frac{1}{10}$ so divide by 10*

Knowing the common fractions and their decimal equivalents also helps.

| *Percentage* | | *Fraction* | | *Decimal* |
|---|---|---|---|---|
| 1% | = | $\frac{1}{100}$ | = | 0.01 |
| 5% | = | $\frac{1}{20}$ | = | 0.05 |
| 10% | = | $\frac{1}{10}$ | = | 0.10 |
| $12\frac{1}{2}$% | = | $\frac{1}{8}$ | = | 0.125 |
| 15% | = | $\frac{3}{20}$ | = | 0.15 |
| 25% | = | $\frac{1}{4}$ | = | 0.25 |
| 30% | = | $\frac{3}{10}$ | = | 0.30 |
| $33\frac{1}{3}$% | = | $\frac{1}{3}$ | = | $0.\dot{3}$ |
| 50% | = | $\frac{1}{2}$ | = | 0.50 |
| 75% | = | $\frac{3}{4}$ | = | 0.75 |

To find, say, $12\frac{1}{2}$% you could multiply by $\frac{1}{8}$ (which means divide by 8) or multiply by 0.125. To find VAT at its current rate of 15% you could multiply by 0.15.

If you have a % button on your calculator read the instructions to find out how to use it.

Usually to find, say, 15% of £3 you would press the buttons [3] [×] [1] [5] [%] [=] to give the result (0.45) on the display which would mean £0.45.

## EXERCISE 7

(Try to use your calculator as little as possible for questions **1** to **15**.)

Find

**1.** 10% of £530 **2.** 3% of £5

**3.** 5% of £700 **4.** 1% of £7050

**5.** 1% of £9 **6.** 25% of £4

**7.** $12\frac{1}{2}$% of £16 **8.** 150% of £6

**9.** 250% of £5 **10.** 15% of £10

**11.** 75% of £12 **12.** 50% of £655

**13.** 10% of £5 **14.** 10% of £1

**15.** $33\frac{1}{3}$% of £66 **16.** 34% of £46

**17.** 17% of £504 **18.** 7.25% of £56.97

**19.** 5% of £6897.

**20.** In a sale all items are reduced by 4%. How much is 4% of £200?

**21.** A discount of 20% is given on a £25 coat. How much is the discount?

**22.** A £32 dress is increased by 25%. What is the increase?

**23.** A surcharge of 10% is made on a £12.50 air ticket. How much is the surcharge?

**24.** A man pays 6% interest per year on a loan of £3000. What is the interest paid per year?

**25.** A washing machine cost £325 but was reduced by 25% because of damage. By how much was the washing machine reduced in price?

**26.** Find the VAT payable at 15% on £18.

# Simple Interest

When you save your money with a building society then extra money, called *interest*, is paid to you while you save.

### Examples

**1.** You save £525 for 2 years at a rate of 12% simple interest per year. How much interest do you earn?

For 1 year the interest would be $(£525 \div 100) \times 12 = £63$.
Therefore for 2 years the interest is $2 \times £63 = £126$.

Notice that with simple interest, if the time is 2 years then the interest is doubled, for 3 years tripled, for 6 months halved and so on.

You could use the simple interest formula

$$I = \frac{P \times R \times T}{100}$$

where $P$ is the principal, which is the amount of the loan in pounds, $R$ is the rate of interest as a percentage, $T$ is the time of the loan in years, and $I$ is the interest in pounds.

**2.** Find the simple interest on £6470 loaned for 3 years at 8% rate of interest.

$P = 6470, \quad R = 8, \quad T = 3$

so, using the simple interest formula

$$I = \frac{6470 \times 8 \times 3}{100}$$

$$= 1552.80$$

The simple interest is £1552.80.

Notice that if you *borrow* money then you have to *pay* interest for the loan.

## EXERCISE 8

Find the simple interest on

**1.** £7250 at 8% for 2 years
**2.** £4100 at 11% for 3 years
**3.** £230 at 10% for 5 years
**4.** £4500 at 12% for 1 year
**5.** £6724 at 5% for 4 years
**6.** £704 at 9% for 6 months
**7.** £915 at 8% for 3 months
**8.** £200 at 7% for 1 year 6 months
**9.** £34.16 at 6% for 2 years
**10.** £282.50 at 15% for 10 years.

# Compound Interest

If the interest is added to your savings at the end of the year so that the interest itself earns interest then this is called *compound interest.*

### Examples

**1. Method 1**
Find the compound interest earned on £1700 invested at 10% for 2 years.

interest at end of 1st year = 10% of £1700 = £170
total amount at end of 1st year
= £1700 + £170 = £1870
(savings plus interest)
interest at end of 2nd year = 10% of £1870 = £187
total amount at end of 2nd year
= £1870 + £187 = £2057
compound interest
= £2057 − £1700 = £357

Notice that the interest on the second year is more than the interest on the first year. Why is this?

**2. Method 2**

Find a) the final amount b) the compound interest on £700 at 9% invested for 4 years.

To find the amount at the end of one year at 9%, add 9% of £700 to £700. You could do this instead by multiplying £700 by $\frac{109}{100}$.

To find the amount (savings plus interest) at the end of the second year again multiply by $\frac{109}{100}$.

So the final amount at the end of four years is

$$£700 \times \frac{109}{100} \times \frac{109}{100} \times \frac{109}{100} \times \frac{109}{100}$$

$$= £700 \times 1.09 \times 1.09 \times 1.09 \times 1.09$$

$$= £988.11 \text{ to the nearest penny}$$

Therefore the final amount is £988.11 and the compound interest is

$$£988.11 - £700 = £288.11$$

## EXERCISE 9

Find a) the final amount b) the compound interest on

**1.** £3000 invested for 2 years at 10%

**2.** £2000 invested for 3 years at 10%

**3.** £600 invested for 2 years at 5%

**4.** £7050 invested for 3 years at 10%

**5.** £25 invested for 3 years at 8%

**6.** £4500 invested for 3 years at 2%

**7.** £1800 invested for 2 years at 9%

**8.** £5000 invested for 3 years at 4%

**9.** £520 invested for 2 years at 5%

**10.** £6750 invested for 4 years at 6%.

# Writing a Number as a Percentage of Another Number

To express a number as a percentage of another number just write as a fraction and then convert to a percentage by multiplying by 100%.

### Example

What is £35 as a percentage of £40?

$$\frac{£35}{£40} = \frac{35}{40}$$

$$\frac{35}{40} = \frac{35}{40} \times 100\%$$

$$= 87.5\%$$

(Either cancel the fraction or use a calculator,

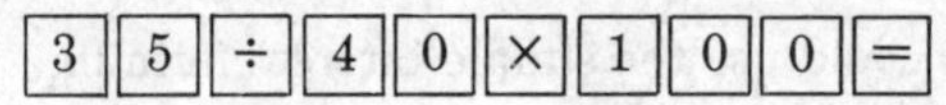


On a calculator with a % key use

)

## EXERCISE 10

Write the first number as a percentage of the second.

**1.** £5, £20 **2.** £250, £500

**3.** £3, £60 **4.** £8, £40

**5.** £5, £50 **6.** £2, £800

**7.** £72, £1440 **8.** £53, £212

**9.** £52, £150 **10.** £704, £800

**11.** 75 p, £1.20 **12.** £1.15, £1.25

**13.** £76.50, £80 **14.** £1.50, £1

**15.** 12 p, 96 p

# Profit and Loss

If goods are bought for £50 and sold for £53 then a £3 *profit* has been made. The £50 is known as the *cost price* and the £53 as the *selling price.*

$$\text{profit} = \text{selling price} - \text{cost price}$$

$$\text{percentage profit} = \frac{\text{selling price} - \text{cost price}}{\text{cost price}} \times 100\%$$

So for the example above,

$$\text{percentage profit} = \frac{£53 - £50}{£50} \times 100\%$$

$$= \frac{3}{50} \times 100\%$$

$$= 6\%$$

## EXERCISE 11

Copy the table and fill in the missing figures.

| *Cost price* | *Selling price* | *Profit* | *Percentage profit (to one decimal place)* |
|---|---|---|---|
| £100 | £104 | a)...... | b)...... |
| £500 | £550 | c)...... | d)...... |
| £250 | £260 | e)...... | f)...... |
| £8.00 | £8.20 | g)...... | h)...... |
| £0.45 | £0.50 | i)...... | j)...... |

When the selling price is less than the cost price then a *loss* has been made.

$$\text{loss} = \text{cost price} - \text{selling price}$$

$$\text{percentage loss} = \frac{\text{cost price} - \text{selling price}}{\text{cost price}} \times 100$$

### Example

A car is bought for £650 and sold for £400. What is a) the loss b) the percentage loss?

a) loss = cost price − selling price

$$= £650 - £400$$

$$= £250$$

b) $$\text{percentage loss} = \frac{£650 - £400}{£650} \times 100\%$$

$$= \frac{250}{650} \times 100\%$$

$$= 38.5\% \quad \text{to 1 decimal place}$$

## EXERCISE 12

Copy the table and fill in the missing figures.

| *Cost price* | *Selling price* | *Loss* | *Percentage loss (to 1 d.p.)* |
|---|---|---|---|
| £100 | £97 | a)...... | b)...... |
| £250 | £200 | c)...... | d)...... |
| £0.50 | £0.40 | e)...... | f)...... |
| £18 | £12.50 | g)...... | h)...... |
| £60 | £50 | i)...... | j)...... |

Suppose you know the percentage profit (or loss) and the cost price and have to find the selling price.

### Example

A suite of furniture is sold at a 5% loss. It was bought for £500. What is the selling price?

the loss is 5% of the cost price

$$= \frac{5}{100} \times £500$$

$$= £25$$

the suite was sold for £500 − £25

$$= £475$$

## EXERCISE 13

Copy the table and fill in the missing figures.

| *Cost price* | *Selling price* | *Profit or loss* | *Percentage profit or loss* |
|---|---|---|---|
| £700 | a)...... | b)...... loss | 5% loss |
| £400 | c)...... | d)...... loss | 10% loss |
| £350 | e)...... | f) ...... profit | 2% profit |
| £600 | g)...... | h)...... profit | 15% profit |
| £225 | i) ...... | j) ...... loss | 9% loss |

You need to use a different method if you know the selling price and the percentage profit (or loss), and need to find the cost price.

## Examples

**1. Profit**

A coat was sold for £520 at 4% profit.
How much did it cost?

The cost price is 100% and the profit is 4% so the selling price is 104%.

$$104\% = £520$$

$$1\% = £520 \div 104$$

$$100\% = £520 \div 104 \times 100$$

$$= £500$$

The coat was bought for £500.

**2. Loss**

Find the cost of a suit sold for £188 at a loss of 6%.

The cost price is 100% and the loss is 6%, so the selling price is 94%.

$$94\% = £188$$

$$1\% = £188 \div 94$$

$$100\% = £188 \div 94 \times 100$$

$$= £200$$

The suit cost £200.

## EXERCISE 14

(This exercise is more difficult than the ones before; use the method shown in the example.) Copy the table below and fill in the missing figures.

| *Cost price* | *Selling price* | *Profit or loss* | *Percentage profit or loss* |
|---|---|---|---|
| a)...... | £115 | b)...... profit | 15% profit |
| c)...... | £78 | d)...... profit | 4% profit |
| e)...... | £36.75 | f) ...... profit | 5% profit |
| g)...... | £47 | h)...... loss | 6% loss |
| i) ...... | £1.96 | j) ...... loss | 2% loss |
| k)...... | £24 | l) ...... loss | 4% loss |

## EXERCISE 15

**1.** A video is sold for £550 and it cost the shop £500 to buy. Give the percentage profit.

**2.** If a table and chairs have a cost price of £200 and a selling price of £175 give the loss as a percentage of the cost price.

**3.** How much profit has been made if a television which cost £250 is sold for £270? Express this as a percentage.

**4.** A 5% profit is made when selling an article which cost £120. What is the selling price?

**5.** Find the price at which a set of cassettes was sold if there was 3% loss on the cost price of £25.

**6.** The selling price of a kettle is £31.50. It is sold at 5% profit. What is the cost price?

**7.** Sets of crockery are offered for sale at £19.50 each. The retailer bought each set for £15. If the retailer sells thirty-five sets find a) the total profit b) the percentage profit.

**8.** If ninety-five books are sold for £215 and cost £237 to buy find a) the loss per book to the nearest penny b) the percentage loss.

**9.** A set of saucepans is sold for £53.04 at a 2% profit. Give the cost price.

**10.** Exercise books, pens and rulers are for sale at 12 p, 18 p and 35 p respectively. 100 exercise books, 25 pens and 10 rulers are sold by mistake for £19. Find a) the actual loss b) the percentage loss.

## EXERCISE 16 (MISCELLANEOUS PERCENTAGE QUESTIONS)

**1.** The bill at a hotel is £32.40 but VAT at a rate of 15% must be added to this. Find the total to be paid.

**2.** If the rate of inflation is 16% by how much would the price of goods costing £500 have risen a) the following year b) after 2 years?

**3.** Income tax is 29% of taxable income. If John earns £8500 and has allowances of £2700 how much tax does he pay?

**4.** Liz borrows £1400 for two years. Calculate a) the simple interest at 16% per annum payable on this loan
b) the monthly repayments if the loan plus interest is paid back over two years.

**5.** A car depreciates by 12% of its value at the beginning of the year each year for 3 years. If it is valued at £5120 at the beginning of 1980 what is its value at the beginning of 1983?

**6.** If a saleswoman earns a salary of £4800 a year plus commission of 20% on the first £3000 of her sales and 12% on the rest, calculate the total amount received in a year if her total sales were £12 000.

**7.** Copy the table and fill in the missing figures.

| *Cost price* | *Selling price* | *Profit or loss* | *Percentage profit or loss* |
|---|---|---|---|
| £50 | £60 | a) ...... | b) ...... |
| £500 | c) ...... | d) ...... profit | 4% profit |
| £105 | e) ...... | £15 profit | f) ...... |
| g) ...... | £300 | h) ...... loss | 20% loss |

**8.** £100 is invested at 10% per annum compound interest. How much will it amount to after a) 1 year b) 2 years?
c) What investment would amount to £605 after 2 years at compound interest?

**9.** A man's income is cut first by 5% and then by a further 2%. If he was originally on a salary of £10 000 calculate his final salary. Explain why this is *not* equal to a 7% wage cut.

**10.** Joan has £1000 to invest for one year. She considers two schemes.
a) a tax-free scheme paying interest of 9.5% per annum
b) a scheme where interest of 15% is paid but tax of 30% has to be paid on the interest
Calculate the interest payable for each of the two schemes and say which is most advantageous.

# Invoices

When goods are supplied to a customer an invoice is also sent. This is a piece of paper showing the quantity of goods supplied together with the cost. It is not a bill.

| INVOICE | | | | | |
|---|---|---|---|---|---|
| S. Peterson<br>7 High St.<br>GOLCAR | | | | | |
| Qty | Description | Unit cost | Gross | Percentage discount | VAT |
| 35 | Men's shirts | 4.99 | £174.65 | 8% | 15% |
| 40 | Ladies' blouses | 2.99 | £119.60 | - | 15% |
| | Total gross | | £294.25 | | |
| | Less discount | | 13.97 | | |
| | | | 280.28 | | |
| | Plus VAT | | 42.04 | | |
| | | | £322.32 | | |

The *quantity* is multiplied by the *unit cost* to give the entry in the column labelled *'gross'*. *Discount* is worked out for each entry by multiplying the percentage discount by the entry in the gross column. The total discount is then taken from the total of the gross column. *VAT* is added to the bill after the discounts have been allowed.

In this example,

discount = 8% of £174.65
= £13.97

VAT = 15% of £280.28
= £42.04

# EXERCISE 17

**1.** Copy and complete this invoice.

| INVOICE | | | | | |
|---|---|---|---|---|---|
| Mr Singh<br>General Stores<br>7 Ashford Rd<br>DONCASTER | | | | | |
| Qty | Description | Unit cost | Gross | Percentage discount | VAT |
| 7 | Sets of cutlery | £9.15 | | 12% | 15% |
| 14 | Glass bowls | £1.85 | | 12% | 15% |
| 15 | Jugs (small) | £0.72 | | – | 15% |
| 25 | Jugs (large) | £1.35 | | – | 15% |
| | Total gross | | | | |
| | Less discount | | | | |
| | | | | | |
| | Plus VAT | | | | |
| | | | | | |

Make similar invoices for the following.

**2.** Mrs C. Clarke (Drapers)
18 Dellfield Close
Appleford
Middx
17 reels cotton (white) @ 32 p each
17 reels cotton (black) @ 32 p each
105 reels cotton (assorted) @ 35 p each
14 pkt bias binding @ 40 p each
16 zips (green) @ 68 p each
26 zips (blue) @ 68 p each
Discount to trade $12\frac{1}{2}$%
VAT at 15%

**3.** Mr W. Patel
15 North Street
Newcastle
23 sheets (single) @ £8.95 each
45 pillow-cases @ £3.50 each
50 duvet covers @ £12.95 each
35 valences @ £5.75 each
Allow a trade discount of 15% and VAT at 15%.

# End Test

**1.** a) Write in words i) £0.23 ii) £4.50.
b) Write using figures
i) fourteen pence
ii) three hundred and five pounds seven pence.

**2.** a) Add £6.37, £42.05 and 95 p.
b) Multiply 87 p by 4.
c) Subtract £9.17 from £23.50.
d) Divide £813.52 by 8.

**3.** List five different ways of making up 21 p in coins.

**4.** I buy five cakes at 19 p each, a loaf costing 42 p and half a dozen rolls at 13 p each. How much does this cost altogether?

**5.** Work out the change from a £20 note given for items costing £8.17, £9.15 and £0.95.

**6.** Joan is paid 3 p per zip fitted. She fits on average 500 zips per day for five days a week. How much does she earn per week?

**7.** A coach is hired for £85. The cost is shared among 17 people. How much do they each pay? If three extra people joined in, how much less would each person pay?

**8.** Find a) 1% of £72 b) 20% of £350
c) $12\frac{1}{2}$% of 72 p d) 15% of £5.12

**9.** Find a) the simple interest at 7% on £806 for i) 2 years ii) 6 months
b) the compound interest at 5% on £7500 for 3 years.

**10.** Write a) £45 as a percentage of £400
b) £3 as a percentage of £36.

**11.** A TV costs the retailer £300 and is sold for £325. Find the percentage profit.

**12.** What is the selling price if a video is bought by a shop for £385 and sold at 8% profit?

**13.** A suite of furniture is sold at 10% profit for £495. What was the cost price? Find the cost price of a TV sold at £364 for a 4% profit.

**14.** Wages fall from £215 to £195. What percentage drop is this?

**15.** Make out an invoice to Mr D. Smith, Hazeldene, Trentham for
125 cakes @ £0.15
150 buns @ £0.12
175 biscuits @ £0.05
with trade discount of 14% and zero-rated VAT.

Supposing VAT had been payable on this at 15% – how much would the VAT have been?

Chapter 5

# Using a Calculator

## The Keys

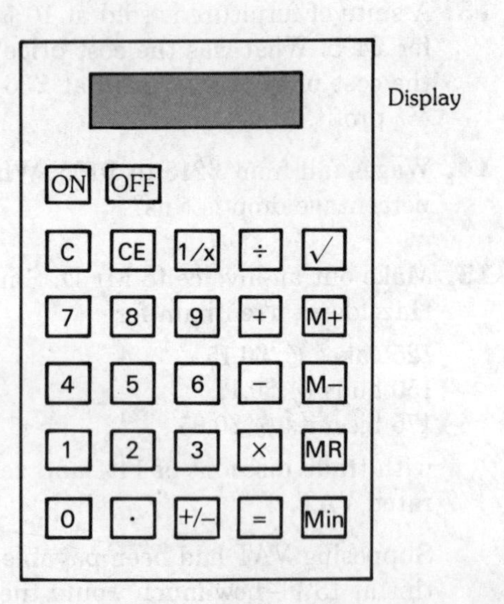


Always read the instructions for your own calculator carefully, as the way that the keys are marked does vary from calculator to calculator. Usually the key marked [C] will clear the calculator and the key which is marked [CE] clears the last entry only.

To work out the cost of, say, 7 items at £6.02 the keys pressed would be

[7] [×] [6] [.] [0] [2] [=] giving (42.14) on the display which stands for £42.14.

## EXERCISE 1

Use your calculator to work out the following pressing the keys in the order given.

**1.** [5] [+] [3] [+] [2] [+] [9] [=]

**2.** [1] [6] [−] [1] [8] [=]

**3.** [.] [4] [×] [.] [2] [=]

**4.** [5] [2] [÷] [8] [=]

**5.** [2] [4] [+] [5] [6] [−] [8] [0] [=]

**6.** [3] [×] [2] [+/−] [=]

**7.** [4] [8] [÷] [6] [÷] [2] [=]

**8.** [5] [+] [3] [×] [4] [=]

**9.** [2] [−] [1] [×] [2] [=]

**10.** [5] [+] [4] [×] [0] [=]

Discuss whether you all got the same answers for the last three questions.

# Order of Operations

Some calculators use BODMAS to work out the order in which they do things. This means that **B**rackets, **D**ivision and **M**ultiplication are done before **A**ddition and **S**ubtraction.

Find out whether your calculator uses BODMAS or not. (Your calculator uses BODMAS if your answer to question **10** is 5.)

Suppose you wish to work out the cost of 7 items at £6.02 plus 9 items at £3.57. With a calculator that uses BODMAS you could simply press

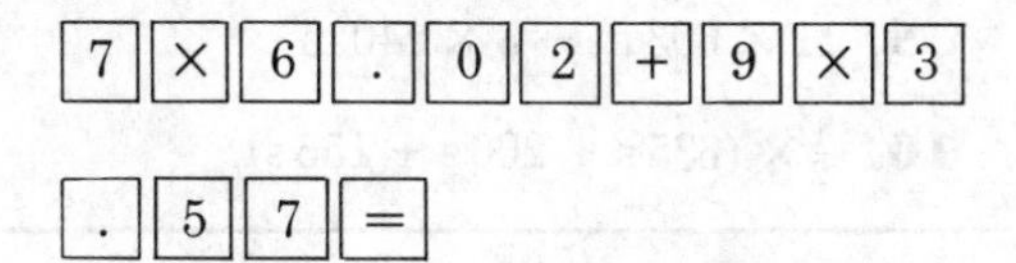


to get the correct answer (74.27) on the display representing £74.27. However, with a simple calculator that does *not* use BODMAS, you would have to press

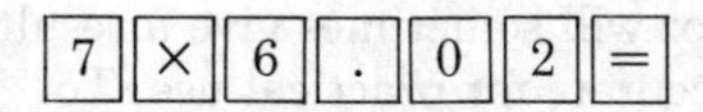


and *write down* or *remember* the result 42.14, then press

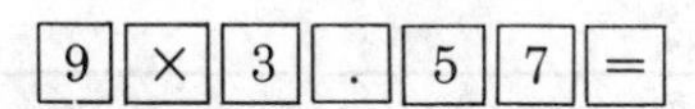


to get 32.13, and then press

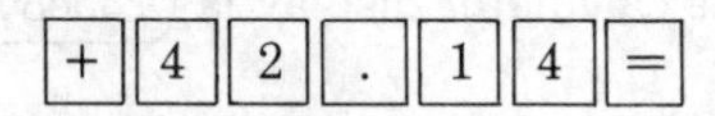


to get the correct answer.

If the calculator has a memory then you need to discover which keys will store a result in the memory, and which will recall it. Suppose these keys are [Min] and [MR]. Then, instead of *writing down* 42.14 you could put it into the memory and recall it when needed by pressing the keys

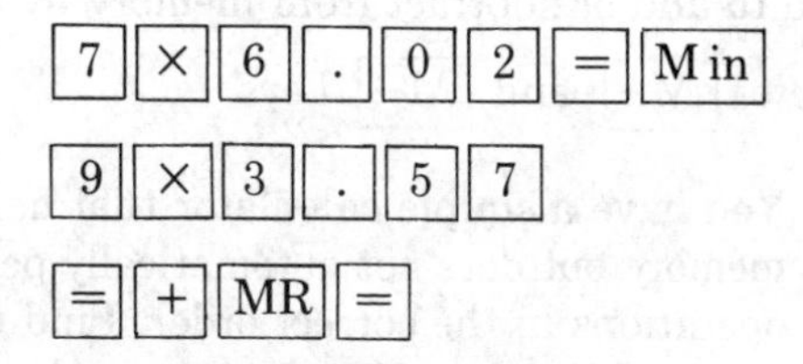


to get the result (74.27) on the display, which represents £74.27. Remember always to clear the memory at the end of a calculation.

## EXERCISE 2

Work out the following using your calculator, and correct the answers to the nearest penny.

**1.** 7 × £85.15 **2.** £186 ÷ 24

**3.** £325.04 + £12.77 **4.** £821 + £105 ÷ 21

**5.** £400 − (£65 + £18)

**6.** 7 × £2.05 + 3 × £1.18

**7.** £84.50 + 3 × £6.19

**8.** 117 × £13 + £28 ÷ 8

**9.** £0.07 + £0.16 + £0.42

**10.** (£19.65 + £3.27) − (£3.18 + £4.50)

If you are working with pounds and pence then work in pounds, entering the pence after the decimal point. For example, 19 p should be entered as £0.19 by pressing the keys

[0] [.] [1] [9]

## Examples

**1.** Add £55.16 + 19 p + £24 + 3 p.

Enter

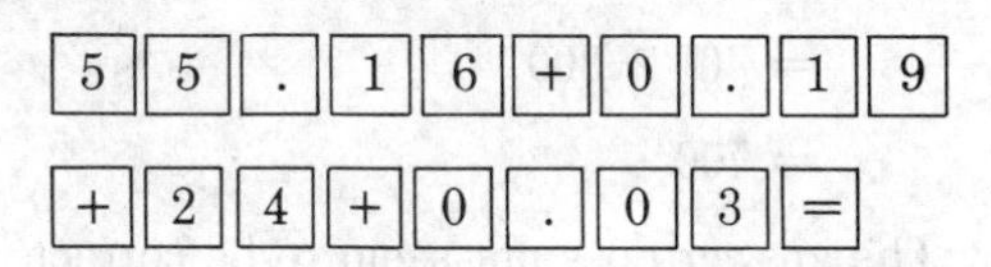


to get the answer (79.38) representing £79.38.

Most calculators with a memory will also allow you to add or subtract from memory by the use of M+ and M− keys.

**2.** You have a simple calculator that has a memory but does not automatically perform operations in the correct order. Find the cost of 8 items at 19 p, 7 at 8 p and 4 at £1.03.

The correct keys to press would be

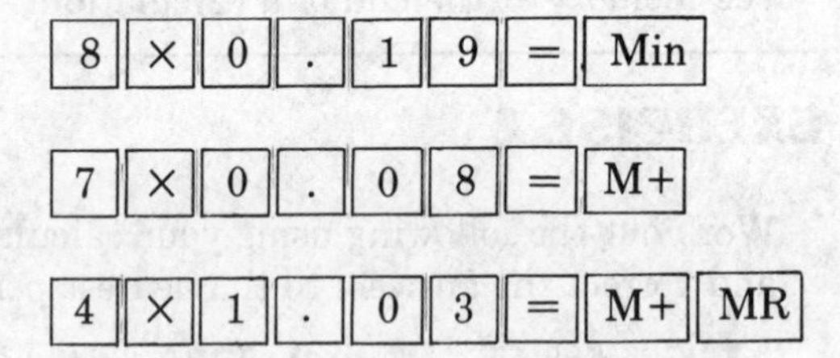


giving the answer 6.2 on the display, representing £6.20.

# Rough Checks

You should always do a rough check to see if the answer on the display is approximately correct. Round to the nearest whole number, ten, fifty or hundred, or whatever is most convenient.

### Example

$17 \times £24.36 + 19 \times £15.50$ has been given as £553.62. Is this correct?

Rough check

$20 \times 20 + 20 \times 15$

$= 400 + 300$

$= 700$

The answer does not seem to be correct, and further effort will reveal that it should have been £708.62.

## EXERCISE 3

Do a rough check then find the answer to

**1.** $9 \times 48$ **2.** $£790 \div 4$

**3.** $516 + 502 + 498$ **4.** $88 + 5 \div 8$

**5.** $(816 + 214) \div (103 + 419)$

**6.** $58.10\,g + 72.15\,g - 35.12\,g$

**7.** $8 \times 3\,p + 19 \times £1.02 + 5 \times 72\,p$

**8.** $(816\,km \div 8) - (711\,km \div 9)$

**9.** $11 \times 603\,m - 3 \times 240\,m$

**10.** $\frac{1}{2} \times (635\,s + 206\,s + 155\,s)$.

# Accuracy and Significant Figures

The calculator will sometimes give a result that is too accurate for practical use. To overcome this you could give a number correct to 2, 3 or 4 significant figures, as shown in the examples below.

### Examples

**1.** On the calculator display is 2356800 .

What is this correct to 2 significant figures?

2 356 800 = 2 400 000 correct to 2 significant figures.

The rule is: count the number of significant figures. Look at the next digit. If it is 5 or greater then increase the previous figure by 1.

**2.** The result on a calculator is (8.571428576) .

Write this correct to 3 significant figures.

The first three figures are 8.57. The number is closer to 8.57 than 8.58. You can see that this is true by examining the next digit. This is a 1 so leave the last figure (7) alone.

8.571 428 576 = 8.57 (correct to 3 s f)

**3.** Write 0.080 976 5 correct to four significant figures.

The first four figures are 0.080 but 8 is the first figure with any value. The first four *significant* figures are 8097. The next digit is 6. This is greater than 5 so increase the 7 by 1.

0.080 976 5 = 0.080 98 (correct to 4 s f)

## EXERCISE 4

Write the following correct to
a) 2 significant figures
b) 3 significant figures
c) 4 significant figures.

**1.** 6.823 612 9

**2.** 13.415 467

**3.** 0.056 962 8

**4.** 54 952 130

**5.** 0.000 819 255

**6.** 9.989 560 312

**7.** 67 089 005 000

**8.** 5607.890 031

# Rounding

Numbers may also be approximated by rounding to the nearest hundred, fifty, whole number etc. See part B chapter 2, page 176 for details of rounding to so many decimal places.

### Example

Round 576.398 to the nearest a) hundred b) fifty c) ten d) 1 decimal place.

567.398 = 600 to the nearest hundred
= 550 to the nearest fifty
= 570 to the nearest ten
= 567.4 correct to 1 decimal place

## EXERCISE 5

**1.** Write 695 to the nearest hundred.

**2.** Round 9946 to the nearest fifty.

**3.** What is 9182 to the nearest ten?

**4.** Put 0.891 correct to 1 decimal place.

**5.** Give 17 517.190 correct to the nearest twenty.

**6.** Write 456.719 89 to 2 decimal places.

**7.** Give 569 080.9 to the nearest thousand.

**8.** What is 67.802 31 to the nearest whole number?

**9.** Put 679 595 correct to the nearest fifty.

**10.** Round 190 030 to the nearest hundred.

**11.** Write 5.671 correct to 1 decimal place.

**12.** Round 56 900 to the nearest thousand.

**13.** Put 0.999 correct to 1 decimal place.

**14.** Give 15 930 correct to the nearest fifty.

**15.** Write 190 854 132 correct to the nearest ten thousand.

# Standard Form Numbers

Very large numbers have to be entered into a calculator in a special way. They are first rewritten in *standard form.*

## Examples

**1.** $500\,000\,000 = 5 \times 100\,000\,000$
$= 5 \times 10^8$

The first number (5) is *between 1 and 10.* The number 8 is called the *exponent*.

Remember that

$10^1 = 10,\ 10^2 = 100,\ 10^3 = 1000$ and so on.

**2.** Write 8 500 000 000 in standard form.

$8\,500\,000\,000 = 8.5 \times 1\,000\,000\,000$

$= 8.5 \times 10^9$

This number would be entered into a calculator using a key, usually marked EXP, by pressing

[8] [.] [5] [EXP] [9]

to give (8.5 09) on the display.

## EXERCISE 6

Write in standard form

**1.** 600 000

**2.** 54 000 000

**3.** 850 000 000

**4.** 400 000 000 000

**5.** 170 000 000 000 000 000

**6.** 30 000 000 000 000 000 000

**7.** 512 000 000 000 000

**8.** 140 000 000 000.

Write as a single number

**9.** $7 \times 10^8$ **10.** $5 \times 10^9$

**11.** $3.6 \times 10^2$ **12.** $5.4 \times 10^4$

**13.** $1.08 \times 10^8$ **14.** $9.73 \times 10^{12}$.

Which keys would you press to enter the following into a calculator with an EXP key?

**15.** $6 \times 10^{12}$ **16.** $1.4 \times 10^{13}$

**17.** $2.35 \times 10^7$

**18.** Suppose the distance from the moon to the earth is $3.84 \times 10^5$ km and the distance from the earth to the sun is $1.5 \times 10^8$ km. Find the distance from the moon to the sun a) when it is directly behind the earth b) when it is in front of the earth and in line with the sun.

**19.** There are about 80 000 000 000 000 000 000 000 molecules of air in a normal sized balloon. Express this number in standard form.

**20.** Using the information from question **19** how many molecules of air would there be in five such balloons? Express your answer in standard form.

How many molecules of air would be in 10 000 such balloons?

## EXERCISE 7 (PRACTICE CALCULATIONS)

**1.** A garden centre orders 500 packets of seeds at 25 p a packet, 800 packets at 20 p and 300 packets at 15 p. Find the total cost.

**2.** Find the average of 8.9 8.7 9.3 9.5 8.8 8.9 9.0 8.7.

**3.** Convert $\frac{5}{16}$ to a decimal by dividing 5 by 16.

**4.** Similarly write $3\frac{11}{16}$ as a decimal, by dividing 11 by 16 and adding 3.

**5.** A man works overtime at time-and-a-quarter. Find how much he earns in 7 hours of overtime if his basic hourly rate is £4.50.

**6.** The area of a circle is given approximately by 3.14 × radius × radius. Work out, correct to 3 significant figures, the area of a circle with radius 9.34 cm.

**7.** The radius of a pond as shown below is 5.46 m and the width of the paving stones is 0.79 m. What is the diameter of the pool plus stones?

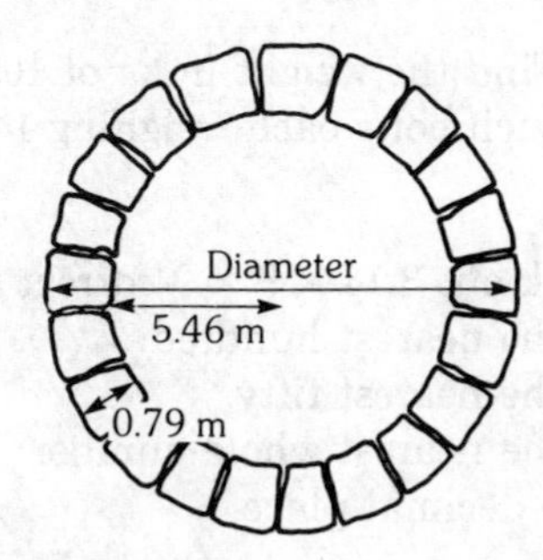


**8.** Twelve 72.5 cm lengths are cut from a cable that is 900 cm long. The remaining length is then cut into three equal pieces. How long is each piece?

**9.** The volume of a packet of cereal is found by multiplying the length by the width by the height. If the length is 21 cm, the width 7.5 cm and the height 33 cm work out the volume of the packet correct to 2 significant figures.

**10.** VAT is charged at 15%. Find the VAT on a bill of £65.72.

**11.** Follow the flow chart through and enter the values in the table.

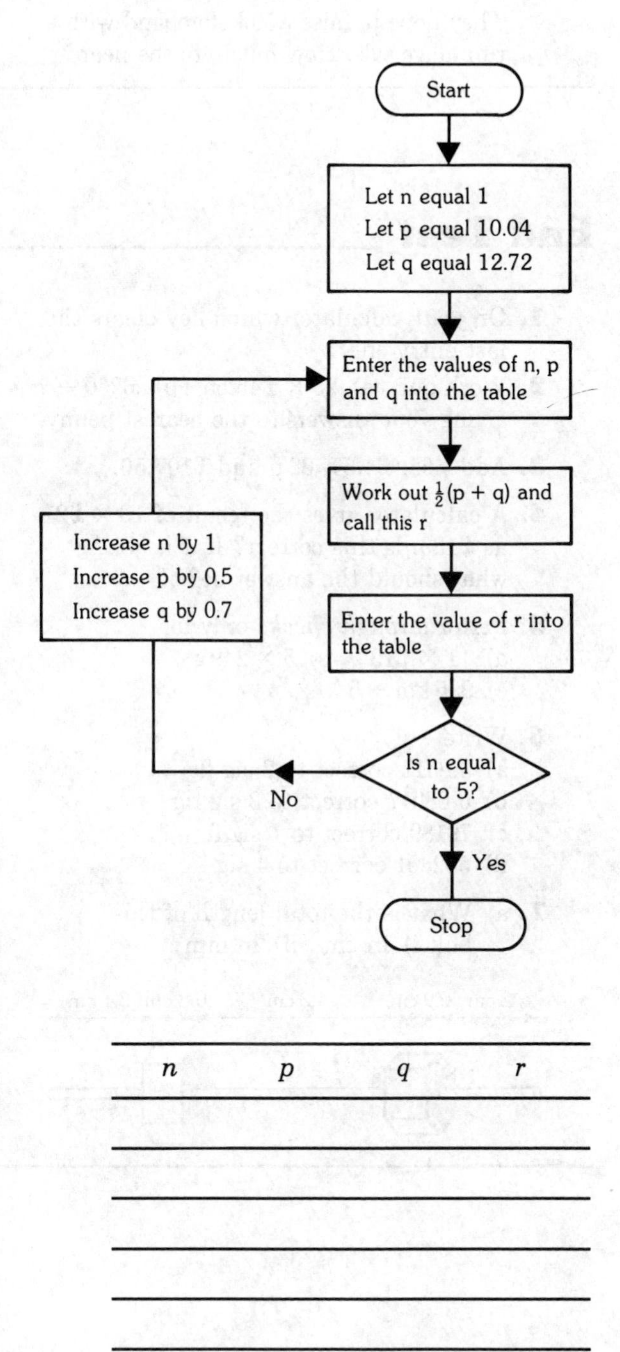


| $n$ | $p$ | $q$ | $r$ |
|---|---|---|---|
| | | | |
| | | | |
| | | | |
| | | | |
| | | | |

**12.** A group of twenty-two people are to share the cost of the entertainment at an old age pensioners' party. They estimate that it will cost £37.50 to provide food, £12.50 for drinks, and £35 for the music.

They hope to raise £50 beforehand with a rummage sale. How much, to the nearest ten pence, should they each contribute?

If they were not prepared to contribute any money how much would they have to charge each pensioner assuming that there were fifty people at the party?

How much are they estimating that the food and drink will cost per person?

# End Test

**1.** On *your* calculator which key clears the last entry only?

**2.** Work out a) $23 \times £45.55$ b) $£250 \div 7$ giving your answers to the nearest penny.

**3.** Add £65, £4.57, 32 p and £107.50.

**4.** A calculator gives the result of $19 \times £21$ as £189. Is this correct? If it is wrong what should the answer be?

**5.** Perform *rough checks* only for
a) $11 \times £19$ b) $5 \times 249$ g
c) $995$ km $\div 5$

**6.** Write
a) 3.8512 correct to 2 sig fig
b) 0.08047 correct to 3 sig fig
c) 79189 correct to 1 sig fig
d) 562301 correct to 4 sig fig

**7.** a) What is the total length of this bolt i) in cm ii) in mm?

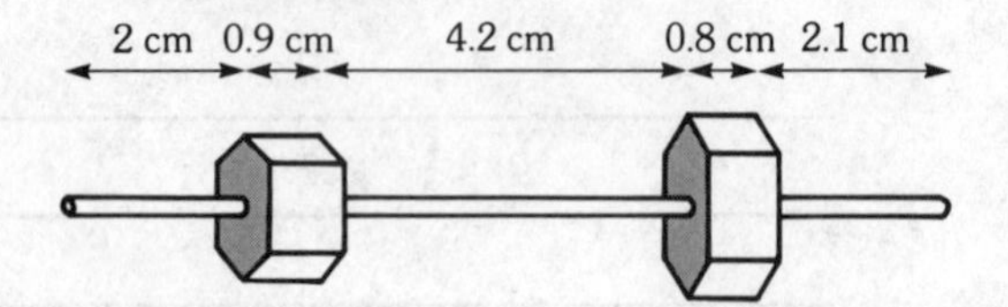


b) Find the weight in kg of 109 such bolts each weighing 14 g.

**8.** Work out $3.14 \times 9 \times 9$ correct to
a) the nearest hundred
b) the nearest fifty
c) the nearest whole number
d) 1 decimal place.

**9.** What would be on the display after pressing [2] [.] [5] [6] [EXP] [3] ?

**10.** Write in standard form
a) 67 000
b) 800 000 000.
c) The distances of the planets Mercury and Neptune from the sun are about $6 \times 10^7$ km and $4.5 \times 10^9$ km. Using a calculator work out how many times farther from the sun Neptune is than Mercury. Write down which keys you would press and the result in the display.

Chapter 6

# Directed Numbers

## Directed Numbers in Practical Situations

Directed numbers are positive and negative numbers, for example 7, −3, +0.2 and so on.

*Positive* numbers have a value greater than zero. *Negative* numbers have a value less than zero. If there is no sign the number is assumed to be positive.

You will have seen negative numbers on graphs.

**Example**

*Temperature Graph for Christmas Week*

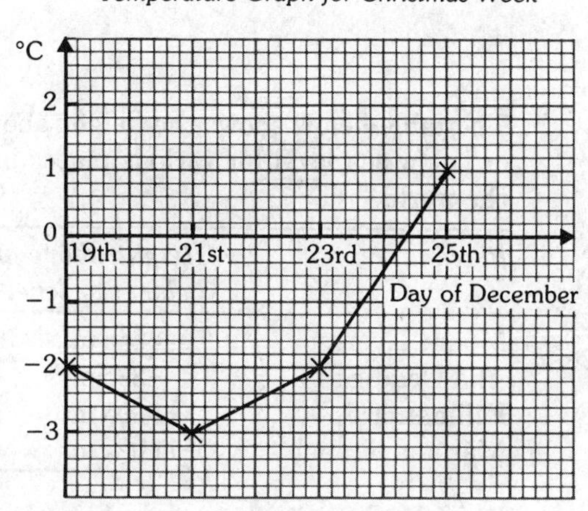


The graph above shows that the temperature rose by 4 °C between 21st and 25th December. The difference between 1 °C and −3 °C is 4 °C.

## EXERCISE 1

**1.** A liquid is cooled from 7 °C to −5 °C in 4 minutes.

*Cooling of a Liquid against Time*

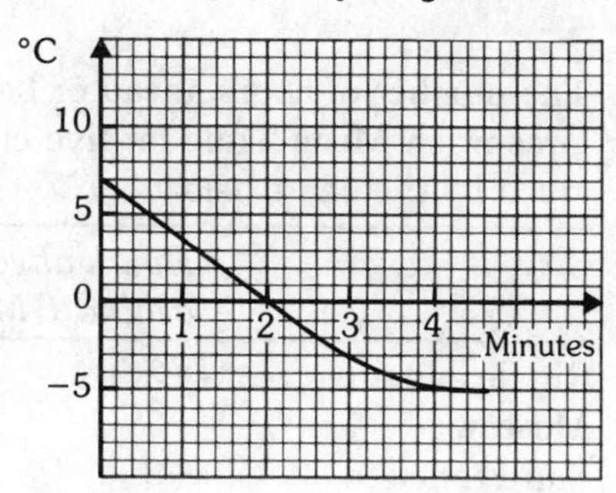


a) By how many degrees has its temperature dropped?
b) How long did it take to cool from 0 °C to −5 °C?
c) What will be its temperature if it cools a further 3 °C after reaching −5 °C?

**2.** Sheila arrived 5 minutes early for her train which was 11 minutes late. How long did she have to wait? Her friend, Janet, arrived 2 minutes after the time at which the train should have departed. How long did Janet have to wait?

**3.** The Grandfather clock is 8 minutes slow and the wall clock is 5 minutes fast. Give the difference in times shown by the two clocks.

**4.** Use the diagram below to answer parts a) and b). (Note that in fact there was no year 0, so your answers will only be approximate!)

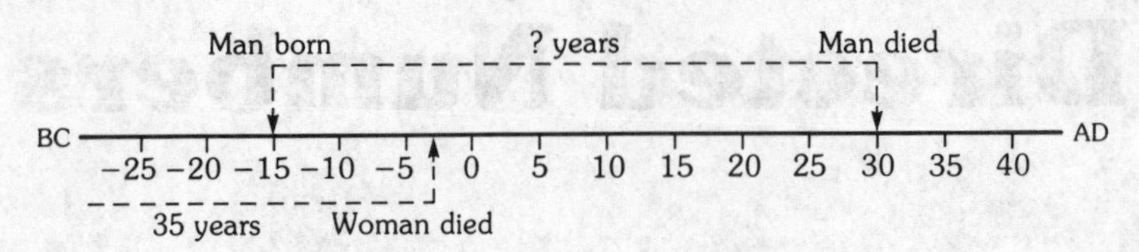


a) How old was a man born in 15 BC who died in AD 30?
b) In which year would a woman who died in 3 BC at an age of 35 have been born?
c) How old were i) Augustus Caesar (63 BC – AD 14) and ii) Mark Anthony (83 BC – 30 BC) when they died?

**5.** The number of hours ahead or behind Greenwich Mean Time for five cities are given by the chart below.

| *City* | *Hours ahead of or behind GMT* |
|---|---|
| Rio de Janeiro | −3 |
| Moscow | +2 |
| San Francisco | −8 |
| Sydney | +10 |
| Mexico City | −7 |

a) At 9 a.m. GMT what is the time in i) Rio de Janiero ii) Sydney?
b) If it is 3 p.m. in Moscow what is the time in Mexico City?
c) How many hours ahead of San Francisco is i) Mexico City ii) Moscow?
d) How many hours behind Sydney is i) Rio de Janeiro ii) San Francisco?
e) On 8th March at 4 p.m. in San Francisco what is the time and date in Sydney?

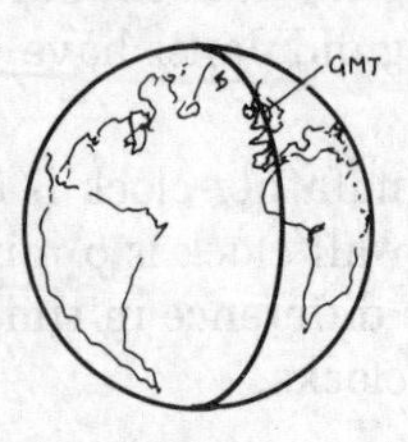


**6.** a) Jeffrey owes the bank £176. He repays £135. How much does he owe now? He pays in a further cheque for £89. By how much will he now be in credit?
b) Copy the bank statement shown below and fill in the right-hand column. Show overdrawn amounts by writing the letters OD after the amount.

**BANK STATEMENT**

| | *Date* | *Payments* | *Receipts* | *Balance* |
|---|---|---|---|---|
| Brought forward | 2 AUG | | | 317.25 |
| | 5 AUG | 320.00 | | 2.25OD |
| | 7 AUG | | 352.25 | 350.00 |
| | 15 AUG | 150.78 | | |
| | 22 AUG | 50.00 | | |
| | 27 AUG | 200.00 | | |
| | 3 SEP | | 348.80 | |

**7.** The table below shows the height above or below sea level for various places in the world.

| *Place* | *Height above or below sea level* |
|---|---|
| Mount Everest | +8848 m |
| The Dead Sea | −395 m |
| Kilimanjaro | +5895 m |
| Mariana Trench | −11 022 m |

Find the difference in height between
a) Kilimanjaro and Mount Everest
b) Mount Everest and the Dead Sea
c) Mount Everest and the Mariana Trench.

**8.** The table below shows the temperature in the first week of January.

| *Temperature in °C* | −5 | −7 | −3 | −1 |
|---|---|---|---|---|
| *Day* | 1st Jan | 2nd Jan | 3rd Jan | 4th Jan |

| *Temperature in °C* | 2 | 3 | 5 |
|---|---|---|---|
| *Day* | 5th Jan | 6th Jan | 7th Jan |

What is the difference in temperature between

a) the coldest and the warmest day

b) 2nd January and 3rd January?

**9.**

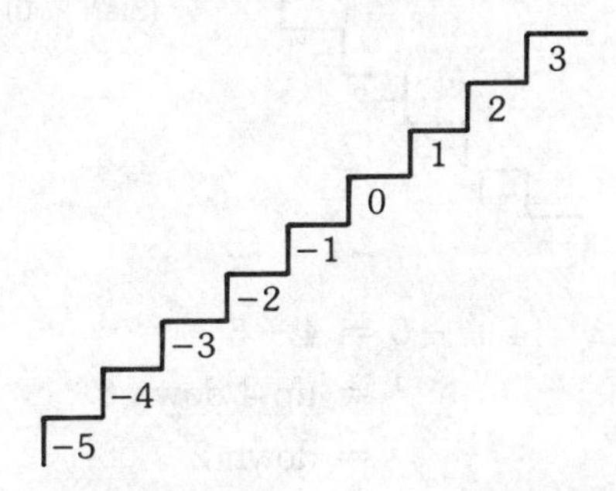


a) Steps on a jetty have numbers painted on them as shown above. If Jo plays a game that entails starting on step 0, going up 1 step and back down 3, then up 1 step and back down 3 again, on which step will Jo now be standing?

b) Taking +2 to mean up 2 steps, and −7 to mean down 7 steps, and so on, which step will Jo be on if she goes +2−7+3 steps starting from 0?

c) Supposing Jo starts on step 0, goes down 2 steps and then down a further 2 steps which number step will she be standing on?

**10.** My watch is 5 minutes slow.

a) If I arrive at the bus station at 9.32 a.m. by my watch what is the actual time?

b) The bus is supposed to leave at 9.40 a.m. but in fact leaves 7 minutes late. How long was I waiting for the bus?

c) The journey is supposed to take 20 minutes but in fact takes 15 minutes. Does it arrive at its destination late or early and by how much?

d) What time does my watch show when I arrive?

# Using a Calculator with Directed Numbers

To make a number negative on a calculator press the +/− key directly *after* entering the number. To enter −3 press keys [3] [+/−].

## Examples

**1.** Calculate $4 + (-2)$.

Press keys [4] [+] [2] [+/−] [=] resulting in (2.) on the display.

**2.** Calculate $-17 + 13 - (-19)$.

Press keys [1] [7] [+/−] [+] [1] [3]

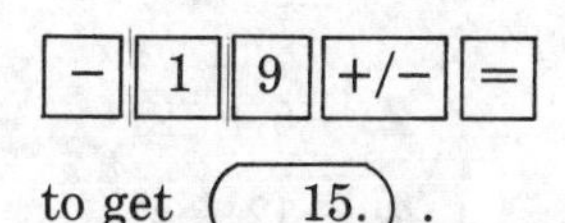


to get (15.).

**3.** Work out $-8 \times -13$.

Press keys [8] [+/−] [×] [1] [3] [+/−] [=] resulting in (104.).

# Order of Operations

Brackets must be done first, then × and ÷ and finally + and −. Otherwise work from left to right. A number outside the bracket means multiply. Sometimes brackets are put round negative numbers for clarity only, and they may be ignored.

## Examples

**1.** Calculate $2(-6 \div 2)$

Press keys [6] [+/−] [÷] [2] [=] [×] [2] [=]

to get (−6.).

If your calculator has brackets you could press

[2] [×] [(] [6] [+/−] [÷] [2] [)] [=].

**2.** Work out $(-5)^2 + 8$.

Press keys [5] [+/−] [$x^2$] [+] [8] [=]

resulting in (33.).

## EXERCISE 2 (ORAL)

Which keys would you press and in which order to work out

**1.** $3 - 11$

**2.** $-5 + 12$

**3.** $-6 \times -8$

**4.** $-12 \div 4$

**5.** $(-9)^2$

**6.** $2(5 - 9)$

**7.** $(-5) \times 2 + 3 \times (-1)$

**8.** $\dfrac{(-12) \times 2}{(-4)}$

**9.** $7(-3) + 2(-4)$

**10.** $\dfrac{5 - (-4)}{3}$ ?

## EXERCISE 3

Repeat exercise 2 using a calculator. This time work out the results.

If no calculator is available you could follow these rules.

**Rule for Addition and Subtraction**
Replace $+-$ and $-+$ by $-$, and replace $--$ and $++$ by $+$, then proceed as if stepping up and down steps.

## Examples

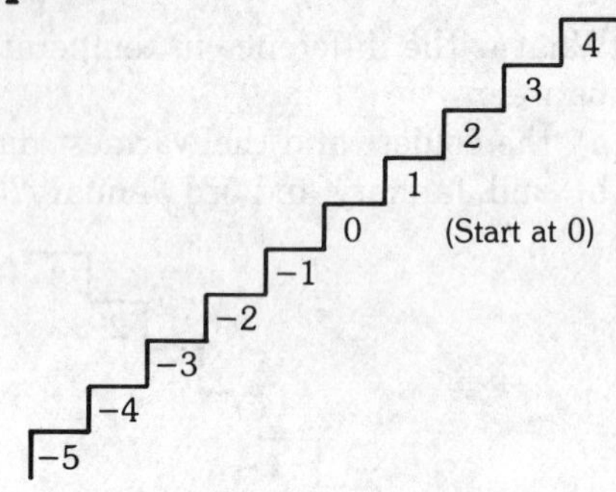


**1.** $4 + -6 = 4 - 6$
$= \text{up 4 down 6}$
$= \text{down 2}$
$= -2$

**2.** $2 - 6 - (-3) = 2 - 6 + 3$
$= \text{up 2 down 6 up 3}$
$= \text{down 1}$
$= -1$

**Rule for Multiplication and Division**
Multiply or divide the numbers as normal. If the signs are the same make the answer positive. If the signs are different make the answer negative.

## Examples

$-4 \times 2 = -8$ $\quad$ $8 \div -2 = -4$

$3 \times -4 = -12$ $\quad$ $-9 \div 3 = -3$

$-2 \times -3 = 6$ $\quad$ $-6 \div -2 = 3$

## EXERCISE 4

**1.** $-2 + -3$

**2.** $12 + -4$

**3.** $-7 + 5$

**4.** $-11 + -9$

**5.** $15 - -5$

**6.** $-8 - 3$

**7.** $(-9) - (-9)$

**8.** $(-8) - (+5)$

**9.** $-7 + 2 + 5$

**10.** $1 - 8 - (-6)$

**11.** $(-7) \times (-5)$

**12.** $8(-3)$

**13.** $(-2) \times 7$

**14.** $(-3)^2$

**15.** $8 \div -4$

**16.** $-35 \div 7$

**17.** $(-4) \div (-2)$

**18.** $\frac{(-15) \div 5}{3}$

**19.** $2(-3 + 2)$

**20.** $2(-3) + 3(-1)$

## End Test

**1.** A metal plate at a temperature of 5 °C is cooled by 12 °C. What is its temperature now?

**2.** A man was born in 28 BC. How old was he when he died in AD 45?

**3.** The time in Cairo is 2 hours ahead of Greenwich Mean Time and the time in Buenos Aires is 5 hours behind Greenwich Mean Time. If it is 4 p.m. in Buenos Aires what time is it in Cairo?

**4.** The average height of land is 840 m above sea level. The average depth of the ocean floor below sea level is 3468 m below the average height of land. What is the average depth of the ocean floor below sea level?

**5.** Which of these are equal to zero?

a) $-8 + 8$ b) $-8 - (-8)$
c) $8 \times (-8)$ d) $(-8) \div 0$
e) $0 \times (-8)$ f) $(-8) \div 8$
g) $0 \div (-8)$ h) $-8 - (+8)$

**6.** Work out

a) $-4 + -7$ b) $4 - (-7)$
c) $(-12) \times 4$ d) $-12 \div -4$
e) $5(-4 + 2)$ f) $3(-1) + 2(-2)$.

**7.**

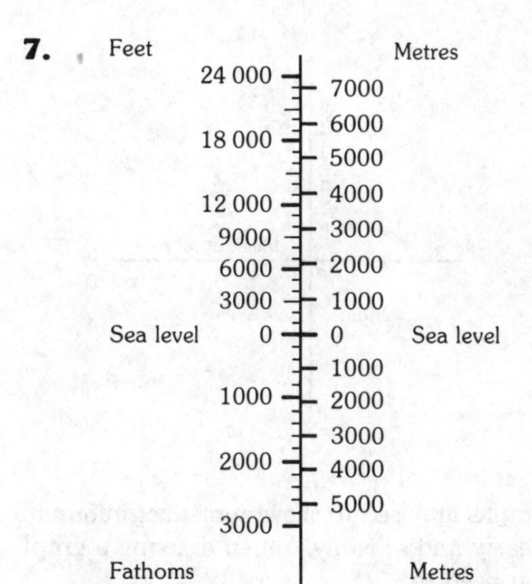


Convert 3000 metres below sea level to fathoms using the above scale. How many feet above sea level is 7000 metres above sea level?

What is the difference in height between 1000 fathoms below sea level and 3000 feet above sea level to the nearest 1000 metres?

**8.** For Montreal the average temperature in January (the coldest month) is −10 °C while the average temperature in July (the warmest month) is 31 °C above this. State the average temperature in July in Montreal.

Chapter 7

# Graphs and Charts

## Axes and Scales

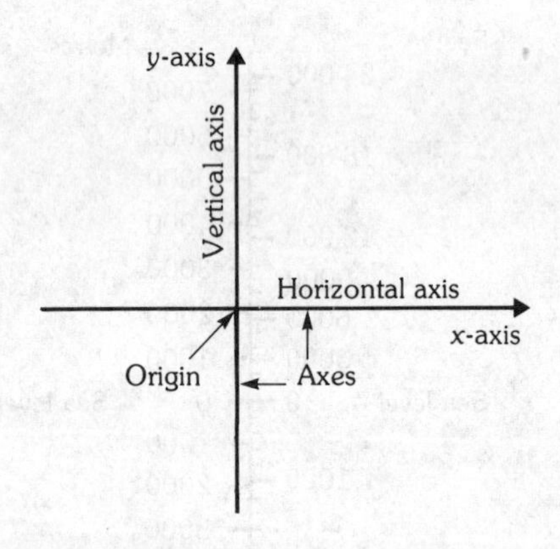


Graphs are used to show numerical information clearly and sensibly. When drawing a graph you need to

(a) choose a reasonable scale

(b) decide how to label the axes

(c) plot the points carefully

(d) decide whether the points should be joined with a straight or curved line

(e) join the points using a sharp pencil.

The scale on the graph is the spacing between the labels on the axes. Number labels on the axes should be spaced evenly. Varying the scale alters the shape of the graph. Try to fill the page without making the graph either too large or too small. Don't forget to put a heading.

### Example

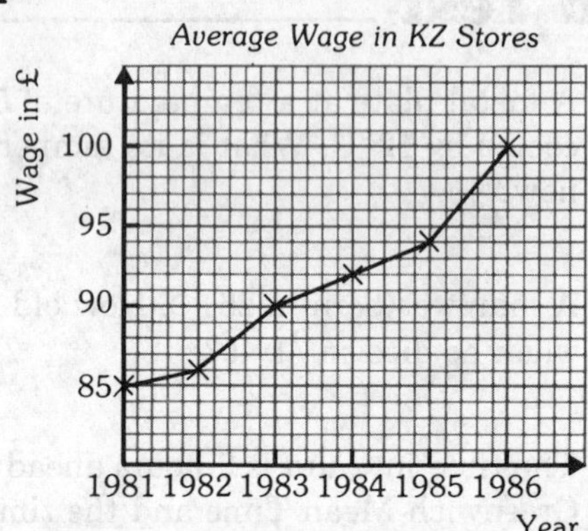


When drawing a graph, labelling the axes is essential. The units used should be clearly shown. The heading needs to be short but descriptive. The scale used on the graph above is 1 cm per year on the horizontal axis and 2 cm per £10 on the vertical axis.

## Coordinates

Coordinates show where things are on a graph or map.

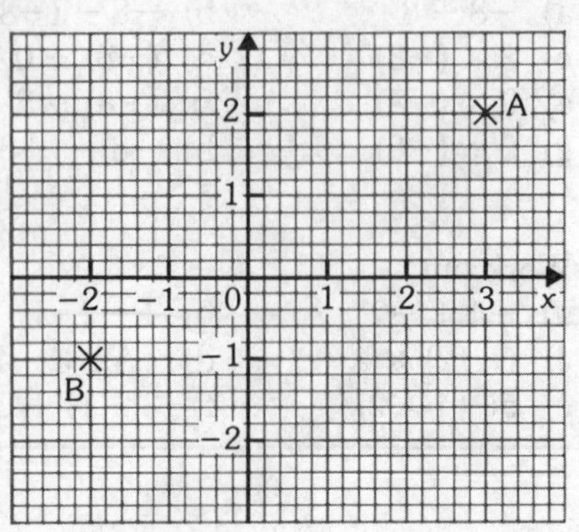

The coordinates of the point A are (3,2). This means A is 3 units in the $x$ direction and 2 units in the $y$ direction from 0. The coordinates of B are $(-2, -1)$.

Negative numbers are to the left and below the origin, O.

Look at the map below.

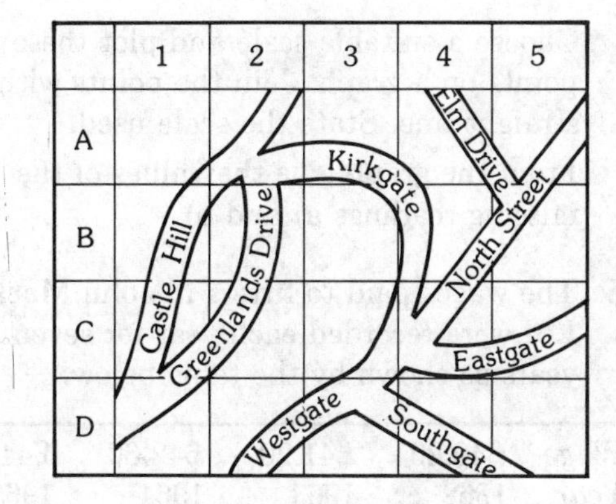


Southgate is in square D4. Each square of the map can be labelled by using a letter and a number.

# Plotting Points

Graphs show a relationship between two sets of numbers.

### Example

A car was tested for petrol consumption at different speeds.

| *Speed* in m.p.h. | 15 | 25 | 40 | 55 | 70 |
|---|---|---|---|---|---|
| *Consumption* in m.p.g. | 27 | 32 | 36 | 38 | 35 |

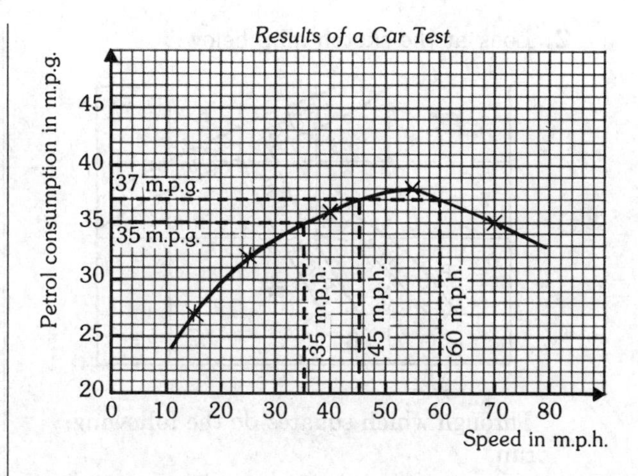


The scale chosen was 1 cm for 10 m.p.h. and 2 cm for 10 m.p.g. The speed in m.p.h. has been marked along the horizontal axis and the consumption in m.p.g. on the vertical axis.

Further information may be obtained from the graph. For example, the petrol consumption at 35 m.p.h. as shown by the dotted line is 35 m.p.g.

The speed when the petrol consumption is above 37 m.p.g. is between 45 and 60 m.p.h.

## EXERCISE 1

**1.** Write down the coordinates of A, B, C, D and E on the graph below.

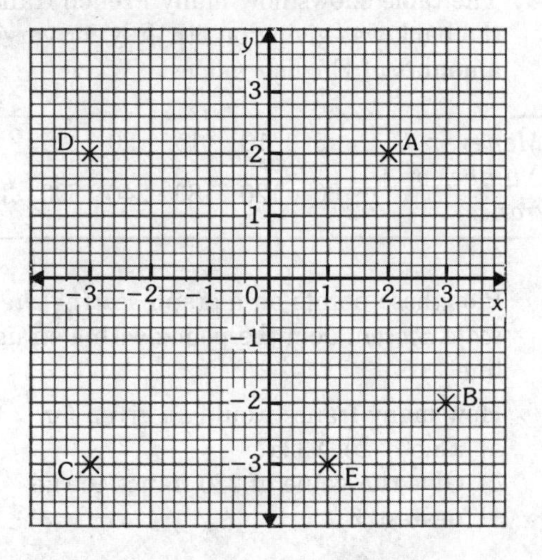

**2.** Look at the sketch map below.

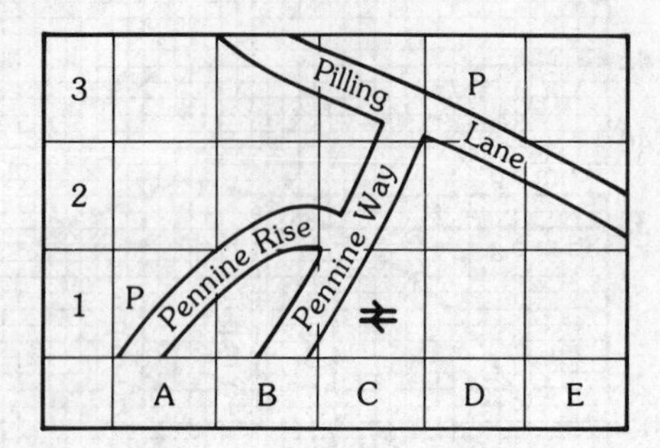


Through which squares do the following run

a) Pilling Lane
b) Pennine Rise
c) Pennine Way?
d) List the squares with
   i) car parks (shown by P)
   ii) a station (shown by ⇌ ).

**3.** Using a scale of 2 cm per unit on both the $x$- and $y$-axes draw a graph and plot the following points.

$(-1,8)$, $(0,3)$, $(1,0)$, $(2,-1)$, $(3,0)$, $(4,3)$

Using a *sharp* pencil join the points with a smooth curve.

What would be the value of $y$ when $x$ has a value of a) 0.5 b) 5?

**4.** The table shows how many French francs the bank was giving on 5th July for varying amounts of English money.

| *Money in £* | 5 | 10 | 15 | 20 | 25 | 30 |
|---|---|---|---|---|---|---|
| *Number of francs* | 60 | 120 | 180 | 240 | 300 | 360 |

Plot these points on a graph *stating the scale chosen.* Join the points with a straight line.

How many francs would be given for
a) £8.50 b) £325?
c) Give the value of 280 francs to the nearest £.

**5.** Various weights were attached to a spring and its length measured. Two readings were missed out. They have been labelled a) and b) in the table below.

| *Weight in lb* | 1 lb | 1 lb 6 oz | 1 lb 9 oz |
|---|---|---|---|
| *Length in cm* | a)...... | 10.2 cm | 10.4 cm |

| *Weight in lb* | b)...... | 2 lb 3 oz |
|---|---|---|
| *Length in cm* | 10.6 cm | 11.0 cm |

Choose a suitable scale and plot these points on a graph. Join the points with a straight line. State the scale used.

From the graph give the values of the missing readings a) and b).

**6.** The wages paid to fitters in John Mackie Ltd were recorded each year for seven years as shown by the table below.

| *Wage* | £40.20 | £41.20 | £42.60 | £44.20 |
|---|---|---|---|---|
| *Year* | 1962 | 1963 | 1964 | 1965 |

| *Wage* | £46 | £48.20 | £50.80 |
|---|---|---|---|
| *Year* | 1966 | 1967 | 1968 |

Use the longest side of the paper for the horizontal axis (year) and the shortest side for the vertical axis (wage).

Plot the above information on a graph, joining the points with as smooth a curve as possible. Use a scale of 2 cm per year on the horizontal axis and 1 cm per £ on the vertical axis.

Draw a second graph of the same information using a scale of 4 cm per year on the horizontal axis and 1 cm per £ on the vertical axis. (It is possible to do this on the same graph paper if you use a coloured pencil.)

Which graph would a manager use to present the information in the best light? Why?

# Bar Charts, Pie Charts and Pictograms

### Bar charts

Some hints for drawing bar charts are given below.

(a) Use a sharp pencil.
(b) Choose a sensible scale.
(c) Draw and label the axes in pencil.
(d) Make sure the bars are of equal width.
(e) Draw the bars in pencil.
(f) Write a heading in ink.
(g) Label the axes in ink.

Use a pencil first so you can correct any mistakes easily. Only when you are *sure* that you have chosen the best scale and that the bar chart shows what you want it to should you label the axes in ink.

## Example

The exam results of 55 students are as follows.

| *Mark* | 0–20 | 21–40 | 41–60 | 61–80 | 81–100 |
|---|---|---|---|---|---|
| *Number of students* | 5 | 10 | 15 | 20 | 5 |

These results are shown on the bar chart below.

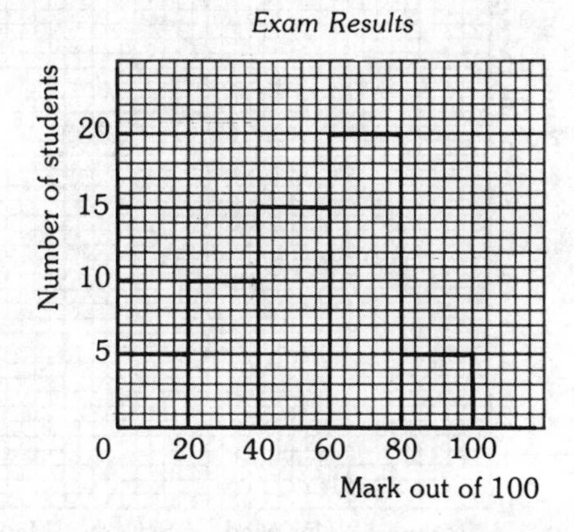


### Pie charts

Some hints for drawing pie charts are given below.

(a) Use a sharp pencil.
(b) Draw a big enough circle.
(c) Calculate the angles to the nearest degree and show your working.
(d) Check the angles add up to 360°, allowing for rounding errors.
(e) Label the sectors of the pie chart in ink.

## Example

Out of a sample of 200 people visiting an exhibition 110 were adults, 50 were girls and 40 were boys. Show this on a pie chart.

total number of people = 200

in the pie chart 360° represents 200 people

$\frac{110}{200} \times 360° = 198°$ represents adults

$\frac{50}{200} \times 360° = 90°$ represents girls

$\frac{40}{200} \times 360° = 72°$ represents boys

total = 360°

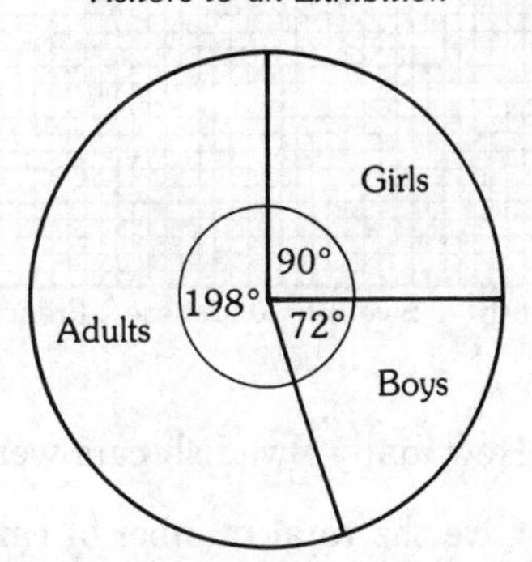

**Pictograms**

Some hints for drawing pictograms are given below.

(a) Choose a good symbol.

(b) Write on the pictogram what the symbol represents.

## Example

The pictogram below shows the sales of cars in February to April one year.

*Sales of cars Feb–Apr*

(car symbol) = 10 000 British Cars

(car symbol) = 10 000 Foreign Cars

## EXERCISE 2

**1.** The results of a survey on the country of origin of cars in a car park are shown in the figures which are given in the table.

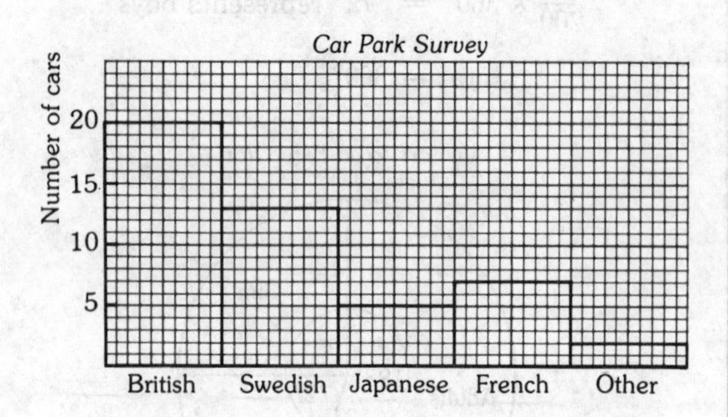


a) How many Swedish cars were there?

b) Give the total number of cars in the car park.

**2.** A dealer has 60 used cars with varying mileages. Draw a bar chart to illustrate the figures which are given below.

| *Number of cars* | 7 | 12 |
|---|---|---|
| *Mileage* | 0–5999 | 6000–11 999 |
| *Number of cars* | 15 | 9 |
| *Mileage* | 12 000–17 999 | 18 000–23 999 |
| *Number of cars* | 13 | 1 |
| *Mileage* | 24 000–29 999 | 30 000–35 999 |
| *Number of cars* | 3 | |
| *Mileage* | 36 000–41 999 | |

What fraction of the cars has a mileage of over 30 000?

**3.** The bar chart shows how many trainees, unskilled, skilled and managerial staff there are at N. Marsden Ltd.

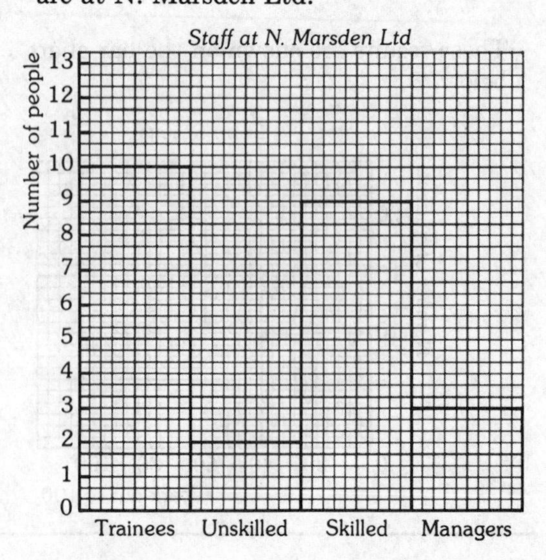

a) How many skilled workers are there at N. Marsden Ltd?
b) Give the number of trainees and unskilled workers as a fraction of the total number of employees.
c) What percentage of staff are managers?

Two of the skilled workers are promoted and become managers. Five trainees leave and four new unskilled workers are recruited. Draw a second bar chart to show the new structure.

**4.** There are 720 people on board a ship. Of these 180 are children, 240 are men, 20 are babies and the rest are women. Show this information on a pie chart, labelling the sectors and angles clearly.

**5.** Look at the pie chart below showing use of land.

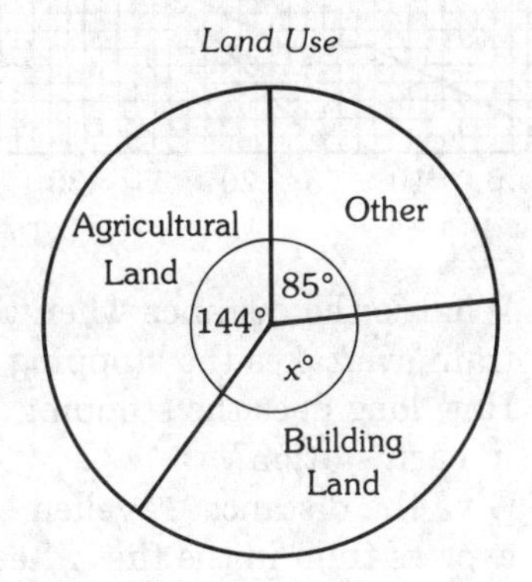


a) What is the value of the angle $x°$?
b) What fraction is agricultural land?

**6.** Designers planning a leaflet advising young people how to budget their spending decide to include a pie chart. Using the information in the table below draw a suitable pie chart to show what they advise.

| | |
|---|---|
| Food | £36 |
| Entertainment and other expenses | £27 |
| Rent and rates | £18 |
| Gas and electricity | £18 |
| Clothes | £9 |
| Total weekly wage | £108 |

a) What fraction of the budget is spent on food?
b) Express this as a percentage.

**7.** The pictogram shows the sales of washing machines for the years 1985 and 1986.

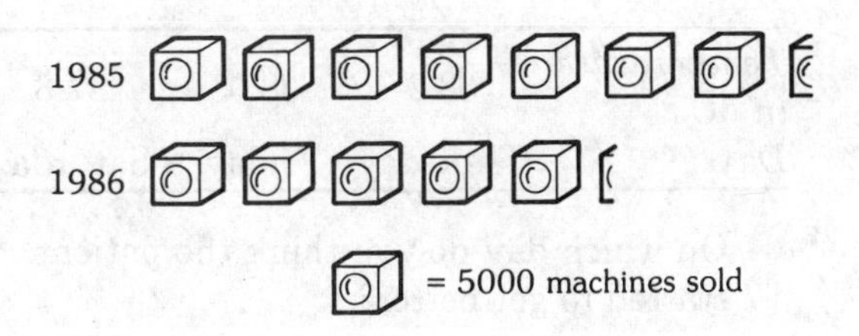


a) Give the number of washing machines to the nearest 10 000 sold in 1985.
b) What is the decrease in sales from 1985 to 1986 to the nearest 5000?
c) Draw a pictogram showing sales of 32 000 electric cookers in 1980 and a 25% increase in sales in 1981, using a suitable symbol.

# End Test

**1.** The temperature chart for a patient in hospital is shown below.

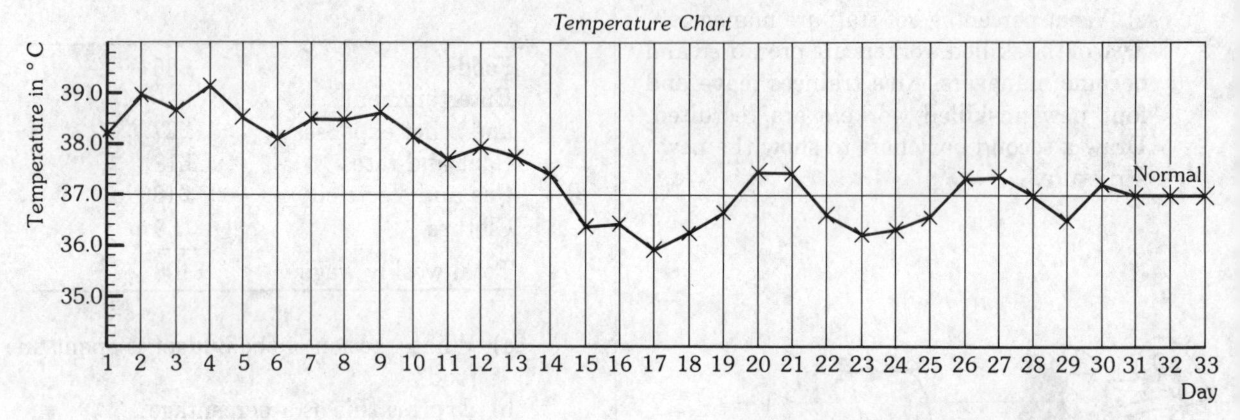


a) What is the normal temperature in °C?

b) For how many days did the patient have an abnormal temperature?

c) For how many days was the temperature above normal?

d) Draw a similar chart for the readings given below.

| *Temperature in °C* | 39.2 | 39.4 | 39.6 |
|---|---|---|---|
| *Day* | Monday | Tuesday | Wednesday |

| *Temperature in °C* | 38.2 | 38.0 | 37.6 |
|---|---|---|---|
| *Day* | Thursday | Friday | Saturday |

On which day do you think the patient started to get better?

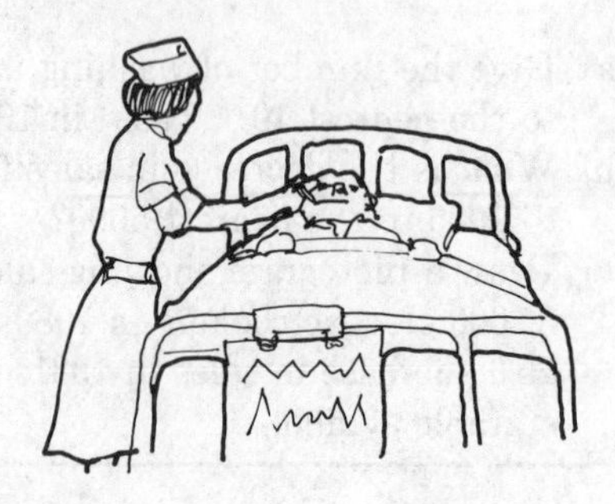

**2.** The graph below shows the journeys of an express train and a local stopping train.

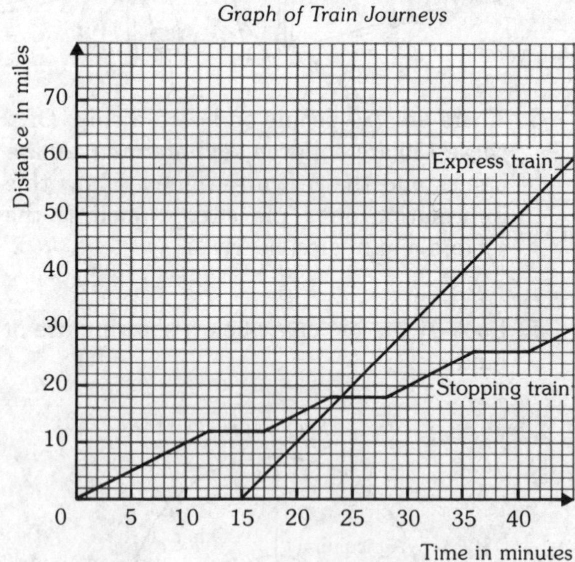


a) What is the distance when the express train overtakes the stopping train?

b) How long does the stopping train wait in each station?

c) Give the distance travelled by the express train in the first fifteen minutes of its journey.

d) What is the average speed of the express train?

**3.** An electricity bill is made up of a fixed charge plus the cost of the units used, as shown by the graph below. The cost if 400 units are used is £21.80, and if 600 units are used is £29.80. Draw an accurate graph to show the fixed charge and the cost per unit.

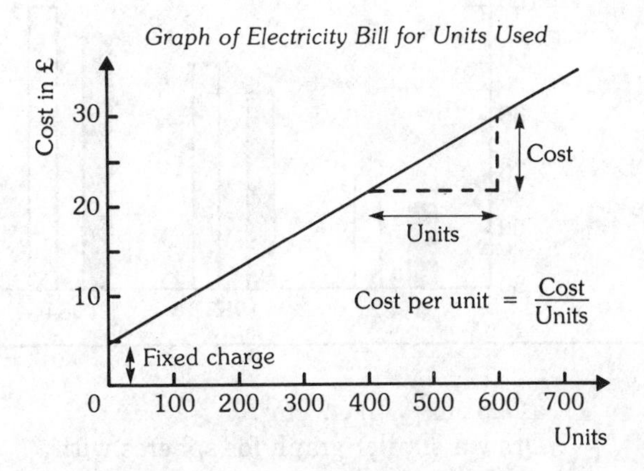


**4.** 200 light bulbs were tested in a factory. The results are shown in the table below.

| | | | |
|---|---|---|---|
| *Number of light bulbs* | 2 | 4 | 22 |
| *Average life in hours* | 0–199 | 200–399 | 400–599 |
| *Number of light bulbs* | 60 | 48 | 32 |
| *Average life in hours* | 600–799 | 800–999 | 1000–1199 |
| *Number of light bulbs* | 26 | 6 | |
| *Average life in hours* | 1200–1399 | 1400–1599 | |

Choose a suitable scale and draw a bar chart to show this information.

Using the data above complete the following table.

| | | | |
|---|---|---|---|
| *Number of light bulbs* | 2 | 6 | 28 |
| *Life span in hours less than* | 200 | 400 | 600 |
| *Number of light bulbs* | 88 | 136 | ? |
| *Life span in hours less than* | 800 | 1000 | 1200 |
| *Number of light bulbs* | 194 | ? | |
| *Life span in hours less than* | 1400 | 1600 | |

Plot this information on a second graph. Using your graph find approximately

a) how many light bulbs had a life span of 700 hours or less

b) the median life span (i.e. the life span for which 100 bulbs lasted a longer and 100 bulbs lasted a shorter time).

**5.** The pie chart shows the proportion of people using electricity, solid fuel, gas and oil for their central heating in a certain town. If there are 18 000 people with central heating in the town and 2000 of them use oil, what is the angle in the sector representing oil?

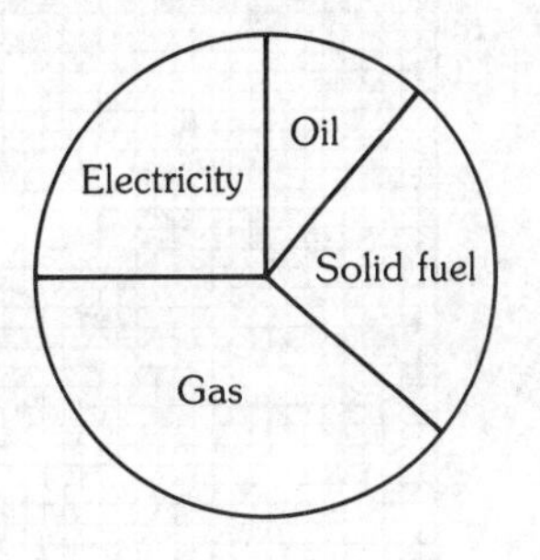


If the angles at the centre are 90° for both electricity and solid fuel how many people use gas?

**6.** The vertical bar charts below show the sales of two types of gas cooker in the years 1982–1984.

*Cooker Sales 1982-4*

a) What is wrong with the scale on the left-hand graph?

b) Explain why the sales look to have increased much more rapidly in the second graph.

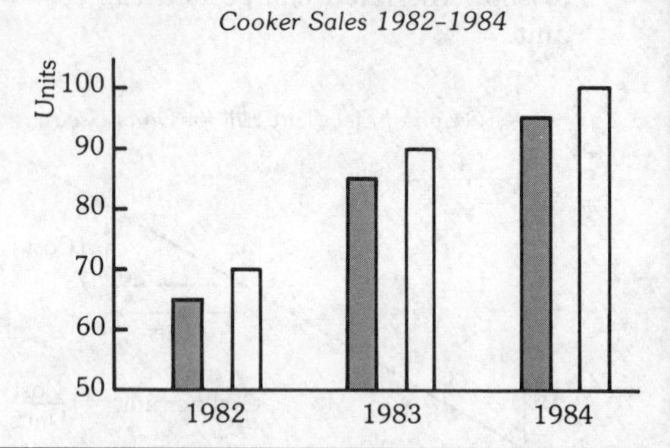


**7.** The graph shows the relationship between the radius and the surface area of a sphere for values of the radius from 1 cm to 1.9 cm.

From the graph estimate to the nearest cm the radius of spheres with surface areas

a) 25 $cm^2$ b) 42 $cm^2$.

Draw a similar graph for spheres with radii from 2 cm to 2.9 cm using the formula

surface area $= 4 \times \pi \times$ radius $\times$ radius

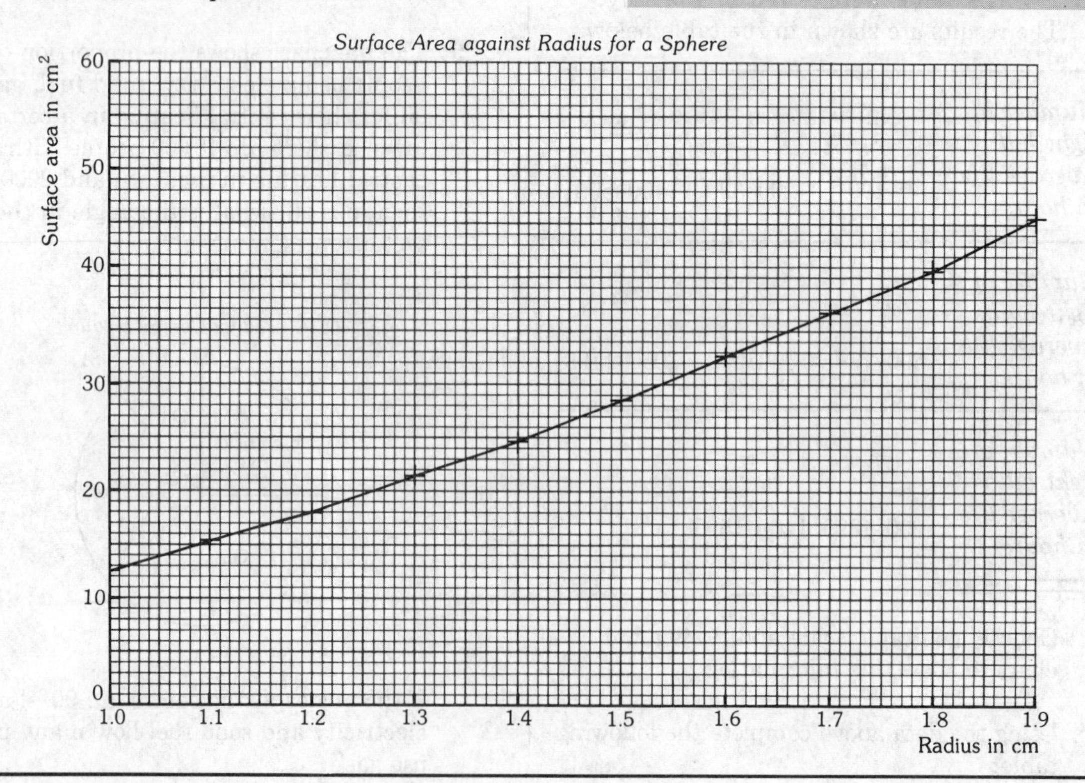

# ANSWERS

# Answers to Part A

## CHAPTER 1

*page 3*

**1.1** 1. a) 8 oz b) 4 oz c) 24 oz d) 22 oz
2. a) $\frac{1}{2}$ pt b) $\frac{1}{4}$ pt c) $\frac{3}{4}$ pt d) $1\frac{1}{2}$ pt
3. a) 112 g b) 168 g c) 224 g d) 675 g e) 336 g
4. a) 7 oz b) 18 oz c) 36 oz d) 54 oz e) 4 oz
5. a) 1200 ml b) 300 ml c) 150 ml d) 120 ml e) 210 ml
6. a) 35 fl oz b) 17 fl oz c) 3 fl oz d) 5 fl oz
7. a) 5 b) 3 c) 9 d) 5 e) 3

**1.2** 1.

| °*F* | °*C* |
|---|---|
| 325 | 163 |
| 350 | 177 |
| 375 | 191 |
| 400 | 204 |
| 425 | 218 |

Apart from the first one the values given in the table in the text are correct to the nearest 10 °C.

2. a) 121 °C b) 149 °C c) 246 °C d) 27 °C e) 16 °C
3. a) 212 °F b) 68 °F c) 79 °F d) 32 °F e) 50 °F

**1.3** 1. a) 4 oz b) 4 oz c) 7 oz
2. a) 10 g b) 60 g c) 70 g d) 10 g
3. a) 150 g b) 150 g c) 225 g d) 375 g
4. a) 160 ml b) 200 ml c) 560 ml d) 360 ml

**1.4** 1.

| *P* | *Q* | *R* | *S* |
|---|---|---|---|
| 120 | 300 | 560 | 700 |

| *T* |
|---|
| 1060 (or 1.06 kg) |

2.

| *X* | *Y* |
|---|---|
| 2 | 11 |

| *Z* | *W* |
|---|---|
| 18 | 27 |
| (or 1 lb 2 oz) | (or 1 lb 11 oz) |

3.

| *A* | *B* | *C* |
|---|---|---|
| 180 | 230 | 85 |

4. a)

| *X* | *Y* | *Z* |
|---|---|---|
| 0.45 | 0.33 | 0.15 |

b)

| *R* | *S* | *T* |
|---|---|---|
| 9 | 17 | 3 |

**2.1** 1. a) £2.00 b) 75 p c) 33 p d) £1.25 e) 67 p
2. a) 11 p b) 27 p c) 16 p d) 65 p
3. a) 17 p b) 51 p c) 13 p d) 19 p
4. a) £5.50 b) 55 p c) £1.65 d) 83 p
5. a) 92 p b) 17 p c) 17 p d) 9 p

**2.2** 1. a) 15 oz b) 25 oz c) 18 oz d) 45 oz (or 2 lb 13 oz)
2. a) 3 lb b) $\frac{3}{4}$ lb c) $3\frac{3}{4}$ lb d) $6\frac{3}{4}$ lb
3. a) $\frac{3}{4}$ lb b) $2\frac{1}{4}$ lb c) 14 oz d) $1\frac{1}{2}$ lb
4. $5\frac{1}{4}$ oz flour (probably use 5 oz)
$\frac{3}{4}$ oz cocoa
3 eggs
6 oz sugar
6 oz margarine

**2.3** 1. a) $2\frac{1}{2}$ fl oz b) 125 ml c) 62.5 (or 65) ml d) 25 ml
2. a) $1\frac{1}{3}$ oz (in practice $1\frac{1}{2}$ oz)
b) 67 (or 70) g
c) $2\frac{2}{3}$ oz (in practice 3 oz)
d) 333 (or 325 g)
3. a) 5 pt b) 2.5 l
4. a) 4 lb grapefruits, 2 lb lemons
b) 2.7 kg grapefruits, 1.3 kg lemons

**Feeding a conference**

*page 9*

1. $7\frac{1}{2}$ lb beef
$3\frac{3}{4}$ lb onions
$52\frac{1}{2}$ oz tomatoes
2. $7\frac{1}{2}$ lb beef
4 lb onions
3 large and 1 small tins of tomatoes
3. 10 lb potatoes
8 lb carrots
8 lb beans
4. 1 lb rice (or 500 g)
1 kg sugar
8 pt milk
8 lb rhubarb
5. 8 slabs

6. 5 pt milk
   20 tea bags (smallest packet)
   200 g coffee
   another 1 kg sugar

# CHAPTER 2

*page 11*

**1.1** 1. a) 2 miles b) 16 miles
c) 40 miles d) 10 miles
2. a) 1 cm
b) i) 20 ii) 50
iii) 80 iv) 200
3. a) 2 km b) 6 km
c) 5 km d) 12.5 km

**2.1** a) 54° b) 82°
c) 124° d) 142°
e) 301° f) 213°

**2.2** 1.

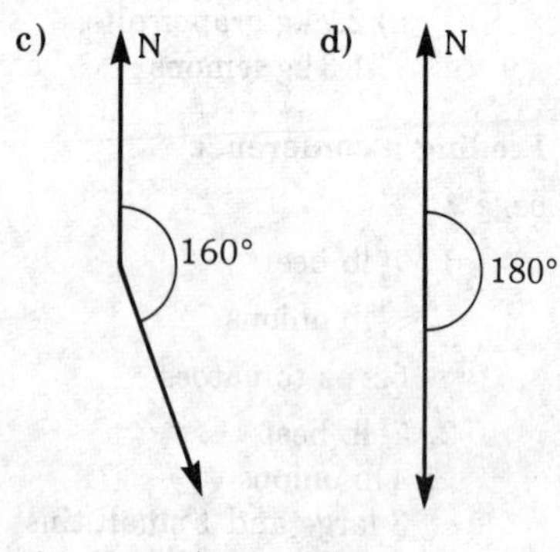


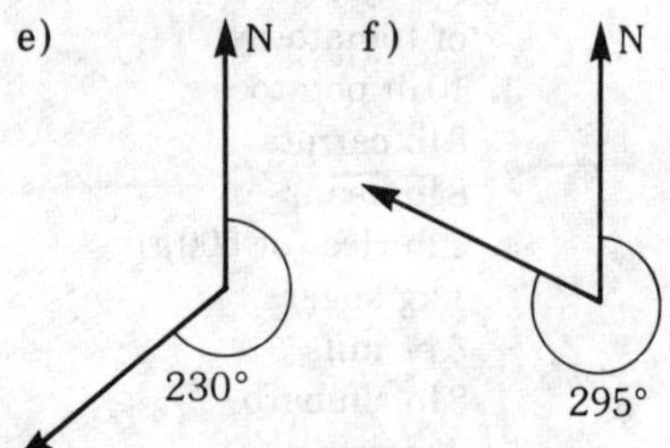


2. a) 053° b) 260°
c) 316° d) 104°

**3.1** 1. a) Barley Mow Fm
b) where A422 crosses old railway
c) Fasmore Ho
d) telephone symbol
2. a) 534337 b) 548395
c) 588406 d) 574388

**3.2** 2. a) E2 b) D3
c) D4 d) E2

**4.1** 1. a) small hill with gentle slope on all sides
b) hill with steep slope on one side, gentle slope on other sides
c) narrow valley with steep hills on both sides
2.

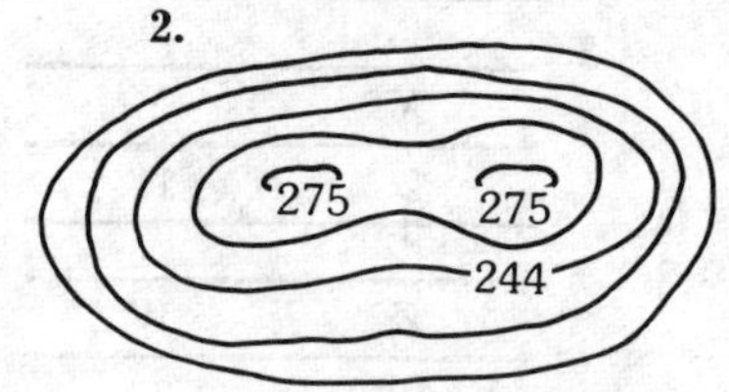


**4.2** 1. a) 1 in 16 b) 1 in 8
c) 1 in 26
2. a) 1 in 20 b) 1 in 10
c) 1 in 30
3. a) 6% b) $12\frac{1}{2}$%
c) 4% d) 5%
e) 10% f) 3%
The higher the percentage the steeper the road.

# CHAPTER 3

*page 20*

**1.1** 1. Thursday
2. Wednesday
3. Tuesday
4. Saturday
5. Monday (but only if posted before mid-day)
6. Monday
7. Tuesday
8. Friday

**1.2** 1. £1.16 2. £1.30
3. £1.80 4. £1.56
5. £6.48 6. £5.10

**1.3** 1. too wide
3. too long
4. acceptable if first class, only acceptable for second class if under 750 g
5. too long

**1.4** 1. 32 p 2. 20 p
3. 30 p 4. 26 p
5. 40 p 6. 40 p
7. 24 p 8. 26 p
9. 18 p 10. 30 p

**2.1** 1. 13
2. a) £6.60 b) £1.65
3. £275 4. 8
5. £201
6. a) £529.55 b) £104.55
7. £167.50

**2.2** 1. taken out
2. demand
3. £94.18
4. £72.42
5. amount in account
6. £87.13

**2.3** 1. £1.50
2. £1060
3. a) £400 b) £800
4. £3630

**3.1** 1. £42.60 2. £56.80
3. £142.00 4. £369.20
5. £170.40
6. a) £85.20 b) £35.50
c) £127.80
7. a) £150.36 b) £1954.68
8. a) £152.44 b) £1981.72

**Premium Bonds versus Savings Accounts**

*page 27*

1. £890.60
2. yes, 8% of £200 is only £16
3. £16.87
4. He gains over the two years.
5. a) £180 interest, £33 tax, net gain £147, £3147 at end of year

b) £240 interest, £72 tax, net gain £168, £3168 at end of year; scheme (b) is most advantageous

# CHAPTER 4

*page 29*

**1.1** 1. a) 160 cm b) 2250 cm c) 80 cm d) 207 cm
2. a) 2.1 m b) 0.85 m c) 35 m d) 68.2 m
3. a) 2 m b) 3.72 m c) 4.8 m d) 1.32 m
4. a) 135 cm b) 3.87 m c) 1.7 m d) 63 cm

**1.2** 1. a) 48 in b) 32 in c) 66 in d) 30 in e) 72 in f) 162 in
2. a) 1 ft 4 in b) 3 ft 8 in c) 2 ft 6 in d) 5 ft
3. a) 1 ft 9 in b) 5 ft 8 in c) 7 ft d) 6 ft 4 in
4. a) 1 ft 4 in b) 1 ft 8 in c) 5 ft 4 in d) 3 ft 5 in

**1.3** 1. a) i) 300 cm ii) 3 m
b) i) 375 cm ii) 3.75 m
c) i) 195 cm ii) 1.95 m
d) i) 160 cm ii) 1.6 m
e) i) 295 cm ii) 2.95 m
f) i) 247.5 m ii) 2.475 m
2. a) i) 80 in ii) 6 ft 8 in
b) i) 140 in ii) 11 ft 8 in
c) i) 32 in ii) 2 ft 8 in
d) i) 168 in ii) 14 ft
e) i) 66 in ii) 5 ft 6 in
f) i) 97 in ii) 8 ft 1 in

# CHAPTER 5

*page 36*

**1.2** 1. 64 $m^2$ 2. 40 $in^2$ 3. $38\frac{1}{4}$ $ft^2$ 4. 40 $cm^2$ 5. 14 $ft^2$ 6. 154 $cm^2$ 7. 350 $in^2$ 8. 32 $ft^2$

**1.3** 1. 32 m 2. 26 in 3. 26 ft 6. 44 cm 7. 66 in 8. 23 ft

**1.4** 1. i) 5 cm ii) 13 cm iii) 8 cm iv) 10.9 cm v) 11.7 cm
2. a) 10 cm b) 25.5 cm
3. a) 24 $cm^2$ b) 128 $cm^2$
4. 11.7 m

**1.5** 1. area 86 $ft^2$, perimeter 40 ft
2. area 11.5 $m^2$, perimeter 15 m
3. area 13.7 $m^2$, perimeter 14.9 m
4. area 17.8 $m^2$, perimeter 20 m
5. area 31.1 $m^2$, perimeter 21.5 m
6. area 103 $ft^2$, perimeter 43 ft

**1.6** 1. 320 $ft^2$ 2. 36 $m^2$ 3. 35.8 $m^2$ 4. 48 $m^2$ 5. 51.6 $m^2$ 6. 344 $ft^2$

**1.7** a) 9 $yd^2$ b) 7.6 $m^2$ c) 8.1 $yd^2$ d) 6.8 $m^2$ e) 225 $ft^2$ f) 21 $m^2$ g) 310 $ft^2$ h) 29 $m^2$ i) 300 $ft^2$ j) 33 $yd^2$

**2.1** 1. a) 4.3 l b) 5.3 l c) 5.3 l d) 7.3 l e) 7.9 l f) 4.7 l
3. a) 1.3 l b) 1.9 l c) 2.3 l d) 3 l e) 5.2 l f) 1.6 l
4. a) 0.5 l b) 0.5 l c) 0.5 l d) 0.6 l e) 0.6 l f) 0.5 l

**2.2** 1. a) 5 b) 7 c) 7 d) 9 e) 10 f) 6

**2.3** 1. a) £100 b) £150 c) £180 d) £230 e) £410 f) £130

**Decorating Harish's Flat**

*page 42*

1. 15 $m^2$ 2. 2.5 l
3. about 10 $m^2$ without doors, 2 l
4. 47 ft, 6 rolls
5. 0.75 l of each
6. 18 m, 1.7 l of each
7. 2.5 l of each
9. £124 10. £27.85

# CHAPTER 6

*page 44*

**1.1** 1. *km* 0 16 32 48 64 80 96 112 128
a) 37.5 miles
b) 62.5 miles
c) 45 miles
2. a) 48 km/h b) 64 km/h c) 80 km/h d) 113 km/h
3. a) 2 h
b) 2 h 48 min
c) 4 h 9 min
d) 60 m.p.h.
e) 60 m.p.h.
f) 60 m.p.h.
g) 150 miles
h) 170 miles
i) 20 miles
4. about 1 h 40 min

**1.2** 1. *l* 0 9.1 18.2 27.3 36.4 45.5 54.6 68.2
a) 4.4 g b) 6.6 g c) 9.9 g
2. a) 39.6 p b) 42.2 p c) 42.7 p d) 44 p
3. a) 33 m.p.g.
b) 43 m.p.g.
c) 36 m.p.g.

**4.**

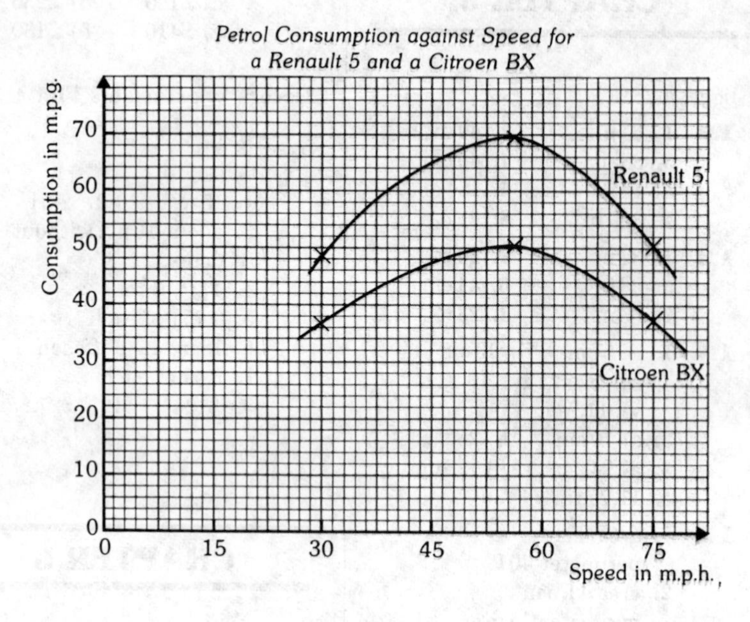


**5.**

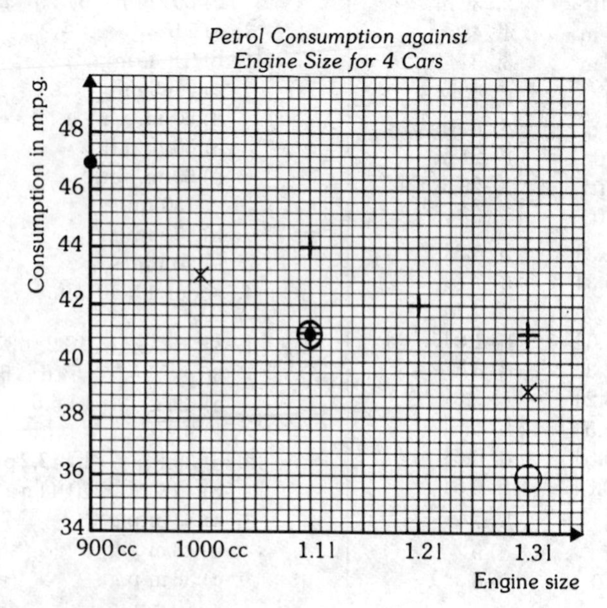


**2.1**

| a) Total easy terms price | b) Extra paid (% of basic price) | c) Extra cost (% of balance) |
|---|---|---|
| £5686.20 | 8 | 10 |
| £6755.84 | 45 | 50 |
| £2668.11 | 34 | 45 |
| £5137.88 | 53 | 65 |
| £6003.68 | 42 | 52 |

**2.2** 1. a) £350 b) £486
c) £625 d) £766
e) £802
2. a) 44 p b) £1.64
c) £4.38 d) £16.41

**2.3** 1. a) £165 b) £332
c) £208 d) £185
2. a) £108 b) £330
c) £84 d) £72
e) £96 f) £684
3.

| | Cost per year | Cost per week |
|---|---|---|
| a) | £933 | £18 |
| b) | £1322 | £25 |
| c) | £952 | £18 |
| d) | £987 | £19 |

**Dave's Travel Costs**
*page 50*

1. £302.40 2. £156
3. £3
4. 2880 miles
5. 2600 miles
6. £400 7. £336
8. £1608.80
9. value £1232, depreciation £376
10. £1110
11. £650 in a year, £13 a week
12. £70

# CHAPTER 7

*page 53*

**1.1** 1. a) £108 b) £133 c) £142.50 d) £112.80 e) £150 f) £107.80 g) £3 h) £1.95 i) £2.50 j) £3.12
2. a) £3 b) £4.50 c) £13.50 d) £24.75 e) £138 f) 46 h
3. a) £3.50 b) £4.38 c) £13.14 d) £24.09 e) £161.04 f) 40 h 48 min

**1.2** 1. £9 2. £14.80 3. £11.10 4. £19.20

**1.3** 1. commission £204, total wage £224
2. commission £213, total wage £253
3. £108

**1.4** 1. £450 2. £500 3. £717 4. £800 5. £900 6. £600 7. £917

**2.1** 1. yes 2. no 3. no 4. no 5. yes

**2.2** 1. a) £97.55 b) £46.70 c) £126.45 d) £107.75 e) £44.35

**3.1**

| | *Taxable income* | *Tax paid per year* |
|---|---|---|
| 1. | £4315 | £1251.35 |
| 2. | £3665 | £1062.85 |
| 3. | £7545 | £2188.05 |
| 4. | £8710 | £2525.90 |
| 5. | £19165 | £5774 |
| 6. | £8665 | £2512.85 |

**3.2**

| | *National insurance* | *Pension* |
|---|---|---|
| 1. | £54 | £30 |
| 2. | £48.60 | £27 |
| 3. | £60 | £33.33 |
| 4. | £82.50 | £45.83 |
| 5. | £78 | £43.33 |

**3.3** 1. £12 099 2. 6% 3. £714.83 4. 29% 5. £668.22

**Family Incomes**

*page 58*

1. £119.25 2. £96
3. £44.90 4. £16.70
5. £68.39 6. £19.83
7. £82.72 8. £68.79
9.

| TAXABLE ALLOWANCES £ | | DEDUCTIONS £ | |
|---|---|---|---|
| | | Income Tax | 98.05 |
| | | Nat. Ins. | 51.00 |
| Total gross pay | 566.67 | NON-TAXABLE ALLOWANCES | |
| Superannuation | 34.00 | | |
| Taxable pay | 532.67 | | |
| | | NET PAY | 373.48 |

10. £31.90 per month extra, yes

# CHAPTER 8

*page 60*

**1.1** 1. a) £936 b) £1170 c) £863.20 d) £1081.60
2. a) £17.82 b) £22.68 c) £24.19 d) £19.55
3. a) 12.5% b) 15% c) 8% d) 12.5%
4. a) £480 b) £442.40 c) £394

**1.2** 1. a) £90.80 b) £154.58 c) £126.48 d) £193.23
2. a) £28 044, £21 792
b) £39 751.20, £27 824.40
c) £50 314.50, £37 944
d) £66 252, £46 375.20

**1.3** 1. a) £304 b) £199.50 c) £250.80 d) £207.10
2. a) 88 p b) 94 p c) 99 p d) £1.04
3. a) £1.52 b) 11.7 p c) £310.58 d) 49.8%

**2.1** 1. a) £2 500 b) £2 290 c) £1 840 d) £4 990
2. a) 3 years 4 months
b) 1 year 48 weeks
c) 2 years 1 month
3. a) £26 250 b) £18 900 c) £17 430 d) £23 625
4.

| After | 1 | 2 years |
|---|---|---|
| a) £ | 18 900 | 19 845 |
| b) 10% | 1 890 | 1 985 |

| After | 3 | 4 years |
|---|---|---|
| a) £ | 20 838 | 21 879 |
| b) 10% | 2 084 | 2 188 |

| After | 5 years |
|---|---|
| a) £ | 22 973 |
| b) 10% | 2 297 |

c) 2 years

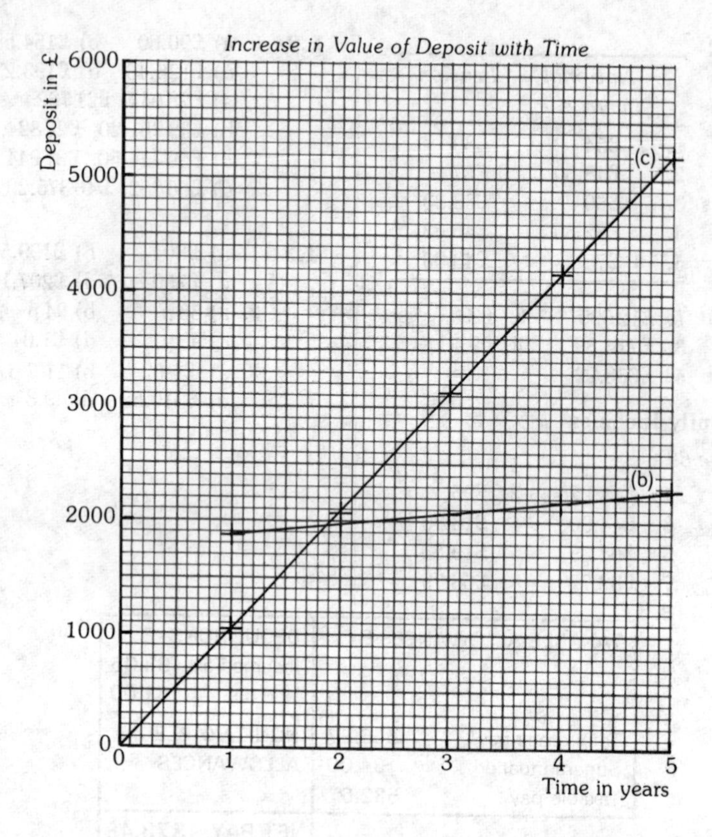


**2.2** 1. £293.87 2. £805.26
3. £1367.63 4. £1108.69

**2.3** 1. £1420 2. £1850
3. a) £780 b) £800
c) £840
4. 3 years 4 months
5. interest £220
6. about £47

**From Renting to Buying**

*page 66*

1. £1690 2. £18 500
3. £1850 4. £2850
5. £70 6. £16 200
7. £136.60 8. £214.60
9. £1853.80 10. £40 980

# CHAPTER 9

*page 68*

**1.1** 1. a) 10.35 a.m.
b) 7.20 a.m.
c) 1.15 p.m.
d) 5.50 p.m.
e) 10.30 p.m.
f) noon
g) 11.59 p.m.
h) midnight
2. a) 0730 b) 1930
c) 1215 d) 0015
e) 2200 f) 1315

**1.2** a) 5 h b) 7 h 10 min
c) 3 h 55 min d) 15 h 15 min
e) 3 h 50 min f) 8.20 p.m.
g) 2.20 a.m. h) 1010
i) 12.35 p.m. j) 0735

**1.3** 1. a) 1 h 45 min
b) 1 h 20 min
c) 1 h 30 min
d) 0900
2. a) 0838
b) 1 h 34 min
c) 1118 d) 30 min
e) 37 min f) 2118

**2.1** 1. a) Bridlington, Scarborough
b) Lincoln or Doncaster
c) York, Leeds, Huddersfield
d) York, Harrogate, Skipton
2. a) 205 direct
b) 206 to Dolgellau, then 211
c) 203 to Chirk, then 205
d) 271 to Dolgellau, 211 to Machynlleth, then 202

**2.2** 1. a) 107 b) 154
c) 275 d) 604
e) 200
2. a) 67 b) 59
c) 74 d) 83

**3.1** 1. a) £16.25
b) i) yes ii) no
2. a) £10.33 b) £41
c) £36.67 d) £24.17

**3.2** 1. a) 4.6 p b) 4.1 p
c) 3.9 p
2. a) £6 b) £10
c) £25 d) £18
3. a) £7 b) £12
c) £30 d) £22
4.

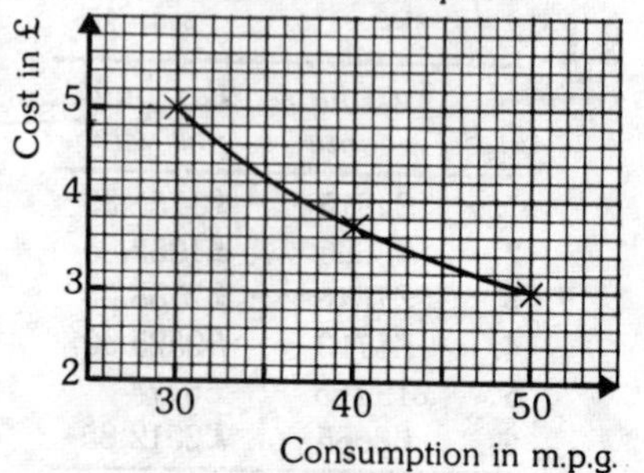

5. a) £18 b) £30
   c) £75 d) £54
6. a) £48 b) £35
   c) £130 d) £122
   e) £240 f) £40

**A Holiday for Dave and Stuart**

*page 76*

1. £9 2. £45
3. £59.50 4. £43.75
5.

| Farm | Hostel |
|---|---|
| £124 | £94 |

| Caravan | Camping |
|---|---|
| £109 | £93 |

6. about 600 miles
7. about £30
8. £24.80
9. a) £149 b) £119
   c) about £150
   d) about £120

# CHAPTER 10

*page 78*

**1.1** 1. a) 6th July
   b) 1st August
   c) 7th September
   d) 3rd March 1984
   e) 4th March 1982
   f) 2nd August
2. a) 16 days
   b) 24 days
   c) 12th September
   d) 19th June

**1.2** 1. a) 4.00 p.m.
   b) 2.30 a.m.
   c) 2330
2. a) 1715
   b) 6.45 a.m.
   c) 10.45 p.m.
3. a) 3 h b) $3\frac{1}{2}$ h
   c) 15 min

**1.3** 1. a) 6 p b) 33 p
   c) £1.30 d) £6.49
2. a) 4 p b) 32 p
   c) £1.61
3. a) 49 p b) £12.20
   c) 1 p

**1.4** See graph on page 228.

**2.1a** 1. a) 625 miles
   b) 470 miles
   c) 620 miles
   d) 895 miles
   e) 725 miles
2. a) 1790 miles
   b) 1480 miles
   c) 1780 miles
   d) 2330 miles
   e) 1990 miles
3. a) £320 b) £270
   c) £320 d) £420
   e) £360

**2.1b** 1. a) 0730 b) 1030
   c) 1200
2. £27
3. a) £44 b) £52
   c) £52
4. £79 5. £164
6. £524

**2.1c** 1. £17.10 2. £32.30
3. £12.80 for an adult, £8.60 for a child
4. £49.30 5. £573.30

**2.2** 1. £378 2. £7
3. £395.50 4. £573
5. £1072

# CHAPTER 11

*page 87*

**1.1** 1. 12 cm$^2$ 2. 12 cm$^2$
3. 22.2 cm$^2$ 4. 19.84 cm$^2$

**1.2** 1. obtuse-angled; 23°, 30°, 127°; angle sum 180°; height 2 cm with 8 cm as base; area 8 cm$^2$
2. right-angled; 37°, 53°, 90°; angle sum 180°; height 4.5 cm with 6 cm as base; area 13.5 cm$^2$
3. isosceles; 71°, 38°, 71°; angle sum 180°; height 5.7 cm with 4 cm base; area 11.4 cm$^2$
4. Acute-angled; 44°, 57°, 79°; angle sum 180°; height 4.1 cm with 7 cm as base; area 14.4 cm$^2$

**1.3** 1. each angle 108°; sum 540°; area 38 cm$^2$
2. each angle 135°; sum 1080°; area 45 cm$^2$
3. each angle 120°; sum 720°; area 65 cm$^2$
4. angles as in **3**; area 65 cm$^2$
5. 131°, 98°, 131°; sum 540°; area 32 cm$^2$
6. a) 10 sides
   b) 12 sides

**2.1** 1. See table on page 228.
2. a)

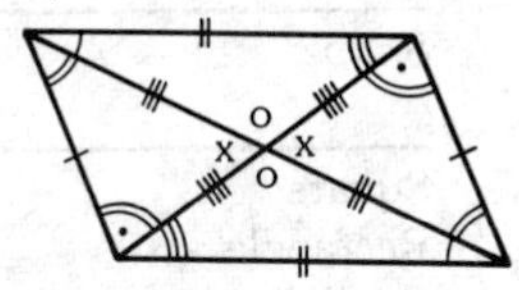

b)

c)

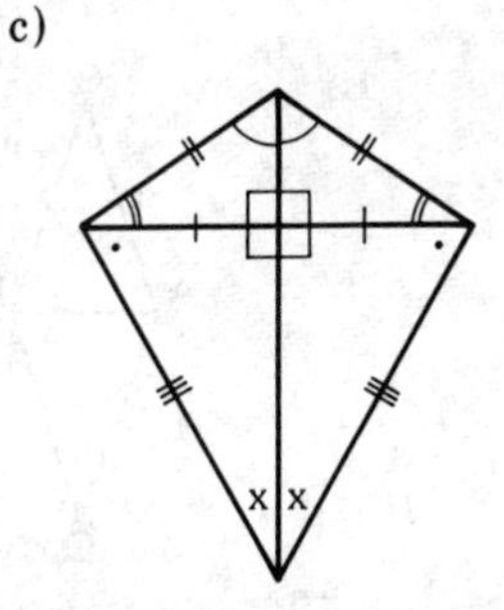

**Chapter 10, 1.4**

1\.

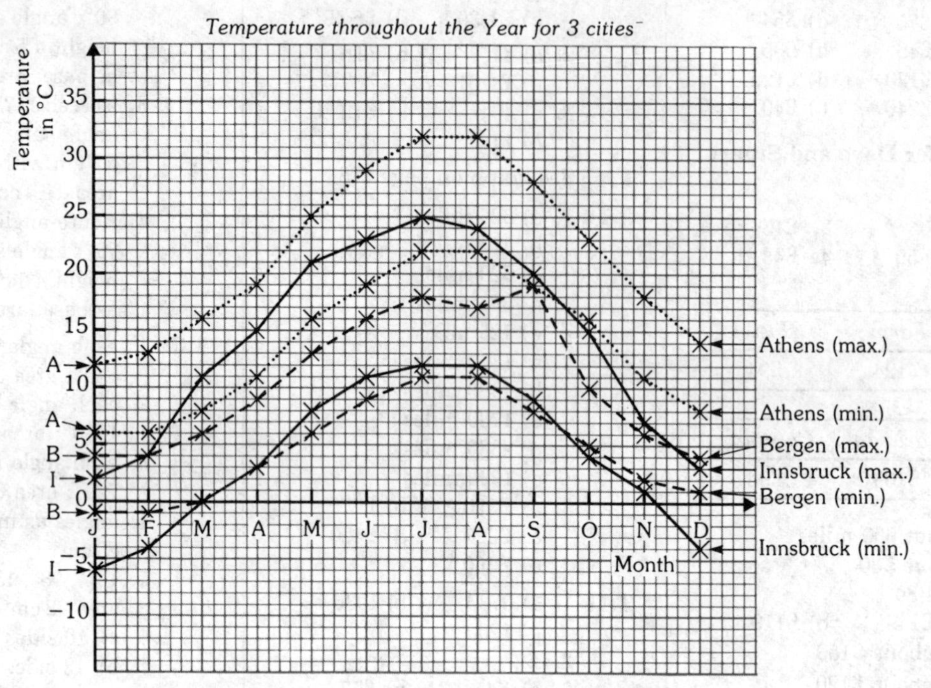


**2.1**

1\.

| *Shape* | *Number of lines of symmetry* | *Order of rotational symmetry* |
|---|---|---|
| Square | 4 | 4 |
| Rectangle | 2 | 2 |
| Parallelogram | none | none |
| Kite | 1 | none |
| Rhombus | 2 | 2 |
| Isosceles triangle | 1 | none |
| Equilateral triangle | 3 | 3 |
| Regular hexagon | 6 | 6 |

**2.2** The answers given here are drawn at a scale of 1:4.

1\.

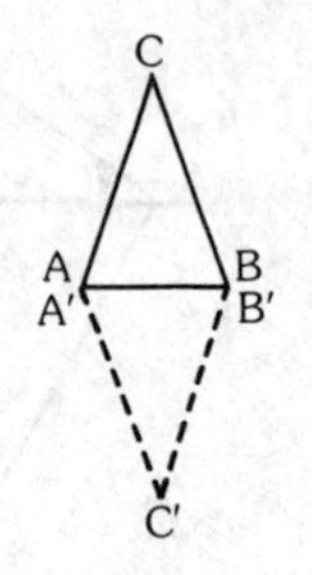


2\.

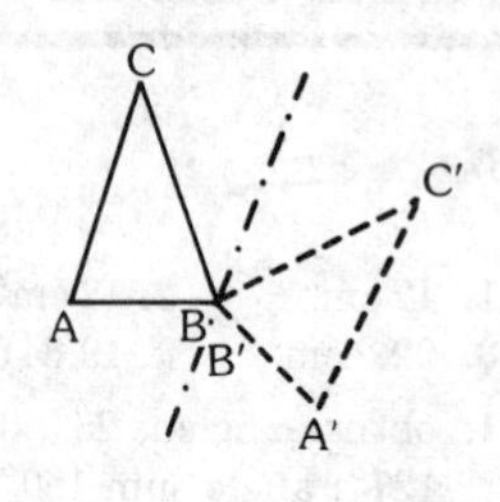


3\.

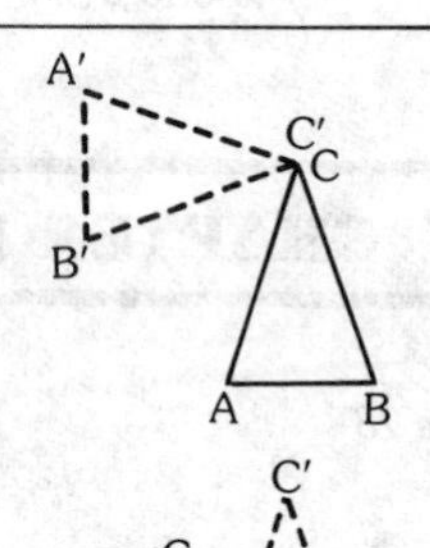


4\.

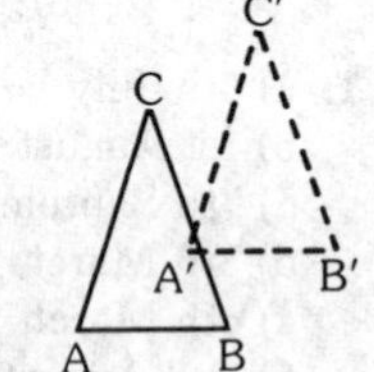


5\.

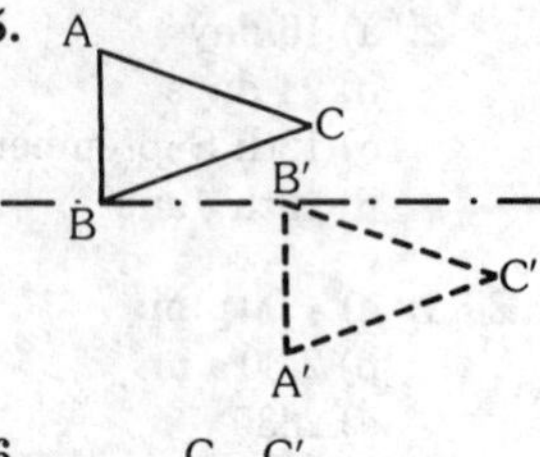


6\.

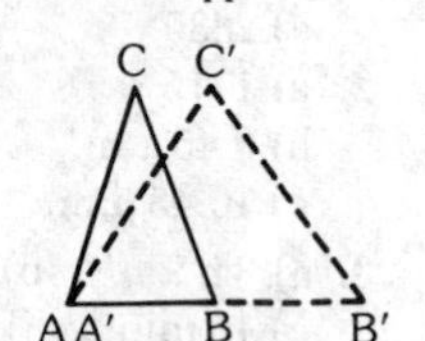

**7.**

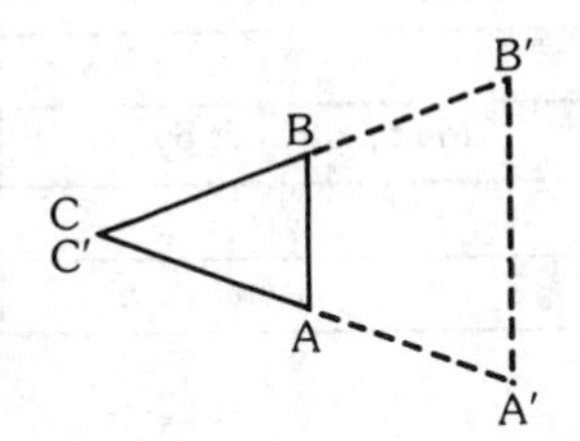


**8.**

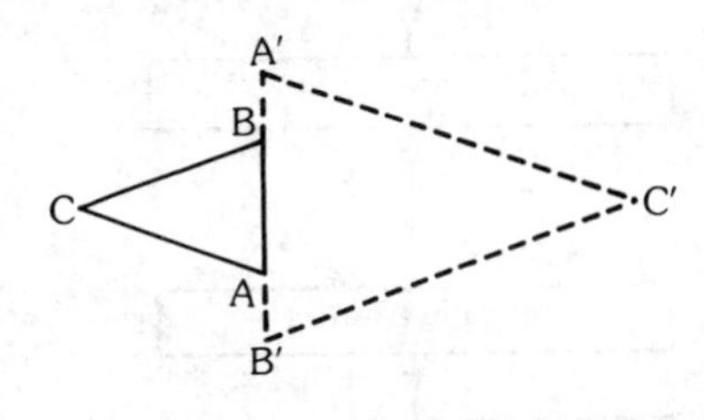


# CHAPTER 12

*page 96*

**1.1** **1.** 8427 **2.** 9099
**3.** 7171 **4.** 9999
**5.** 5000 **6.** 6512
**7.** 4919 **8.** 5022
**9.** 0769

**1.2** **1.** and **2.**

| *Reading at end* | *Reading at beginning* | *Cubic feet used (100s)* | *Therms used* | *Cost at 35 p per therm* | *Total bill* |
|---|---|---|---|---|---|
| 2798 | 2512 | 286 | 296 | £103.60 | £112.20 |
| 3112 | 2975 | 137 | 142 | £49.63 | £58.23 |
| 4889 | 4772 | 117 | 121 | £42.38 | £50.98 |
| 3002 | 2775 | 227 | 235 | £82.23 | £90.83 |
| 0972 | 0638 | 334 | 346 | £120.99 | £129.59 |
| 5827 | 5598 | 229 | 237 | £82.96 | £91.56 |

**4.**

| *Present reading* | *Previous reading* | *Units used* | *Cost of units* | *Total bill* |
|---|---|---|---|---|
| 76513 | 75924 | 589 | £29.98 | £38.48 |
| 84155 | 81179 | 2976 | £151.48 | £159.98 |
| 62623 | 59871 | 2752 | £140.08 | £148.58 |
| 70064 | 67872 | 2192 | £111.57 | £120.07 |
| 81021 | 78863 | 2158 | £109.84 | £118.34 |

**1.3** **1.** £315 **2.** £440
**3.** £325 **4.** £450

**1.4** **1.** a) £340 b) £350
c) £390 d) £250

**2.** a) fridge 36 p,
washing machine 46 p,
twin tub 61 p,
freezer 76 p
b) fridge £19,
washing machine £24,
twin tub £32,
freezer £40

**3.** a) 70 p b) £37

**Colin's and Cheryl's Energy Costs**

*page 101*

**1.** 273 h **2.** 27
**4.** 845 h **5.** 338
**10.** 375, more **11.** 273
**13.** 169

# CHAPTER 13

*page 103* (Note – examples given here. Other versions may also be correct.)

**1.1**

**1.**

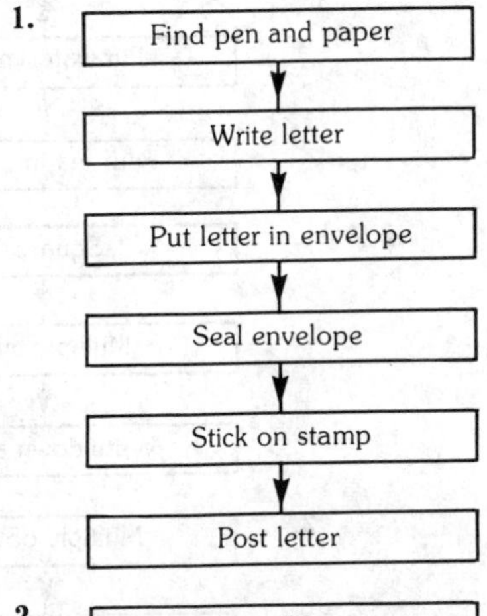


**2.**

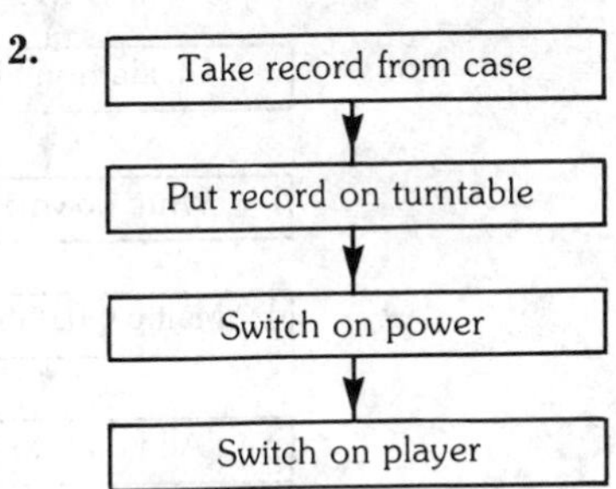


**3.**

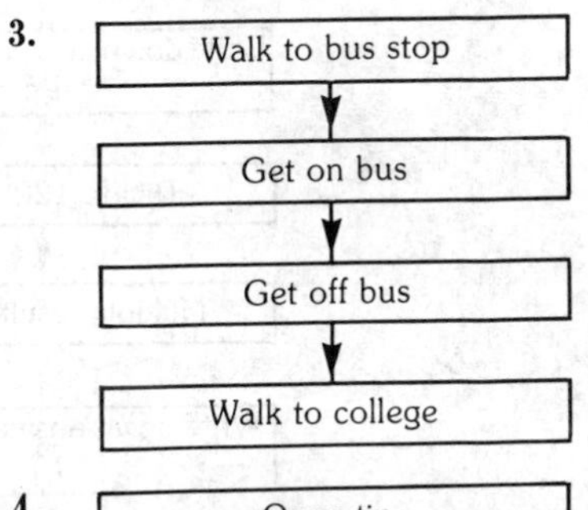


**4.**

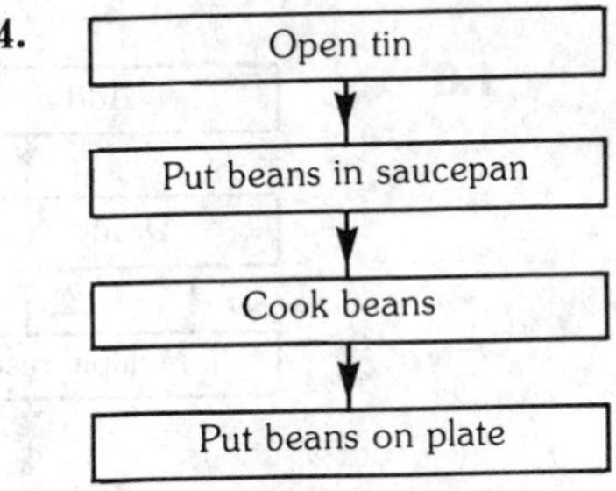

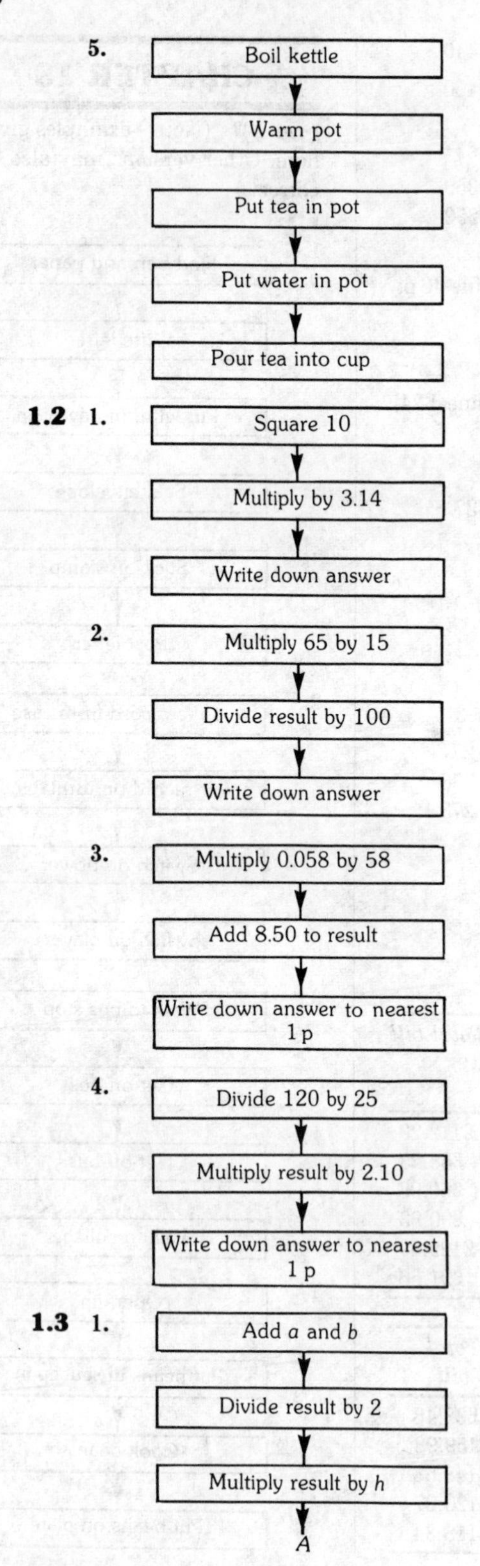


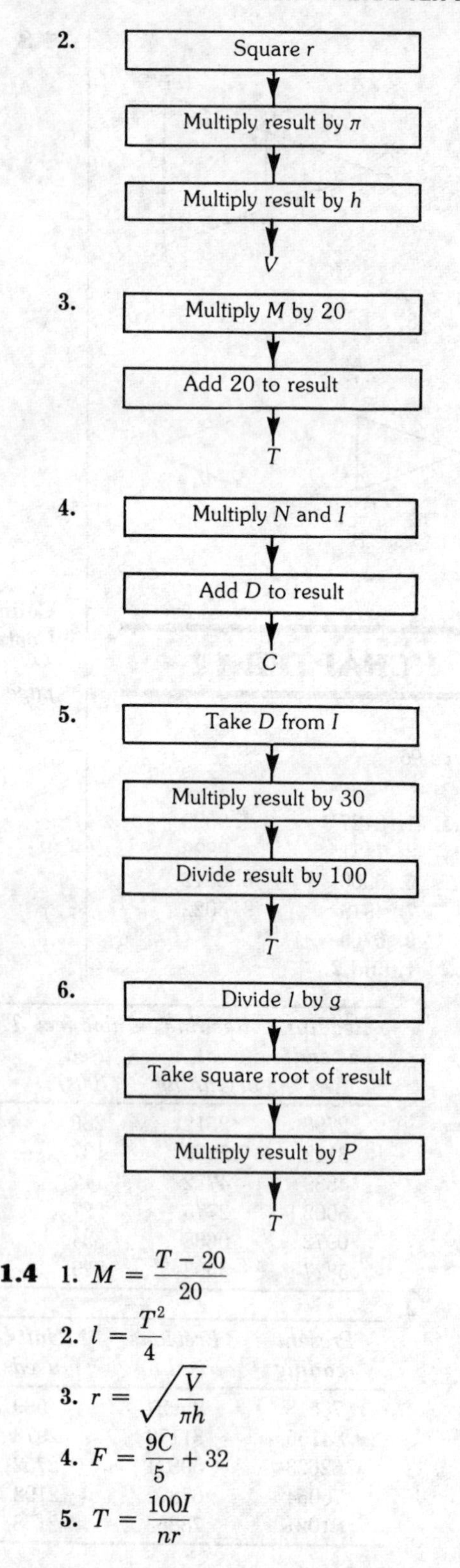


**1.4** 1. $M = \dfrac{T - 20}{20}$

2. $l = \dfrac{T^2}{4}$

3. $r = \sqrt{\dfrac{V}{\pi h}}$

4. $F = \dfrac{9C}{5} + 32$

5. $T = \dfrac{100I}{nr}$

**2.1** 1.

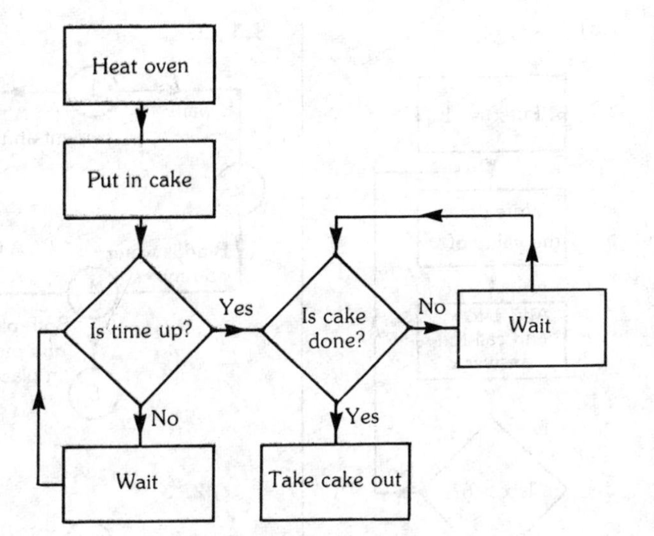


3.

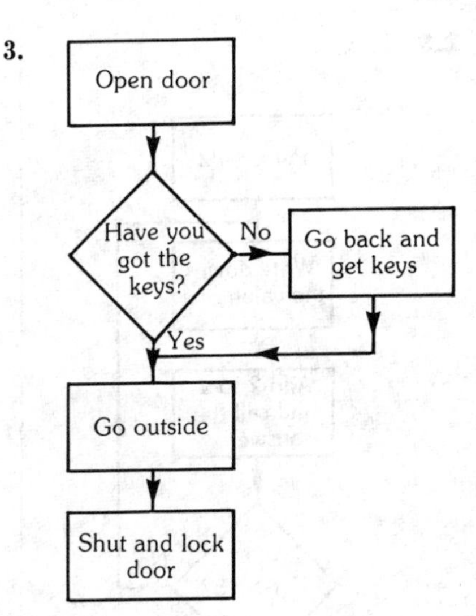


2.

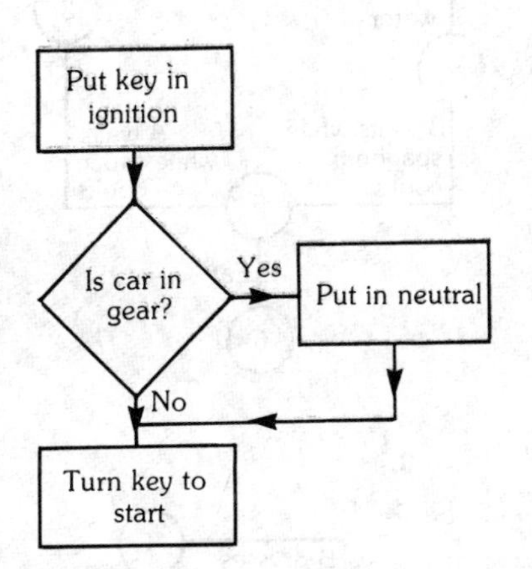


**Strictly you should also check the handbrake and pull out the choke.**

**2.2**

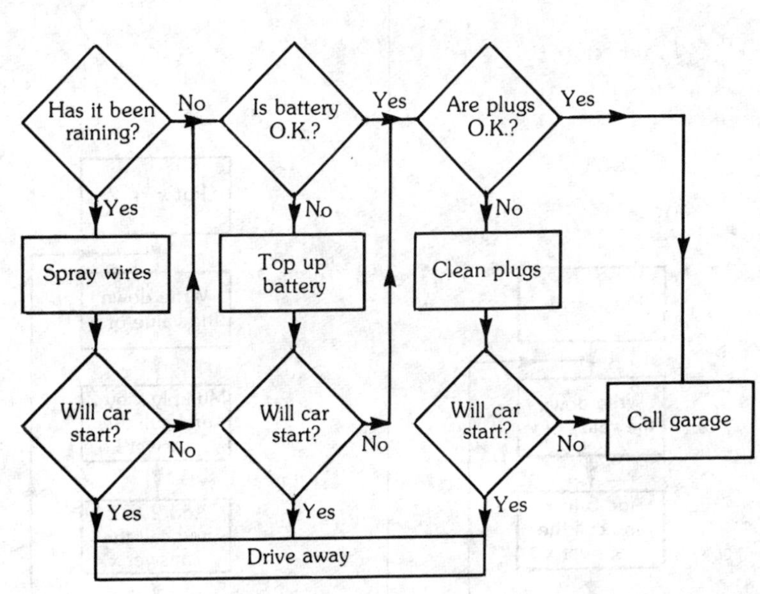

**2.3** 2. a)

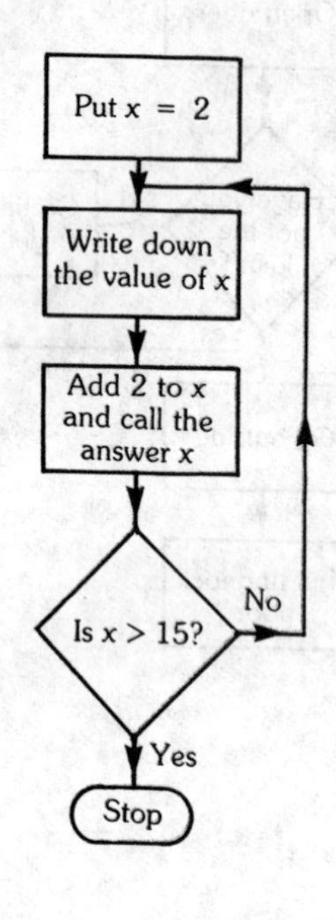


b)

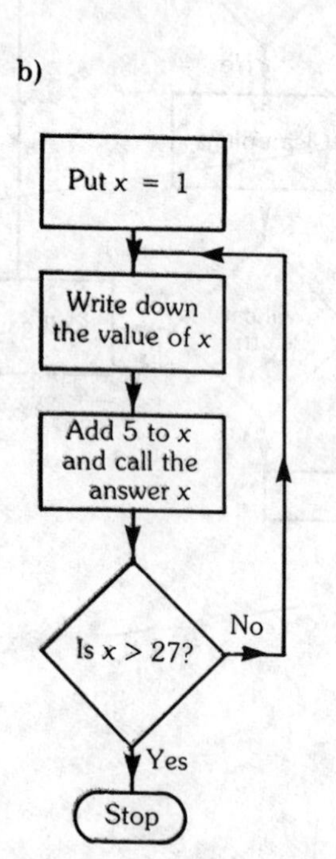


c)

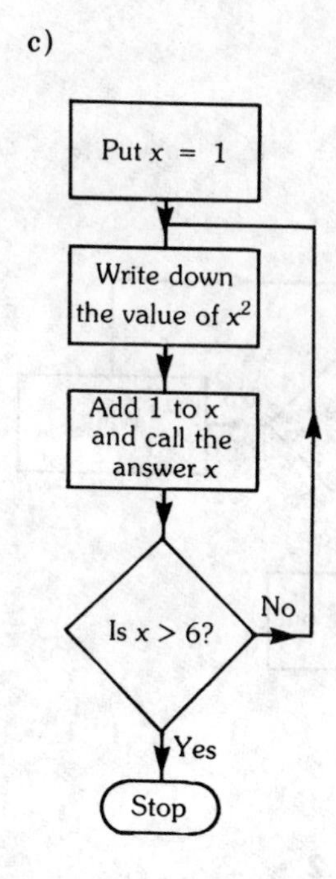


d)

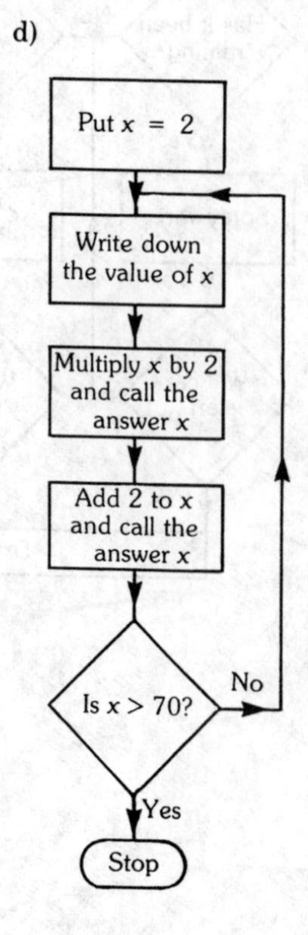


**3.1** 1.

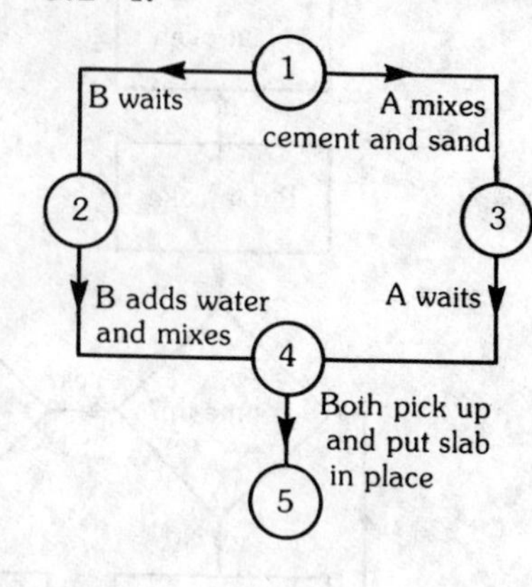


2.

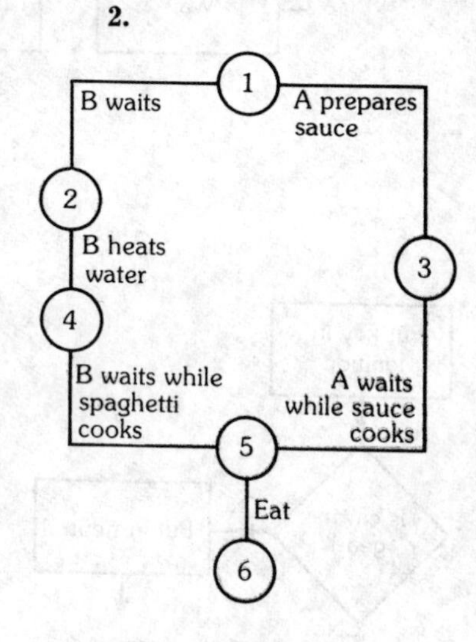


3.

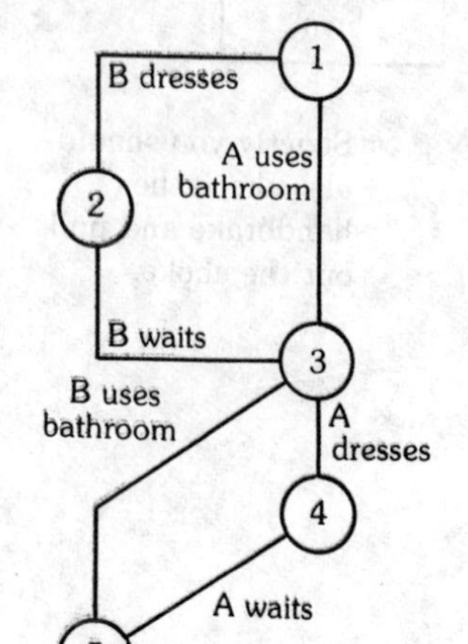

4.

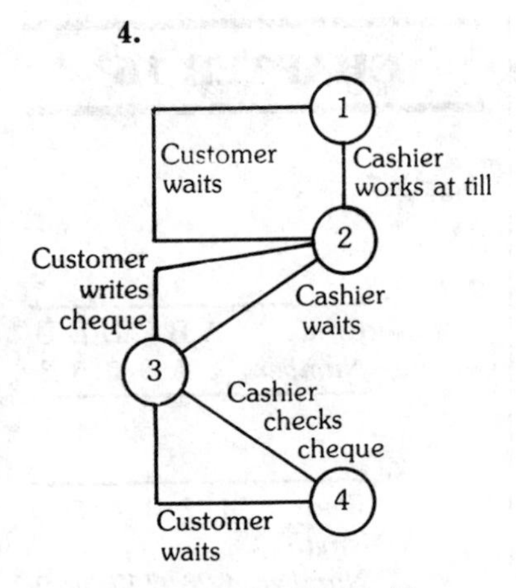


**3.2** 1. times 18, 15, 13, 10, critical path 1, 2, 5
2. times 50, 54, 58, 70, critical path 1, 4, 3, 5, 6, 7

# CHAPTER 14

*page 112*

**1.2** 1. a) 177 $ft^3$ b) 241 $ft^3$
c) 396 $ft^3$
2. a) 1.27 $m^3$ b) 2.88 $m^3$
c) 1.40 $m^3$

**1.3** 1.

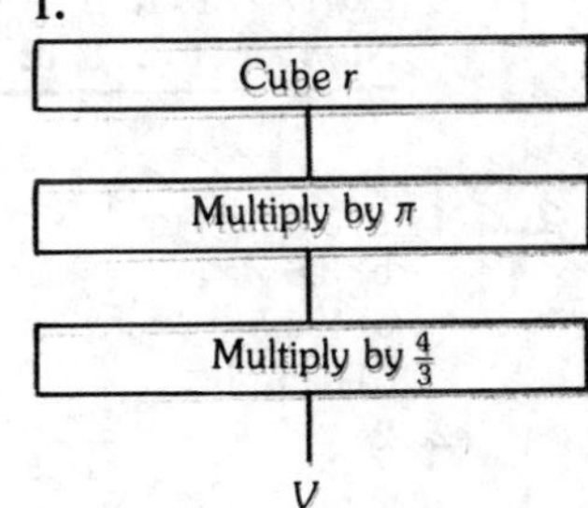


2. 512 $in^3$ 3. 603 $cm^3$
4. 2.4 $m^3$ 5. 50.2 $cm^3$
6. 91 400 000 $ft^3$

**1.4** 1. 420 $cm^3$, 0.95 $g/cm^3$
2. 439 $cm^3$, 1.0 $g/cm^3$
3. 236 $cm^3$, 0.90 $g/cm^3$
4. 169 $cm^3$, 1.0 $g/cm^3$
5. 4030 $cm^3$, 0.12 $g/cm^3$

**2.1** 1. 213 $cm^2$

2. 2076 $cm^2$

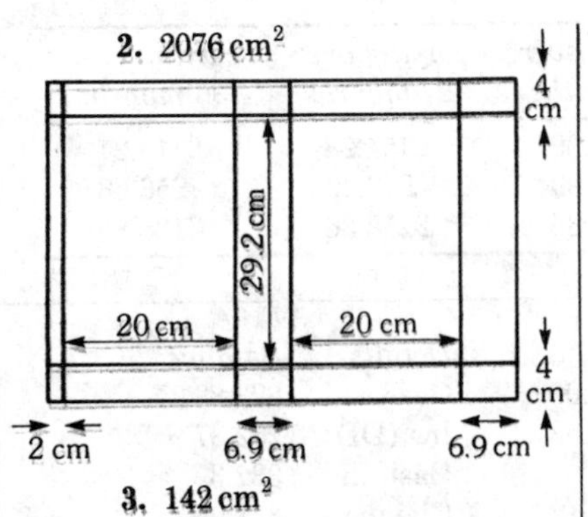


3. 142 $cm^2$



4. 184 $cm^2$

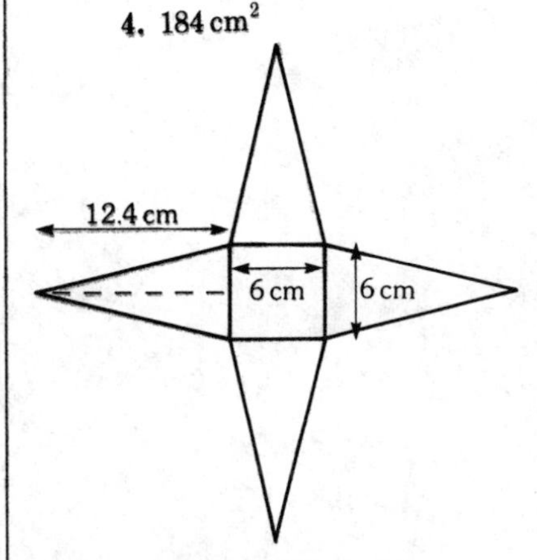


5. no, 483 $cm^2$

**2.2** 1. a) i) 25 ii) 125
b) i) 100 ii) 1000
c) i) 10 000
ii) 1 000 000
d) i) $\frac{1}{4}$ ii) $\frac{1}{8}$
2. 44 oz 3. 15.4 cm

**2.3** 1. 402 $cm^2$, 326 $cm^2$, 319 $cm^2$, 329 $cm^2$, 346 $cm^2$

**3** 1. 0.085 p/g, 0.086 p/g
2. £4.70/l, £3.75/l
3. 0.088 p/g, 0.083 p/g
4. 0.085 p/g, 0.079 p/g
5. 0.136 p/g, 0.132 p/g
6. 1.35 p/g, 1.31 p/g

# CHAPTER 15

*page 120*

**1.1** 1. a) £17 b) 1 p
2. a) £90.06 b) 7 p
3. a) £8.60 b) 10 p
4. a) £606 b) £106.01
5. a) £600.02 or £727
b) 3 p or £127.01
6. 21.2%, 21.2%

**1.2** 1. a) £120 b) £192
c) £240 d) £360
e) £600
2. a) £194.56
b) £188.99, £183.29, £177.45, £171.47
d) £171.47

**1.3** 1. See table overleaf.
2. a) £1119.96 b) £1239.84
c) £1359.72

**2.1** 1. a) £8 b) £14
c) £9 d) £2
e) £4 f) £2
2. a) £6 b) £3
c) £4
3. £200 4. £210

**2.2** 1. £3062 2. £3215
3. a) £40 b) £45
c) £70 d) £61
4. £93
5. about £275

**2.3** 1. 5%, $2\frac{1}{2}$%
2. £403.86, £40.39
3. £56.48

**3** 1. a) £14.29 b) £35.71
c) £214.29 d) £714.29
2. a) £13.89 b) £15.15
3. £148.57

**1.3** 1.

| | Amount of loan (£) | Paying over 12 months | Paying over 24 months | Paying over 36 months |
|---|---|---|---|---|
| a) | 3000 | £279.99 | £154.98 | £113.31 |
| b) | 1500 | £139.99 | £77.49 | £56.66 |
| c) | 5000 | £466.65 | £258.30 | £188.85 |

**4** 2.

| Date | Credits | Debits | Details | Balance |
|---|---|---|---|---|
| a) 6/5/86 | | 50.00 | Cash | 273.89 |
| b) 8/5/86 | | 21.52 | Ins (DD) | 252.37 |
| c) 11/5/86 | 80.00 | | Cash in | 332.37 |
| d) 12/5/86 | | 11.12 | Cheque | 321.25 |
| e) 16/5/86 | | 116.21 | Credit Account | 205.04 |
| f) 22/5/86 | | 60.00 | Cash | 145.04 |

**Family Finances**

*page 125*

**1.** a) £61 b) £464
**2.** £4370 **3.** £155.09
**4.** £67
**5.** about £1400
**6.** about £440
**9.** £94.43

# CHAPTER 16

*page 127*

**2.1** 1.

| *Grade* | A | B | C | D | E | U |
|---|---|---|---|---|---|---|
| *Number* | 3 | 5 | 9 | 5 | 5 | 3 |

2.

| *Word length* | 1 | 2 | 3 | 4 | 5 | 6 |
|---|---|---|---|---|---|---|
| *Number* | 6 | 28 | 22 | 17 | 7 | 9 |

| *Word length* | 7 | 8 | 9 | 10 | 11 |
|---|---|---|---|---|---|
| *Number* | 9 | 6 | 5 | 1 | 1 |

3.

| *Letter* | A | B | C | D | E | F |
|---|---|---|---|---|---|---|
| *Number* | 4 | 2 | 2 | 2 | 15 | 3 |

| *Letter* | H | I | L | M | N | O |
|---|---|---|---|---|---|---|
| *Number* | 6 | 5 | 4 | 3 | 5 | 3 |

| *Letter* | P | R | S | T | U | Y |
|---|---|---|---|---|---|---|
| *Number* | 1 | 2 | 8 | 12 | 3 | 2 |

**2.2** 1. a) 5.5
b) 3.6
c) 105.5
d) 0.36

2. 2

**2.3** 1. a) 1, 6
b) 2, 3
c) 5, 5
d) no mode, 4.5
e) 1, 2

**2.4** 1. a) 9
b) 6
c) 7

# CHAPTER 17

*page 134*

**1.1** 1.

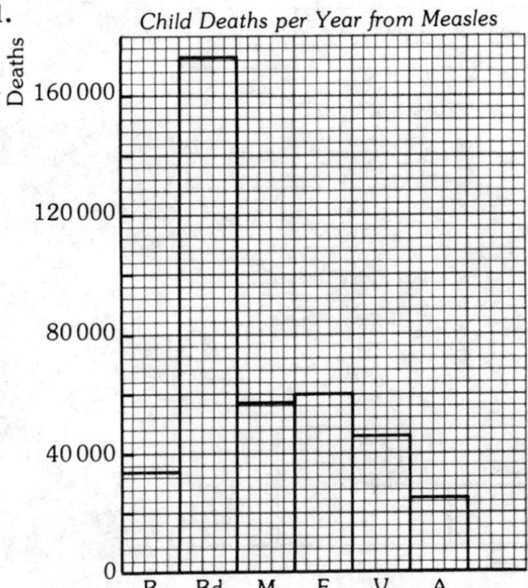


2.

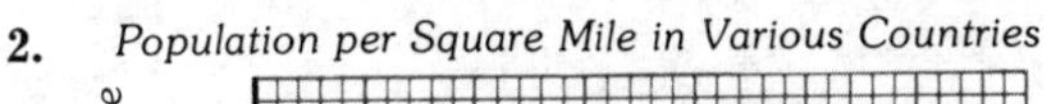


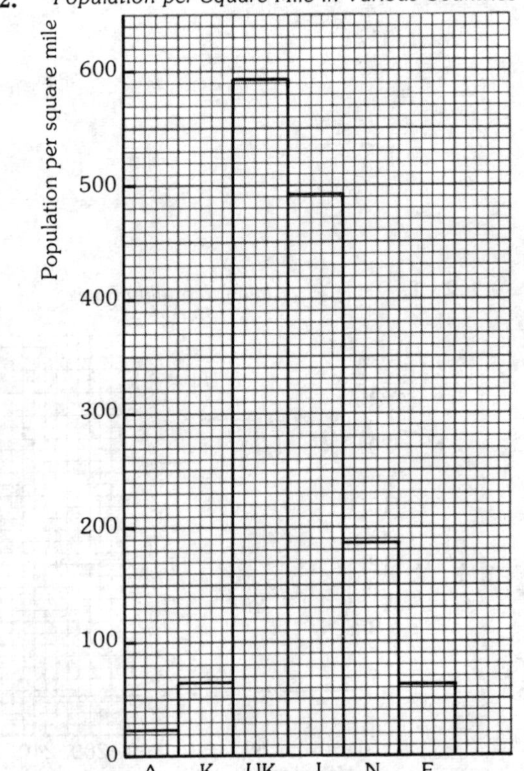


3.

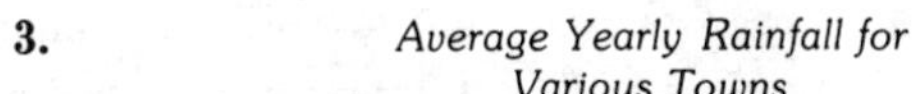


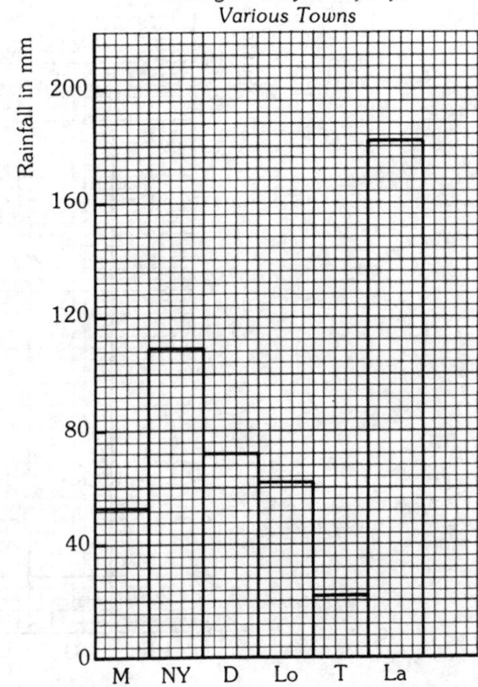


4.

*Life Expectancy at Birth for Various Countries*

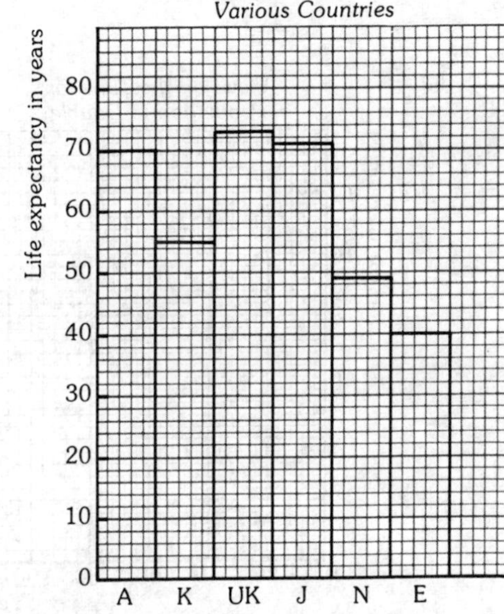

**5.**

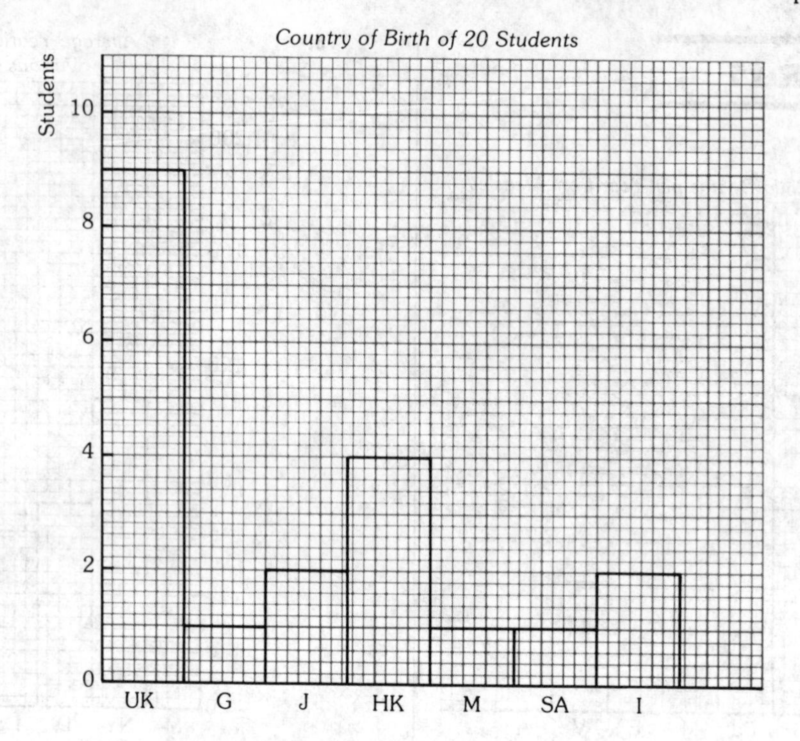
Country of Birth of 20 Students
Students
10
8
6
4
2
0
UK
G
J
HK
M
SA
I

**6.**

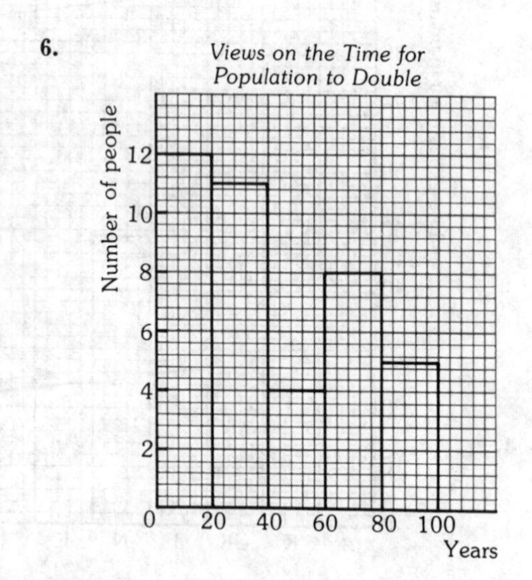
Views on the Time for
Population to Double
Number of people
12
10
8
6
4
2
0
20
40
60
80
100
Years

**1.2** **1.**

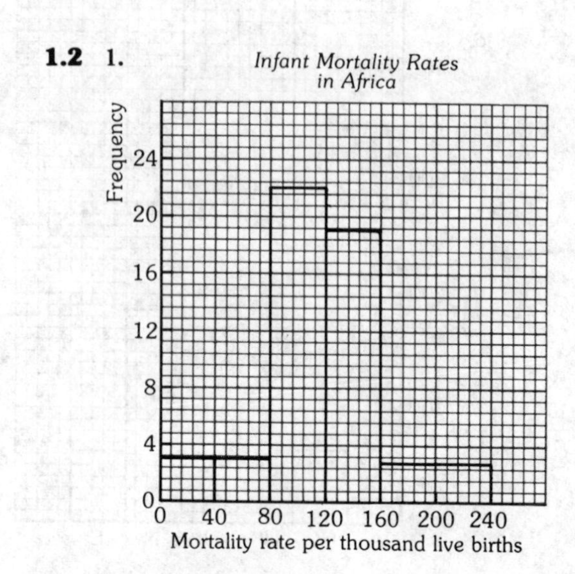
Infant Mortality Rates
in Africa
Frequency
24
20
16
12
8
4
0
40
80
120
160
200
240
Mortality rate per thousand live births

**2.**

Population of Countries in Africa

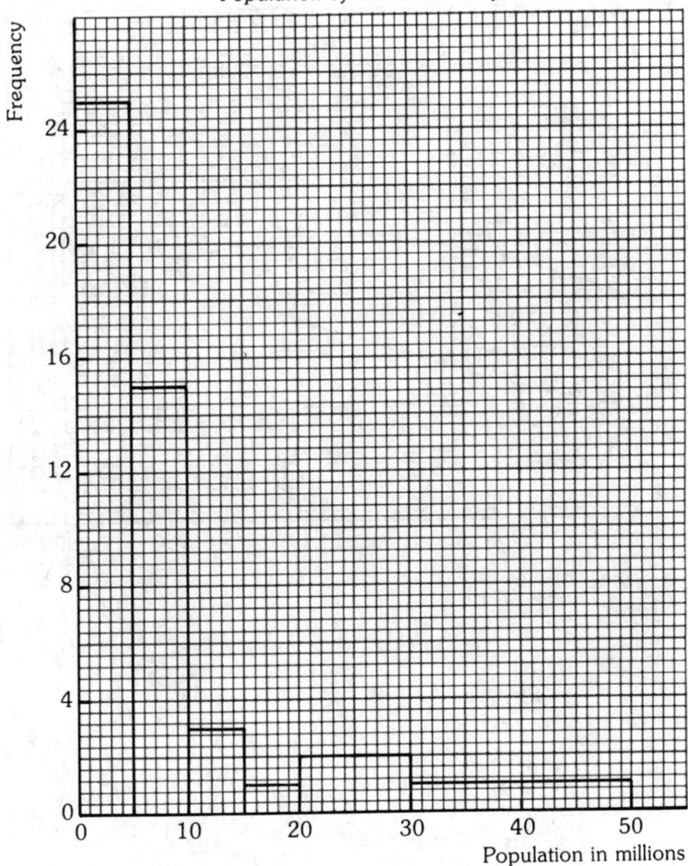
Frequency
24
20
16
12
8
4
0
0
10
20
30
40
50
Population in millions

**1.3** 1.

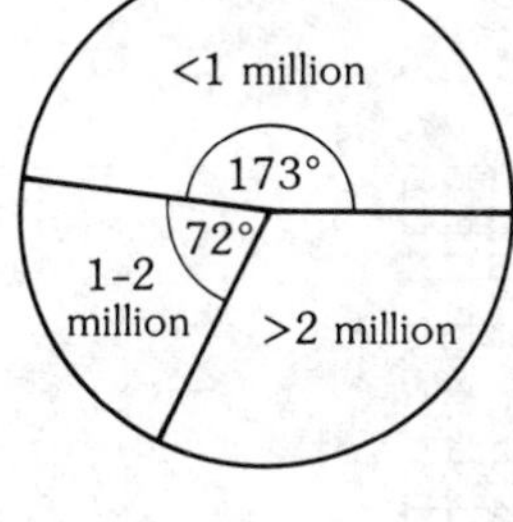
<1 million
173°
72°
1-2
million
>2 million

2.

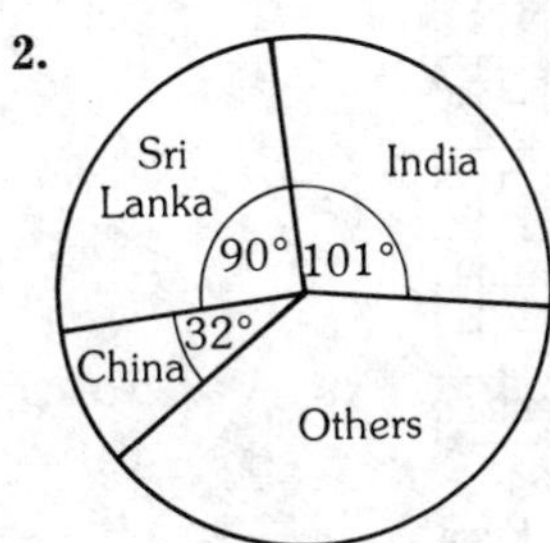
Sri
Lanka
India
90°
101°
32°
China
Others

3.

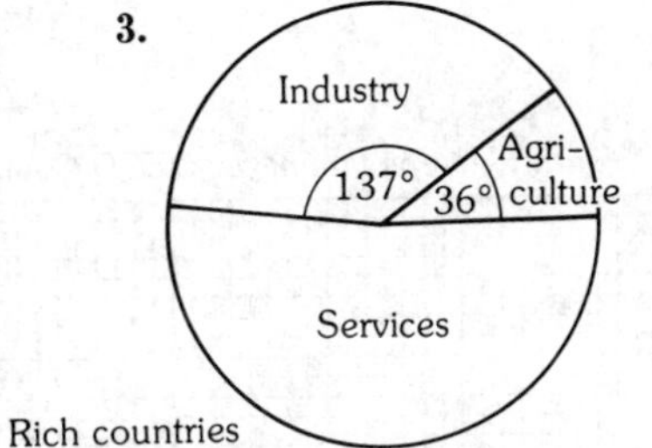
Industry
137°
36°
Agri-
culture
Services
Rich countries

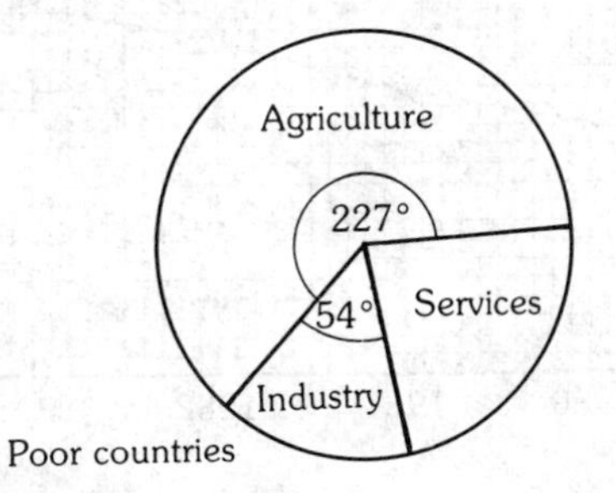
Agriculture
227°
54°
Services
Industry
Poor countries

4.

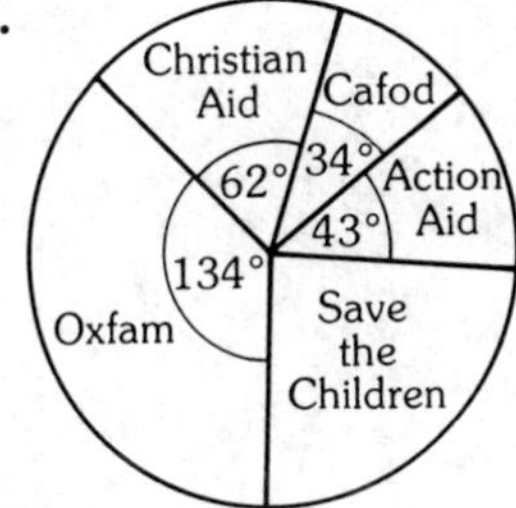
Christian
Aid
Cafod
62°
34°
Action
Aid
43°
134°
Oxfam
Save
the
Children

**2.1** 1. and 2.

*Population over 55 for N. Europe and Japan*

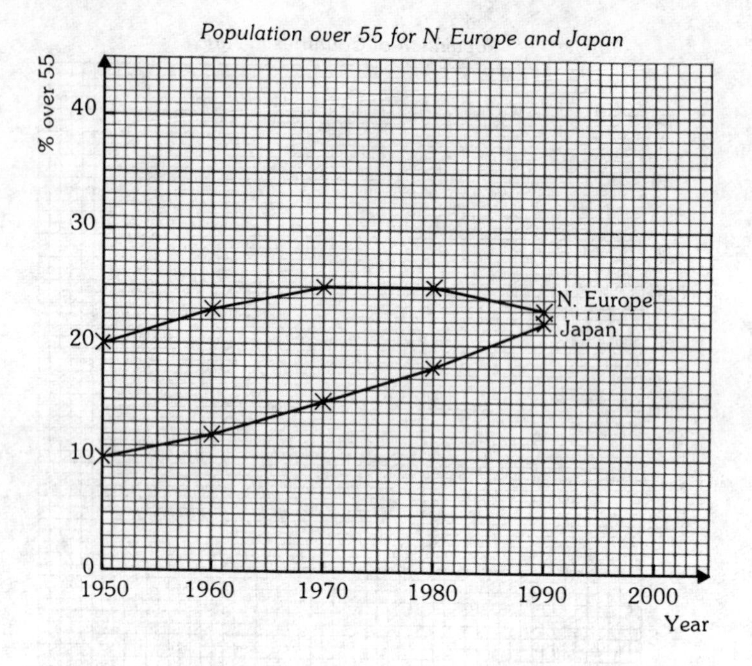


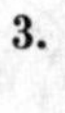

**3.**

*Youth Unemployment for OECD Countries*

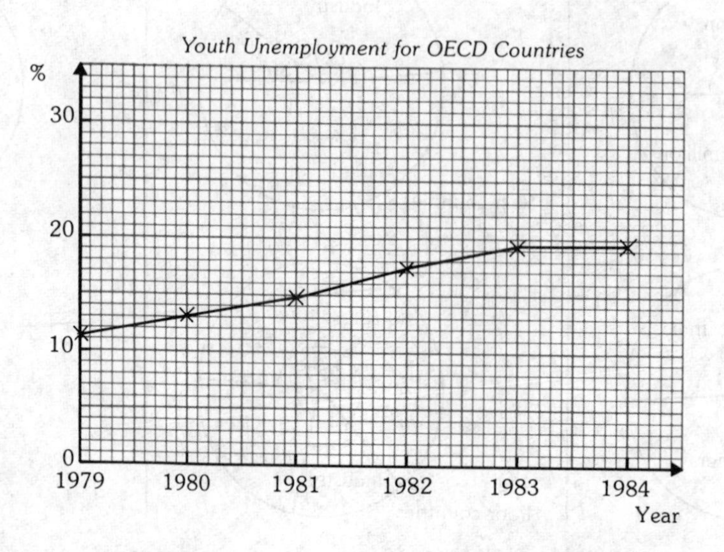

**4.**

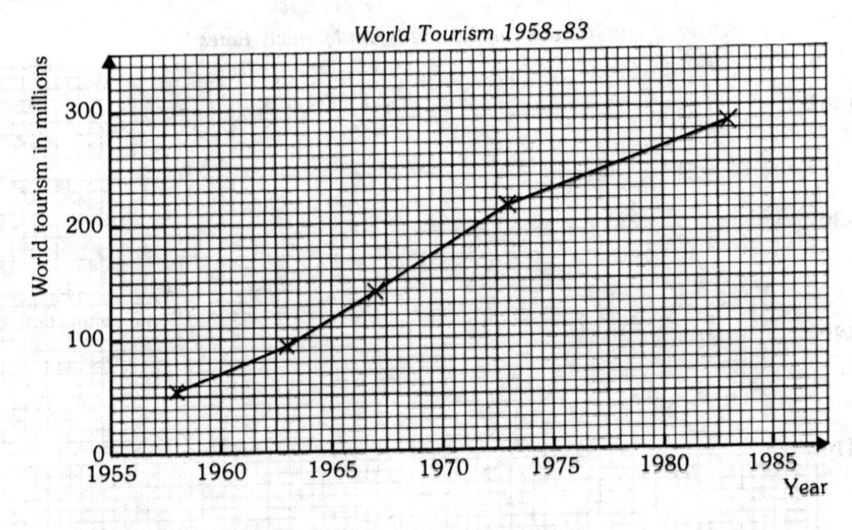
World Tourism 1958-83
World tourism in millions
300
200
100
0
1955
1960
1965
1970
1975
1980
1985
Year

**2.2** **1.**

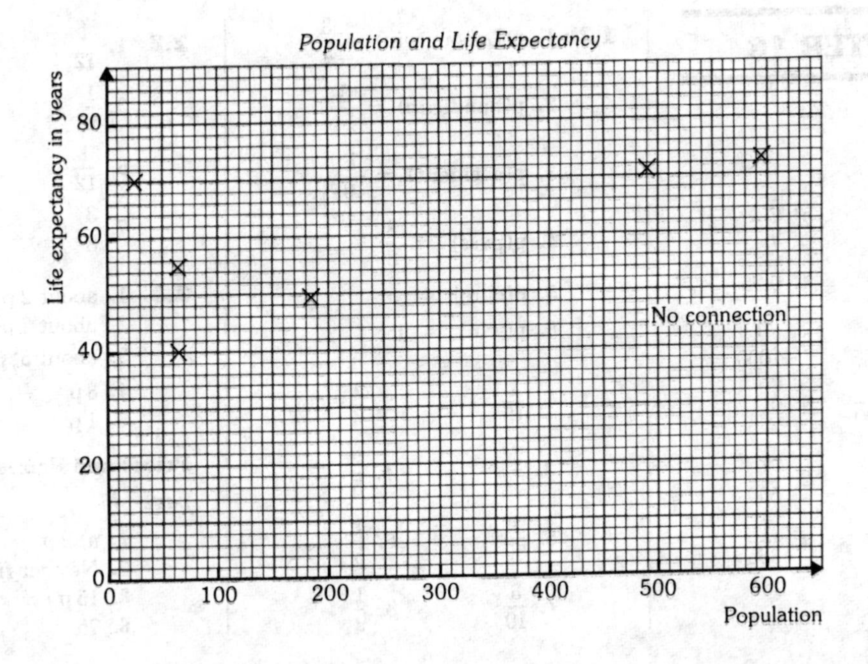
Population and Life Expectancy
Life expectancy in years
80
60
40
20
0
0
100
200
300
400
500
600
Population
No connection

**2.**

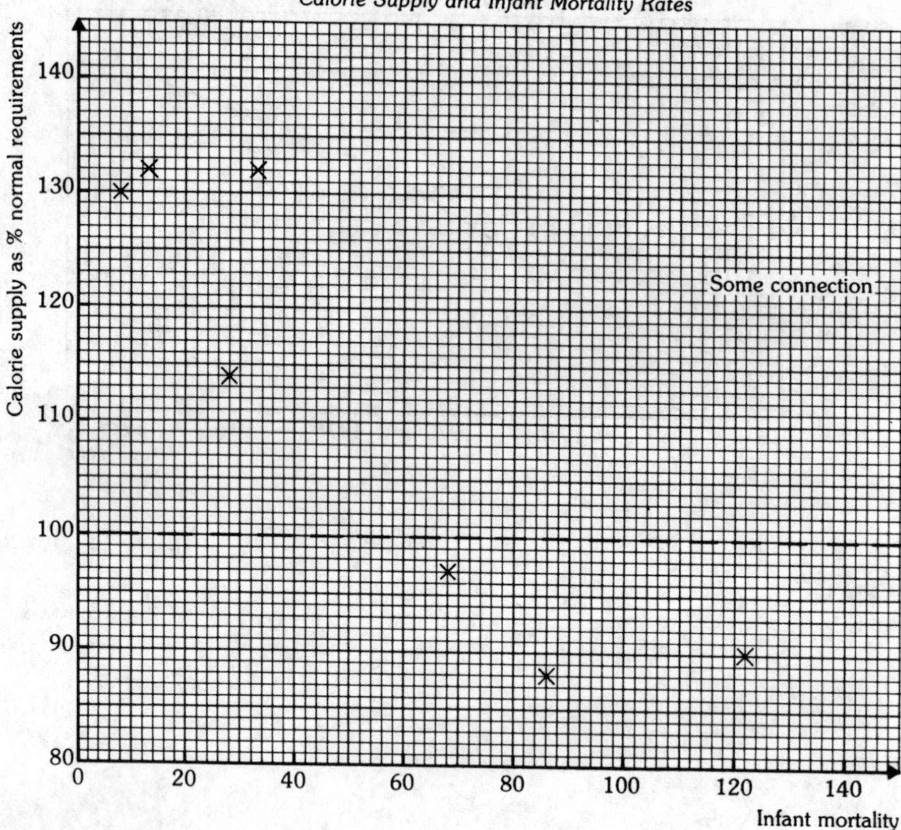


# CHAPTER 18

*page 141*

**1.1** 1. $\frac{2}{13}$ 2. $\frac{5}{9}$

3. $\frac{1}{21}$ 4. 0.95

5. $\frac{1}{4}$ 6. $\frac{1}{1000}$

7. $\frac{1}{1\,000\,000}$ 8. 0

**1.2a** 1. $\frac{1}{28}$ 2. $\frac{1}{500}$

3. $\frac{1}{10}$ 4. $\frac{1}{365}$

5. $\frac{1}{36}$

**1.2b** 1. p(Labour win) $= \frac{3}{4}$

2. p(handicap) $= \frac{3}{14}$

3. p(accident) $= \frac{7}{10}$

4. p(pass) $= \frac{3}{8}$

5. p(cure) $= 0$

6. p(win) $= 1$

**2.1** 1. $\frac{1}{13}$ 2. $\frac{1}{4}$

3. $\frac{2}{3}$ 4. $\frac{2}{5}$

5. $\frac{1}{50}$ 6. $\frac{1}{5}$

7. $\frac{3}{10}$ 8. $\frac{1}{9}$

**2.2** 1. $\frac{1}{12}$ 2. $\frac{1}{18}$

3. $\frac{1}{6}$ 4. $\frac{11}{36}$

5. $\frac{1}{12}$ 6. $\frac{7}{12}$

7. $\frac{3}{8}$ 8. $\frac{1}{8}$

**2.3** 1. about 2 p

2. about 2 p

3. about $5\frac{1}{2}$ p

4. 3 p 5. 5 p

6. 1 p

**Prices and Prizes**

*page 146*

1. a) 3 p b) 6 p
2. Neither fits.
3. 15 p 4. 30 p
6. 75

# CHAPTER 19

*page 147*

**1.1** 1.

| *in* | 20 | 25 | 30 |
|---|---|---|---|
| *cm* | 50.8 | 63.5 | 76.2 |

| *in* | 35 | 40 | 45 |
|---|---|---|---|
| *cm* | 88.9 | 101.6 | 114.3 |

| *in* | 50 | 55 | 60 |
|---|---|---|---|
| *cm* | 127.0 | 139.7 | 152.4 |

3. a) i) 81 cm ii) 97 cm
iii) 117 cm
b) i) 39 in ii) 47 in
iii) 54 in

**2.2** 4. Dave 32, Alan 48

# Answers to Part B

## CHAPTER 1

### Exercise 1

*page 161*

**1.** 30 **2.** 49
**3.** 862 **4.** 326
**5.** 683 **6.** 69
**7.** 519 **8.** 945
**9.** 6524 **10.** 10 780
**11.** **A** 15, 12, 11, 12, 14, 10, 8, 11, 13, 12
**B** 26, 15, 21, 25, 26, 22, 19, 20, 25, 21
**C** 30, 31, 33, 32, 30, 35, 38, 25, 30, 34
**D** 125, 75, 100, 50, 100, 310, 440, 500, 180, 415
**12.** a) 10251 b) 1 566 652
c) 21 757 d) 4757
**13.** a) 29 b) 49
c) 141 d) 128
**14.** 550 000 **15.** 1151
**16.** 3 540 000 **17.** 510 billion
**18.** 13 945 11 779 5100 7756
12 491 51 071 horizontally
20 153 10 683 17 970 2265
51 071 vertically
Totals are all wrong by 35.
**19.** 93 mm, 107 mm, 248 mm
**20.** £1125
**21.** a) 210 b) 1350
c) 25 d) 49
e) 184, 193
**22.** 8300

### Exercise 2

*page 163*

**1.** **A** 2, 5, 4, 5, 4, 4, 2, 2, 3, 3
**B** 7, 6, 4, 1, 6, 6, 3, 4, 1, 2
**C** 3, 6, 8, 7, 8, 8, 6, 2, 8, 3
**D** 3, 9, 9, 9, 7, 7, 2, 8, 7, 8
**2.** 15 **3.** 19
**4.** 403 **5.** 6473
**6.** 2278 **7.** 646
**8.** 3966 **9.** 4982
**10.** 4089 **11.** 1449
**12.** 450 000
**13.** 999 500 000
**14.** five hundred and sixty-six
**15.** b) 3158 c) 3609
d) 61 817 e) 6969
**16.** a) 301(9) b) 18(3)9
c) (3)9991 d) 84(5)31
**17.** 258 **18.** 17 miles
**19.** a) 1073 b) 1452
**20.** 19 000 **21.** 2040

### Exercise 3

*page 165*

**1.** **A** 20, 18, 15, 54, 21, 56, 36, 48, 35, 81
**B** 24, 54, 30, 49, 27, 10, 45, 42, 16, 63
**C** 64, 21, 18, 27, 36, 14, 25, 24, 72, 40
**D** 32, 0, 12, 42, 63, 32, 45, 56, 72, 48
**2.** 156 **3.** 230
**4.** 256 **5.** 3395
**6.** 7408 **7.** 915
**8.** 4644 **9.** 3066
**10.** 48 064 **11.** 88 803
**12.** a) 12, 36 p b) 18, 54 p
c) 30, 90 p
**13.** a) 16 152 b) 12 288
c) 390 880 d) 859 574
e) 474 281 f) 1 267 296
**14.** a) 25 b) 81
c) 27 d) 343
**15.** 54 p
**16.** £5088
**17.** £99 890
**18.** £196 250
**19.** £5600
**20.** 1530 pairs
**21.** yes, 25 cm
**22.** a) i) 70 ii) 700
b) i) Write a zero after the number.
ii) Write two zeros after the number.
**23.** a) 2, 6, 12
b) 9, 15, 19, 23, 25
c) 6, 9, 12, 15
d) 15, 25,
e) 9, 25
**24.** a) 630 b) 8120
**25.** a) 700 b) 62 500

### Exercise 4

*page 166*

**1.** **A** 3, 5, 3, 5, 7, 7, 9, 7, 4, 6
**B** 8, 2, 3, 9, 8, 4, 8, 8, 7, 5
**C** 5, 7, 5, 6, 9, 9, 6, 8, 2, 6
**D** 4, 9, 4, 9, 8, 3, 6, 6, 6, 8
**2.** a) 7 b) 8
c) no
d) 1, 3, 7, 9, 21, 63
e) no
**3.** a) 8 b) 36
c) 5 d) 2
e) 1, 2, 3, 4, 6, 12
**4.** a) 26 b) 13 r 3
c) 7 r 1 d) 12 r 4
e) 78 r 6
**5.** a) 25 r 6 b) 172
c) 154 r 1 d) 1000 r 3
e) 982
**6.** a) 40 b) 50
c) 128 d) 20
e) 39
**7.** a) 93 r 3 b) 700
c) 890 r 1 d) 978 r 7
**8.** 98 cm **9.** £25
**10.** 160 cm **11.** 35
**12.** 14 p, no **13.** 12, 3 cm
**14.** a) 16 b) 3
c) 65 r 10 d) 15 r 2
**15.** £3 **16.** 35
**17.** 32

**Exercise 5**

*page 168*

**1.** 18 **2.** 6
**3.** 45 **4.** 9
**5.** 49 **6.** 4
**7.** 28 **8.** 38
**9.** 14 **10.** 45
**11.** 4 **12.** 3
**13.** £373 **14.** 40 kg
**15.** £174
**16.** 80 min = 1 h 20 min
**17.** 102 p or £1.02
**18.** £12 **19.** 10 °C
**20.** 140 min = 2 h 20 min

**Exercise 6 Newspaper Cuttings**

*page 168*

**1.** 115 miles, 5 miles
**2.** 16
**3.** £90, £36, yes
**4.** 50 p **5.** £625
**6.** £624 700 **7.** £3
**8.** £15 699
**9.** 245, 236, 552
**10.** £68
**11.** 1330, 1415, 37
**12.** £19

**Exercise 7**

*page 170*

**1.** a) 68 b) 231
c) 84 d) 113
**2.** a) 319 b) 60
c) 63 d) 8
**3.** a) 16 b) 7
c) 28 d) 9
**4.** a) 9 b) 1
**5.** a) 2 out of 1, 2, 3, 6, 12 b) no
**6.** a) 600 b) 450
**7.** a) yes b) 3, 11 p
**8.** 4 **9.** 15
**10.** a) 70 b) 20
c) 27 d) 81
**11.** a) 6952 b) 2854
c) 2760 d) 25 500
**12.** fifty-two thousand three hundred and four
**13.** a) 48 b) 7
**14.** a) 5 b) 20

**End Test**

*page 170*

**1.** 128 **2.** 37
**3.** 64 **4.** 19
**5.** 56 **6.** 36
**7.** 8 **8.** 946
**9.** two of 1, 3, 7, 21
**10.** 80 **11.** yes
**12.** a) 27 b) 16
c) 64 d) 216
**13.** 78 **14.** 3
**15.** a) 100 b) 38
c) 138 d) 32
**16.** 5 **17.** yes
**18.** 42 **19.** 1432
**20.** a) 75 802 b) 666 685
c) 6020 d) 18 952
**21.** a) 12 b) 10
c) 13 d) 8
e) 19 f) 14
**22.** a) 37 r 8 b) 1500
**23.** 480 000 **24.** 4 g
**25.** 7 **26.** £169
**27.** sixty thousand and forty
**28.** 6 **29.** 668
**30.** a) 19 b) 40
c) 28
d) 16, 24, 28, 34,40
e) 16 f) 24
g) 34 h) 1

# CHAPTER 2

**Exercise 1**

*page 172*

**1.** 62.28 **2.** 9.7
**3.** 57.67 **4.** 130.5
**5.** 53.85 **6.** 149.87
**7.** 204.6 **8.** 739.76
**9.** 122.98 **10.** 2.01
**11.** 46.01 **12.** 24.2
**13.** 195.6 **14.** 60.4
**15.** 4.62 tonnes **16.** 8.25 kg
**17.** £26.53 **18.** 23.6 cm
**19.** 12.75 l **20.** £30.33
**21.** 158.36, 137.67, 126.6, 422.63 horizontally,
169.12, 186.79, 66.72, 422.63 vertically
**22.** 16.4 km, 35.2 km, yes

**Exercise 2**

*page 173*

**1.** 62.1 **2.** 25.2
**3.** 29.48 **4.** 0.557
**5.** 43.91 **6.** 4.45
**7.** 28.38 **8.** 0.176
**9.** 1.05 **10.** 98.4
**11.** 22.3 **12.** 0.21
**13.** 0.75 **14.** 21.75
**15.** 3.13 **16.** 2.9
**17.** 39.4 **18.** £7.64
**19.** 15.9 m **20.** 0.62 tonnes
**21.** 2.35 m
**22.** 0.425 m by 0.075 m
**23.** 0.09 kg
**24.** 3.2 m, 2.15 m

**Exercise 3**

*page 174*

**1.** 6.8 **2.** 6.6
**3.** 4.84 **4.** 13.64
**5.** 7.5 **6.** 0.32
**7.** 0.08 **8.** 0.028
**9.** 24 **10.** 0.006
**11.** 0.1917 **12.** 2.4
**13.** 7.8 **14.** 0.045
**15.** 1.62 **16.** 0.36
**17.** 0.856
**18.** a) 6.82 b) 453
c) 50 d) 0.456
e) 12.6
Move the decimal point one place to the right.
**19.** a) 67.8 b) 450
c) 3200 d) 9.82
e) 158.7
Move the decimal point two places to the right, or add two noughts for whole numbers.
**20.** 0.21 **21.** 2.16
**22.** 53 **23.** 0.69
**24.** 1 **25.** 3
**26.** 5.4 **27.** 2.2
**28.** 11.99 **29.** 268.8
**30.** 608.4 **31.** 34.56
**32.** £28.90 **33.** 392 kg
**34.** £906.96 **35.** 651 l
**36.** 275.6 p

### Exercise 4

*page 175*

**1.** 2.4 **2.** 4.1
**3.** 0.6 **4.** 2.4
**5.** 0.6 **6.** 4
**7.** 270 **8.** 210
**9.** 2.4 **10.** 2900
**11.** a) 1.42 b) 0.05
c) 8 d) 0.7
Move the decimal point one place to the left.
**12.** a) 0.06 b) 0.008
c) 0.123 d) 0.9
Move the decimal point two places to the left.
**13.** 1950 **14.** 15
**15.** £2.12 **16.** £4.80
**17.** 0.48 m **18.** 0.5 min
**19.** 0.25 mm **20.** £92.50

### Exercise 5

*page 176*

**1.** 0.929 **2.** 0.107
**3.** 6.667 **4.** 5.692
**5.** 71.727 **6.** 1.333
**7.** 17.143 **8.** 18.333
**9.** 35.000 **10.** 0.078

### Exercise 6

*page 177*

**1.** $0.\dot{6}$ **2.** $0.\dot{1}$
**3.** $0.\dot{2}\dot{7}$ **4.** $0.\dot{7}1428\dot{5}$
**5.** $0.\dot{4}\dot{5}$ **6.** $0.\dot{4}$
**7.** $0.\dot{6}\dot{3}$ **8.** $0.\dot{8}$
**9.** $0.\dot{4}2857\dot{1}$ **10.** $0.\dot{3}8461\dot{5}$

### End Test

*page 177*

**1.** 67.5 **2.** 0.05
**3.** 2.97 **4.** 21.2
**5.** 2.2
**6.** a) 42.03 b) 18.63
**7.** a) 0.1 b) 3
**8.** 0.833 **9.** $0.\dot{8}5714\dot{2}$
**10.** 8 tenths and 1 thousandth
**11.** 7.2
**12.** 0.2753, by 0.1503
**13.** 0.0125, 0.125, 0.375, 0.5
**14.** 2.25 **15.** 1.03
**16.** 2.8 **17.** 1.296
**18.** a) 117.11 b) $2.6\dot{4}$
**19.** 0.4, 0.375, 0.25, $0.\dot{2}$
**20.** £6.93 **21.** 21.6 kg
**22.** £10.91 **23.** 0.3 m
**24.** 31.3 kg **25.** 4
**26.** 13 nuts, 13 bolts
**27.** 3.46 cm **28.** £3.56
**29.** a) 56.25 $cm^2$ b) 0.3 $m^2$
**30.** 187.5, 220

# CHAPTER 3

### Exercise 1

*page 179*

**1.** 5 **2.** 12
**3.** 10 **4.** 4
**5.** 8 **6.** 3
**7.** 24 **8.** 20
**9.** 375 **10.** 28
**11** 32 **12.** 28

### Exercise 2

*page 179*

**1.** $\frac{1}{4}$ **2.** $\frac{1}{4}$
**3.** $\frac{2}{5}$ **4.** $\frac{1}{2}$
**5.** $\frac{1}{4}$ **6.** $\frac{1}{6}$
**7.** $\frac{1}{2}$ **8.** $\frac{3}{4}$
**9.** $\frac{8}{9}$ **10.** $\frac{1}{3}$
**11.** $\frac{3}{10}$ **12.** $\frac{9}{125}$

### Exercise 3

*page 180*

**1.** $2\frac{1}{5}$ **2.** $1\frac{3}{10}$
**3.** $1\frac{5}{7}$ **4.** $3\frac{1}{3}$
**5.** $3\frac{3}{5}$ **6.** $5\frac{7}{8}$
**7.** $1\frac{13}{16}$ **8.** $2\frac{11}{64}$
**9.** $1\frac{23}{100}$ **10.** $3\frac{3}{1000}$

### Exercise 4

*page 180*

**1.** $\frac{7}{3}$ **2.** $\frac{13}{4}$
**3.** $\frac{3}{2}$ **4.** $\frac{11}{4}$
**5.** $\frac{9}{2}$ **6.** $\frac{34}{9}$
**7.** $\frac{75}{8}$ **8.** $\frac{133}{16}$
**9.** $\frac{201}{100}$ **10.** $\frac{4273}{1000}$

### Exercise 5

*page 181*

**1.** $\frac{5}{6}$ **2.** $\frac{13}{15}$
**3.** $\frac{11}{12}$ **4.** $1\frac{7}{10}$
**5.** $1\frac{13}{16}$ **6.** $4\frac{7}{8}$
**7.** $4\frac{1}{2}$ **8.** $4\frac{1}{4}$
**9.** $12\frac{1}{4}$ **10.** $5\frac{5}{16}$
**11.** $4\frac{5}{22}$ **12.** $4\frac{9}{32}$

### Exercise 6

*page 181*

**1.** $2\frac{1}{4}$ **2.** $\frac{1}{2}$
**3.** $\frac{1}{2}$ **4.** $\frac{7}{8}$
**5.** $2\frac{5}{12}$ **6.** $\frac{5}{8}$
**7.** $2\frac{9}{16}$ **8.** $\frac{5}{8}$
**9.** $5\frac{3}{16}$ **10.** $5\frac{9}{10}$
**11.** $1\frac{21}{64}$ **12.** $6\frac{43}{100}$

### Exercise 7

*page 182*

**1.** $\frac{3}{8}$ **2.** $1\frac{1}{3}$
**3.** $\frac{2}{9}$ **4.** $1\frac{1}{2}$
**5.** $1\frac{1}{2}$ **6.** $\frac{17}{64}$
**7.** 2 **8.** 6
**9.** $\frac{1}{5}$ **10.** $11\frac{1}{9}$
**11.** $10\frac{1}{4}$ **12.** $1\frac{1}{6}$

### Exercise 8

*page 182*

**1.** $3\frac{3}{4}$ **2.** $2\frac{1}{4}$
**3.** $\frac{3}{8}$ **4.** $1\frac{1}{2}$
**5.** $2\frac{4}{5}$ **6.** 2
**7.** 32 **8.** $\frac{1}{2}$

### Exercise 9

*page 183*

**1.** $1\frac{3}{8}$ **2.** $1\frac{5}{8}$
**3.** $3\frac{7}{8}$ **4.** $4\frac{1}{2}$
**5.** $1\frac{35}{64}$ **6.** 21
**7.** 4. **8.** $67\frac{1}{2}$

### Exercise 10

*page 183*

**1.** $\frac{1}{2}$ **2.** $1\frac{1}{4}$
**3.** $3\frac{3}{4}$ **4.** $\frac{1}{8}$
**5.** $8\frac{3}{100}$ **6.** $\frac{7}{10}$
**7.** $\frac{27}{40}$ **8.** $4\frac{1}{5}$
**9.** $1\frac{3}{8}$ **10.** $\frac{1}{16}$

### Exercise 11

*page 183*

**1.** 0.5 **2.** 0.75
**3.** 2.333 **4.** 1.2
**5.** 0.125 **6.** 4.375
**7.** 8.167 **8.** 0.667
**9.** 2.6 **10.** 3.143

### Exercise 12

*page 184*

**1.** $\frac{1}{10}$ **2.** $\frac{1}{4}$
**3.** $\frac{1}{2}$ **4.** $\frac{3}{4}$
**5.** $\frac{3}{10}$ **6.** $\frac{7}{10}$
**7.** $\frac{1}{25}$ **8.** $1\frac{1}{2}$
**9.** $\frac{1}{8}$ **10.** $\frac{1}{3}$
**11.** 2 **12.** $\frac{3}{8}$

### Exercise 13

*page 184*

**1.** 25% **2.** 50%
**3.** 75% **4.** $66\frac{2}{3}$%
**5.** 60% **6.** 70%
**7.** $62\frac{1}{2}$% **8.** 65%
**9.** 125% **10.** 3%
**11.** 250% **12.** 310%
**13.** $\frac{1}{4}$% **14.** $1\frac{1}{2}$%

### Exercise 14

*page 184*

**1.** $\frac{1}{2} = \frac{6}{12}, \frac{2}{3} = \frac{8}{12}, \frac{5}{12}, \frac{3}{4} = \frac{9}{12};$
$\frac{5}{12}, \frac{1}{2}, \frac{2}{3}, \frac{3}{4}$
**2.** $\frac{7}{20}, \frac{1}{5} = \frac{4}{20}, \frac{3}{10} = \frac{6}{20}, \frac{1}{4} = \frac{5}{20};$
$\frac{1}{5}, \frac{1}{4}, \frac{3}{10}, \frac{7}{20}$
**3.** $\frac{3}{8} = \frac{6}{16}, \frac{9}{16}, \frac{1}{4} = \frac{4}{16},$
$\frac{5}{8} = \frac{10}{16}, \frac{1}{2} = \frac{8}{16}; \frac{1}{4}, \frac{3}{8}, \frac{1}{2}, \frac{9}{16}, \frac{5}{8}$
**4.** $\frac{3}{4} = 0.75, \frac{5}{8} = 0.625, \frac{1}{2} = 0.5$
$\frac{7}{20} = 0.35; \frac{3}{4}, \frac{5}{8}, \frac{1}{2}, \frac{7}{20}$
**5.** $\frac{1}{2} = 0.5, \frac{5}{6} = 0.8\dot{3}, \frac{3}{5} = 0.6,$
$\frac{2}{3} = 0.\dot{6}, \frac{1}{3} = 0.\dot{3}; \frac{5}{6}, \frac{2}{3}, \frac{3}{5}, \frac{1}{2}, \frac{1}{3}$
**6.** $\frac{3}{4} = 0.75, \frac{25}{32} = 0.781,$
$\frac{47}{64} = 0.734, \frac{13}{20} = 0.65,$
$\frac{5}{8} = 0.625; \frac{25}{32}, \frac{3}{4}, \frac{47}{64}, \frac{13}{20}, \frac{5}{8}$

### Exercise 15 (Miscellaneous)

*page 185*

**1.** 22 lb **2.** 6 ft $5\frac{1}{8}$ in
**3.** $4\frac{5}{8}$ yd **4.** $5\frac{3}{8}$ in
**5.** $3\frac{3}{8}$ ft **6.** $9\frac{9}{16}$ in
**7.** £268.83 **8.** £4500
**9.** $\frac{1}{3}$, $33\frac{1}{3}$% **10.** 25 l
**11.** $\frac{4}{7}$, 60 **12.** 90

### End Test

*page 186*

**1.** a) 4 b) 12
c) 56 d) 75
**2.** a) $\frac{1}{3}$ b) $\frac{3}{10}$
c) $\frac{1}{4}$ d) $\frac{15}{16}$
**3.** a) $4\frac{1}{5}$ b) $3\frac{1}{6}$
c) $2\frac{3}{16}$ d) $4\frac{9}{10}$
**4.** a) $\frac{7}{2}$ b) $\frac{8}{3}$
c) $\frac{61}{8}$ d) $\frac{81}{16}$
**5.** a) $1\frac{3}{10}$ b) $5\frac{7}{8}$
c) $11\frac{5}{16}$
**6.** a) $1\frac{1}{4}$ b) $1\frac{13}{16}$
**7.** a) $\frac{2}{3}$ b) $2\frac{1}{3}$
c) $7\frac{1}{5}$
**8.** a) $1\frac{1}{3}$ b) $12\frac{1}{2}$
c) 24 d) $1\frac{5}{8}$
**9.** a) $2\frac{1}{2}$ b) $1\frac{9}{10}$
c) 11
**10.** a) 0.55 b) 1.75
c) 5.4375
**11.** a) $\frac{3}{5}$ b) $4\frac{1}{4}$
c) $13\frac{1}{8}$
**12.** a) $\frac{1}{5}$ b) $\frac{2}{3}$
c) $1\frac{2}{5}$
**13.** a) 80% b) $33\frac{1}{3}$%
c) 10% d) $\frac{1}{2}$%
**14.** $\frac{4}{25}, \frac{1}{5}, \frac{3}{10}, \frac{7}{20}$
**15.** about 4
**16.** $\frac{21}{32}$ in
**17.** $6\frac{1}{2}$ h
**18.** $93\frac{1}{3}$ yd, £28.57, $7\frac{1}{6}$ yd
**19.** £6800
**20.** 315 g, $\frac{135}{500} = \frac{27}{100}$

# CHAPTER 4

### Exercise 1

*page 187*

**1.** forty-five pence
**2.** two pounds seventy pence
**3.** three pounds and eight pence
**4.** four pounds fifty-six pence
**5.** ninety-seven pounds sixty pence
**6.** 12 p **7.** £1.10
**8.** 75 p **9.** 6 p
**10.** £205.50 **11.** £0.24
**12.** £0.03 **13.** £0.20
**14.** £0.17 **15.** £0.08

### Exercise 2

*page 187*

**1.** £13.96 **2.** 85 p
**3.** £2.10 **4.** £1.60

5. £16.30 6. £6.55
7. £1.13 8. £16.20
9. £1.05 10. £7.01

**Exercise 3 (Speed Practice)**

*page 187*

1. £3.79 2. £13.07
3. £20.19 4. £15.18
5. £10 6. £344.50
7. £1.74 8. £10.13
9. £10.10 10. £1.90
11. £2.93 12. £39.20
13. £20.70 14. £25.40
15. £154 16. £21.85
17. £432.74 18. £1093.50
19. £4.02 20. £0.83
21. £20.75 22. £1.71
23. £1.94 24. £31.15

**Exercise 4**

*page 188*

Various answers are possible.
Some examples are given below.

1. a) 50 p, 2 p, 1 p; 20 p, 20 p, 10 p, 1 p, 1 p, 1 p
   b) 4 × £1, 50 p, 2 p, 1 p; 4 × £1, 5 × 10 p, 3 × 1 p
   c) £5, 4 × £1, 2 × 20 p, 10 p, 2 p, 1 p; 9 × £1, 50 p, 2 p, 1 p
2. a) 7 × 10 p, 5 p; 3 × 20 p, 10 p, 5 p
   b) 4 × £1, 50 p, 20 p, 5 p; 3 × £1, 3 × 50 p, 5 × 5 p
   c) £5, 4 × £1, 50 p, 20 p, 5 p; £5, 4 × £1, 7 × 10 p, 2 × 2 p, 1 p
3. a) 20 p, 5 p, 2 p; 4 × 5 p, 5 p, 2 p; 2 × 10 p, 7 × 1 p
   b) £20, £20, £10; £50; 5 × £10

**Exercise 5 (Miscellaneous)**

*page 188*

1. £1.17, £2.42
2. £1.68, 43 p
3. £16.20, 30 p
4. £2.08, £1.52
5. £0.80, £10.80
6. a) £10.25 b) £14.35 c) £18.45; a) £9.75 b) £5.65 c) £1.55
7. a) £193.50 b) £223.60
8. £194.80 9. £5296.50
10. £263.75 11. £3.52
12. £63

**Exercise 6**

*page 189*

1. £15 2. £70
3. £45 4. £80
5. £180 6. £30
7. £95 8. £490
9. £320 10. £150

**Exercise 7**

*page 190*

1. £53 2. 15 p
3. £35 4. £70.50
5. 9 p 6. £1
7. £2 8. £9
9. £12.50 10. £1.50
11. £9 12. £327.50
13. 50 p 14. 10 p
15. £22 16. £15.64
17. £85.68 18. £4.13
19. £344.85 20. £8
21. £5 22. £8
23. £1.25 24. £180
25. £81.25 26. £2.70

**Exercise 8**

*page 191*

1. £1160 2. £1353
3. £115 4. £540
5. £1344.80 6. £31.68
7. £18.30 8. £21
9. £4.10 10. £423.75

**Exercise 9**

*page 192*

1. a) £3630 b) £630
2. a) £2662 b) £662
3. a) £661.50 b) £61.50
4. a) £9383.55 b) £2333.55
5. a) £31.49 b) £6.49
6. a) £4775.44 b) 275.44
7. a) £2138.58 b) £338.58
8. a) £5624.32 b) £624.32
9. a) £573.30 b) £53.30
10. a) £8521.72 b) £1771.72

**Exercise 10**

*page 192*

1. 25% 2. 50%
3. 5% 4. 20%
5. 10% 6. 0.25%
7. 5% 8. 25%
9. 34.7% 10. 88%
11. 62.5% 12. 92%
13. 95.6% 14. 150%
15. 12.5%

**Exercise 11**

*page 193*

a) £4 b) 4%
c) £50 d) 10%
e) £10 f) 4%
g) £0.20 h) 2.5%
i) £0.05 j) 11.1%

**Exercise 12**

*page 193*

a) £3 b) 3%
c) £50 d) 20%
e) £0.10 f) 20%
g) £5.50 h) 30.6%
i) £10 j) 16.7%

**Exercise 13**

*page 194*

a) £665 b) £35
c) £360 d) £40
e) £357 f) £7
g) £690 h) £90
i) £204.75 j) £20.25

**Exercise 14**

*page 194*

a) £100 b) £15
c) £75 d) £3

e) £35 f) £1.75
g) £50 h) £3
i) £2.00 j) £0.04
k) £25 l) £1

**Exercise 15**

*page 194*

1. 10% 2. 12.5%
3. £20, 8% 4. £126
5. £24.25 6. £30
7. a) £157.50 b) 30%
8. a) 23 p b) 9.3%
9. £52
10. a) £1 b) 5%

**Exercise 16 (Miscellaneous Percentage Questions)**

*page 195*

1. £37.26
2. a) £80 b) £172.80
3. £1682
4. a) £448 b) £77
5. £3430 6. £6480
7. a) £10 profit
   b) 20% profit
   c) £520 d) £20
   e) £120
   f) 14.3% profit
   g) £375 h) £75
8. a) £110 b) £121
   c) £500
9. £9310
10. a) £95
   b) £105, scheme b)

**Exercise 17**

*page 196*

1.

| | Gross |
|---|---|
| | £64.05 |
| | £25.90 |
| | £10.80 |
| | £33.75 |
| Total gross | £134.50 |
| Less discount | £10.79 |
| | £123.71 |
| Plus VAT | £18.56 |
| | £142.27 |

2.

**INVOICE**

Mrs. C. Clarke (Drapers)
18 Dellfield Close
Appleford
Middx

| Qty | Description | Unit cost | Gross | Percentage discount | VAT |
|---|---|---|---|---|---|
| 17 | Reels cotton (white) | 32 p | £5.44 | $12\frac{1}{2}$% | 15% |
| 17 | Reels cotton (black) | 32 p | £5.44 | $12\frac{1}{2}$% | 15% |
| 105 | Reels cotton (asstd) | 35 p | £36.75 | $12\frac{1}{2}$% | 15% |
| 14 | Bias binding (pkt) | 40 p | £5.60 | $12\frac{1}{2}$% | 15% |
| 16 | Zips (green) | 68 p | £10.88 | $12\frac{1}{2}$% | 15% |
| 26 | Zips (blue) | 68 p | £17.68 | $12\frac{1}{2}$% | 15% |
| | Total gross | | £81.79 | | |
| | Less discount | | £10.22 | | |
| | | | £71.57 | | |
| | Plus VAT | | £10.74 | | |
| | | | £82.31 | | |

3.

**INVOICE**

Mr. W. Patel
15 North Street
Newcastle

| Qty | Description | Unit cost | Gross | Percentage discount | VAT |
|---|---|---|---|---|---|
| 23 | Sheets (single) | £8.95 | £205.85 | 15% | 15% |
| 45 | Pillow-cases | £3.50 | £157.50 | 15% | 15% |
| 50 | Duvet covers | £12.95 | £647.50 | 15% | 15% |
| 35 | Valences | £5.75 | £201.25 | 15% | 15% |
| | Total gross | | £1212.10 | | |
| | Less discount | | £181.82 | | |
| | | | £1030.28 | | |
| | Plus VAT | | £154.54 | | |
| | | | £1184.82 | | |

**End Test**

*page 197*

1. a) i) twenty-three pence
   ii) four pounds fifty pence
   b) i) 14 p ii) £305.07
2. a) £49.37 b) £3.48
   c) £14.33 d) £101.69
3. 10 p, 10 p, 1 p; 20 p, 1 p; 4 × 5 p, 1 p; 10 p, 2 × 5 p, 1 p; 10 p, 11 × 1 p (other ways are possible)
4. £2.15
5. £1.73
6. £75
7. £5, 75 p
8. a) 72 p b) £70
   c) 9 p d) £0.77
9. a) i) £112.84 ii) £28.21
   b) £1182.19
10. a) 11.25% b) $8.\dot{3}\%$
11. $8.\dot{3}\%$
12. £415.80
13. £450, £350
14. 9.3%
15.

| INVOICE | | | | | | |
|---|---|---|---|---|---|---|
| Mr. D. Smith<br>Hazeldene<br>Trentham | | | | | | |
| | Qty | Description | Unit cost | Gross | Percentage discount | VAT |
| | 125 | Cakes | £0.15 | £18.75 | 14% | — |
| | 150 | Buns | £0.12 | £18.00 | 14% | — |
| | 175 | Biscuits | £0.05 | £8.75 | 14% | — |
| | | | Total gross | £45.50 | | |
| | | | Less discount | £6.37 | | |
| | | | | £39.13 | | |
| | | | Plus VAT | — — | | |
| | | | | £39.13 | | |

VAT would have been £5.87.

# CHAPTER 5

**Exercise 1**

*page 198*

1. 19
2. −2
3. 0.08
4. 6.5
5. 0
6. −6
7. 4
8. 32 or 17 (BODMAS)
9. 2 or 0 (BODMAS)
10. 0 or 5 (BODMAS)

**Exercise 2**

*page 199*

1. £596.05
2. £7.75
3. £337.81
4. £826
5. £317
6. £17.89
7. £103.07
8. £1524.50
9. £0.65
10. £15.24

**Exercise 3**

*page 200*

1. 500, 6432
2. £200, £197.50
3. 1500, 1516
4. $11\frac{1}{2}$, 11.625
5. 2, 1.97318
6. 100 g, 95.13 g
7. £25, £23.22
8. 20 km, 23 km
9. 5900 m, 5913 m
10. 500 s, 498 s

**Exercise 4**

*page 201*

1. a) 6.8 b) 6.82
   c) 6.824
2. a) 13 b) 13.4
   c) 13.42
3. a) 0.057 b) 0.0570
   c) 0.056 96
4. a) 55 000 000 b) 55 000 000
   c) 54 950 000
5. a) 0.000 82 b) 0.000 819
   c) 0.000 819 3
6. a) 10 b) 9.99
   c) 9.990
7. a) 67 000 000 000
   b) 67 100 000 000
   c) 67 090 000 000
8. a) 5600 b) 5610
   c) 5608

**Exercise 5**

*page 201*

1. 700
2. 9950
3. 9180
4. 0.9
5. 17 520
6. 456.72
7. 569 000
8. 68
9. 679 600
10. 190 000
11. 5.7
12. 57 000
13. 1.0
14. 15 950
15. 190 850 000

**Exercise 6**

*page 202*

1. $6 \times 10^5$
2. $5.4 \times 10^7$
3. $8.5 \times 10^8$
4. $4 \times 10^{11}$
5. $1.7 \times 10^{17}$
6. $3 \times 10^{19}$

7. $5.12 \times 10^{14}$
8. $1.4 \times 10^{11}$
9. 700 000 000
10. 5 000 000 000
11. 360
12. 54 000
13. 108 000 000
14. 9 730 000 000 000
15. [6] [EXP] [1] [2]
16. [1] [.] [4] [EXP] [1] [3]
17. [2] [.] [3] [5] [EXP] [7]
18. a) 150 384 000 $= 1.50384 \times 10^8$
    b) 149 616 000 $= 1.49616 \times 10^8$
19. $8 \times 10^{22}$
20. $4 \times 10^{23}$, $8 \times 10^{26}$

**Exercise 7 (Practice Calculations)**

*page 202*

1. £330
2. 8.975
3. 0.3125
4. 3.6875
5. £39.38
6. 274 cm$^2$
7. 7.04 m
8. 10 cm
9. 5200 cm$^3$
10. £9.86
11.

| $n$ | $p$ | $q$ | $r$ |
|---|---|---|---|
| 1 | 10.04 | 12.72 | 11.38 |
| 2 | 10.54 | 13.42 | 11.98 |
| 3 | 11.04 | 14.12 | 12.58 |
| 4 | 11.54 | 14.82 | 13.18 |
| 5 | 12.04 | 15.52 | 13.78 |

12. £1.60, £0.70 with a rummage sale raising £50, £1

**End Test**

*page 204*

2. a) £1047.65 b) £35.71
3. £177.39
4. no, £399
5. a) £200 b) 1250 g c) 200 km
6. a) 3.9 b) 0.0805 c) 80 000 d) 562 300
7. a) i) 10 cm ii) 100 mm b) 1.526 kg
8. a) 300 b) 250 c) 254 d) 254.3
9. (2.56 03)
10. a) $6.7 \times 10^4$
    b) $8 \times 10^8$
    c) 75, [4] [.] [5] [EXP] [9] [÷] [6] [EXP] [7] [=], (75.)

# CHAPTER 6

**Exercise 1**

*page 205*

1. a) 12 °C b) 2 min c) −8 °C
2. 16 min, 9 min
3. 13 min
4. a) 45 years b) 38 BC c) 77 years d) 53 years
5. a) i) 6. a.m. ii) 7 p.m.
   b) 6 a.m.
   c) i) 1 ii) 10
   d) i) 13 ii) 18
   e) 10 a.m., 9th March
6. a) £41, £48

b)

| BANK STATEMENT | | | | |
|---|---|---|---|---|
| | Date | Payments | Receipts | Balance |
| Brought forward | 2 AUG | | | 317.25 |
| | 5 AUG | 320.00 | | 2.25OD |
| | 7 AUG | | 352.25 | 350.00 |
| | 15 AUG | 150.78 | | 199.22 |
| | 22 AUG | 50.00 | | 149.22 |
| | 27 AUG | 200.00 | | 50.78OD |
| | 3 SEP | | 348.80 | 298.02 |

7. a) 2953 m b) 9243 m c) 19 870 m
8. a) 12 °C b) 4 °C
9. a) −4 b) −2 c) −4
10. a) 9.37 a.m. b) 10 min **c)** 2 min late d) 9.57 a.m.

**Exercise 2 (Oral)**

*page 208*

1. [3] [−] [1] [1] [=]
2. [5] [+/−] [+] [1] [2] [=]
3. [6] [+/−] [×] [8] [+/−] [=]
4. [1] [2] [+/−] [÷] [4] [=]
5. [9] [+/−] [$x^2$] [=]
6. [5] [−] [9] [=] [×] [2] [=]
7. [5] [+/−] [×] [2] [+] [3] [×] [1] [+/−] [=]
   or for a calculator that does not automatically do × before +,
   [5] [+/−] [×] [2] [=]
   (write down 10)
   [3] [×] [1] [+/−] [+] [1] [0] [=]
8. [1] [2] [+/−] [×] [2] [=] [÷] [4] [+/−] [=]

9. [7] [×] [3] [+/−] [+] [2]
[×] [4] [+/−] [=]
or
[7] [×] [3] [+/−] [=]
(write down −21)
[2] [×] [4] [+/−]
[−] [2] [1] [=]

10. [5] [−] [4] [+/−]
[=] [÷] [3] [=]

**Exercise 3**

*page 208*

1. −8
2. 7
3. 48
4. −3
5. 81
6. −8
7. −13
8. 6
9. −29
10. 3

**Exercise 4**

*page 208*

1. −5
2. 8
3. −2
4. −20
5. 20
6. −11
7. 0
8. −13
9. 0
10. −1
11. 35
12. −24
13. −14
14. 9
15. −2
16. −5
17. 2
18. −1
19. −2
20. −9

**End Test**

*page 209*

1. −7 °C
2. 73 years
3. 11 p.m.
4. 2628 m
5. a), b), e), g)
6. a) −11 b) 11
   c) −48 d) 3
   e) −10 f) −7
7. 1600 fathoms below sea level, 23 000 ft above sea level, 3000 m
8. 21 °C

# CHAPTER 7

**Exercise 1**

*page 211*

1. A(2, 2), B(3, −2), C(−3, −3), D(−3, 2), E(1, −3)
2. a) B3, C3, D3, D2, E2
   b) A1, B1, B2, C2
   c) B1, C1, C2, C3
   d) i) A1, D3 ii) C1
3. a) 1.25 b) 8
4. a) 102 b) 3900
   c) £23
5. a) 9.8 cm b) 1 lb 12 oz
6. the first graph, because it looks steeper

**Exercise 2**

*page 214*

1. 13, 47
2. $\frac{1}{15}$
3. a) 9 b) $\frac{1}{2}$
   c) $12\frac{1}{2}\%$
4. the angles are 90° (children), 120° (men), 10° (babies), 140° (women).
5. a) 131° b) $\frac{2}{5}$
6. a) $\frac{1}{3}$ b) $33\frac{1}{3}\%$
7. a) 40 000 b) 10 000
   c) sales in 1981 are 40 000.
   Could use = 4000.

**End Test**

*page 216*

1. a) 37 °C b) 29 days
   c) 19 days d) Thursday
2. a) 18 miles b) 5 minutes
   c) 30 miles d) 120 m.p.h.
3. £5.80, 4 p
4. 168, 200
   a) 54 b) 850 hours
5. 40°, 7000
6. uneven scale, scale does not start at zero
7. a) 1.4 cm b) 1.8 cm